COMPUTATIONAL MATHEMATICS AND ANALYSIS SERIES

GEOMETRY OF QUANTUM COMPUTATION

Computational Mathematics and Analysis Series

Additional books in this series can be found on Nova's website
under the Series tab.

Additional e-books in this series can be found on Nova's website
under the e-book tab.

Mathematics Research Developments

Additional books in this series can be found on Nova's website
under the Series tab.

Additional e-books in this series can be found on Nova's website
under the e-book tab.

COMPUTATIONAL MATHEMATICS AND ANALYSIS SERIES

GEOMETRY OF QUANTUM COMPUTATION

GIA GIORGADZE

nova publishers
New York

Library of Congress Cataloging-in-Publication Data

Giorgadze, Gia, author.
Geometry of quantum computations / Gia Giorgadze (Department of Mathematics, Faculty of Exact and Natural Sciences, Ivane Javakhishvili Tbilisi State University 2, Tbilisi, Georgia).
pages cm
Includes bibliographical references and index.
ISBN 978-1-62257-325-7 (hardcover)
1. Quantum computers. 2. Geometry--Data processing. 3. Quantum theory--Mathematics. I. Title.
QA448.D38G56 2012
530.12--dc23

2012022859

Published by Nova Science Publishers, Inc. † New York

Contents

Preface

The research in this book is concerned with the geometric theory of computation. We shall use a known paradigm for our computation that involves a combination of an algorithm and a computation process. The latter process will be performed through the use of a dynamical system. In this dynamical system computation, we have chosen a configuration space with a nontrivial geometry or topology. We then discuss the geometric character of this computation. This book examines the known holonomic and topological models of quantum computation and the new monodromic model of quantum computation in detail.

Monodromic quantum computing is a geometric model for quantum computations which is based on the ideas developed by this author and outlined in this book. According to this research, the elementary gates necessary for performing quantum computations may be realized using monodromy matrices induced by differential equations for Fuchs type systems. It is known that when using quantum computation for a holonomic model (holonomic quantum computing) the space of quantum bits (qubits) is a finite dimensional vector space over a field of complex numbers, where the holonomy provides the transformation of qubits from one state to another. On the other hand, quantum computations for the topological model is based on finding such representations such as braid groups, in a group of nondegenerate complex matrices, whose images would form dense subgroups everywhere.

In monodromic quantum computations, the fiber of the vector bundle induced by a Fuchs system is considered as the space of qubit states, whose dynamics is realized by the monodromy operators. From the analytic theory of differential equations, a necessary and sufficient condition for obtaining a universal system of quantum gates is known; whereby the solvability of the Riemann-Hilbert monodromy problem ensures the existence of a Fuchs system that has a corresponding arbitrary system of quantum gates.

Applications that utlize analytic theory for differential equations and geometric control theory in quantum physics for these models has a rather short history. Today, current research in quantum physics provides promising new approaches in this field. Experiments show that the nanostructures do not have trivial geometry, so their study includes the role of geometry as an important ingredient.

The theory of classical computation was put on a firm foundation by proofs that dealt with Turing machines, when possibly supplemented by random-number generation, can efficiently simulate classical systems. Thus, in particular, can include all other proposed models of universal classical computation. Thus, computational complexity results obtained within the Turing machine model, or any of the equivalent models, are universally applicable, up to polynomial factors. The field of quantum computing originated with Feynman's observation that quantum-mechanical systems represent an apparent exception to this efficient mutual simulatability: simulating quantum-mechanical systems with n degrees of freedom on a classical computer requires time scaling exponentially in n. Feynman conjectured that this exception was an indication of two classes of computation: classical and quantum; with all quantum models able to simulate each other efficiently, and all classical models able to simulate each other but not quantum models, efficiently. Deutsch defined quantum Turing machines and proposed them as a concrete model for a universal quantum computer able to efficiently simulate all others. Subsequently proposed models of quantum computation, such as the quantum circuit model, continuous and discrete quantum walks, topological quantum computers, and adiabatic quantum computers, have all been proven to be polynomially equivalent to the quantum Turing machine model. Today, the model of quantum computation most widely used by theorists is the quantum circuit model. Feynman's original question about whether a universal quantum computer can efficiently simulate not only other models of quantum computation but also naturally occurring quantum systems have also been addressed: quantum circuits have been designed to simulate the evolution for time t of systems with n nonrelativistic particles using $poly(n,t)$ gates.

The first chapter in this book contains all the necessary definitions and basic knowledge about the general theory of quantum computation. In chapter 2, a time-dependent periodic Hamiltonian admitting exact solutions is applied to construct a set of universal gates for quantum computation. The approach is based on the transformation of soluble time-independent equations into time-dependent ones by employing a set of special time-dependent transformation operators. A class of periodic time-dependent Hamiltonians with cyclic solutions is constructed in a closed analytic form and the nonadiabatic geometric phase is determined in terms of the obtained solutions.

In chapter 3, a three-level atom is considered that has energy levels $|i>$, $|g>$, and $|e>$ in the so-called Lambda which use Ladder and Vee configurations. It is assumed that the direct transition between $|g>$ and $|e>$ is forbidden and the dipole moments of the two transitions are perpendicular. The atom interacts with a two-mode quantum electromagnetic field. The coupled system is investigated using the rotating wave approximation. In each configuration, the model obtained is similar to the Jaynes-Cummings model. In this case, we study the quantum description for a two-level atom when the time evolution for the system is a semi-classical approximation.

In chapter 4, applications for geometric control theory methods on Lie groups and homogeneous spaces in the theory of quantum computations are investigated. These methods are shown to be very useful for the problem of constructing a universal set of gates for quantum computations: a well-known result has the set of all one-bit gates together with almost any one two-bit gate is universal as considered from the control theory viewpoint.

Differential geometric structures such as the principal bundle for the canonical vector bundle on a complex Grassmann manifold, the canonical connection form on this bundle, the canonical symplectic form on a complex Grassmann manifold, and the corresponding dynamical systems are investigated. The Grassmann manifold is considered as an orbit of the co-adjoint action and the symplectic form is described as the restriction of the canonical Poisson structure on a Lie coalgebra. The holonomy of the connection on the principal bundle over Grassmannian and its relation with the Berry phase is considered and investigated for trajectories within Hamiltonian dynamical systems.

In chapter 5, a model for quantum computation is considered based on the monodromy representation of a Fuchsian system. The rôle of local and entangling operators in monodromic quantum computing is shown when monodromy matrices have this connection with logarithmic singularities that act on the fiber of a holomorphic vector bundle in the space of qubits. The prominent theme is the construction problem for a set of universal gates with monodromy operators induced by a connection with a logarithmic singularity. In the formal scheme developed in this chapter, this can be incorporated already by the known models-topological and holonomic. Discussed in further detail are the possible applications for $1-D$ direct and inverse scattering problem which use the design of universal quantum gates for quantum computation. The potentials for generating some universal gates are also described.

The book is based on materials generated through a research program on the theory of quantum computing which has been developed over several years at the Georgian National Academy of Sciences. Then later, as a group effort while at the Laboratory of Information Technologies at the Joint Institute for Nuclear Research in Dubna, together with Z. Giunashvili, M. Jibladze, G. Khimshiashvili, Z. Melikishvili, A. Suzko, and R. Tevzadze. This book includes the joint efforts by these esteemed researchers. The author is grateful to all for their long-standing fruitful collaboration.

In the aim of geometrizing some topics in function theory, the author frequently benefitted from the advice of Prof. B. Bojarski. Long-standing scientific interaction with Prof. B. Bojarski enabled an adequate adjustment for the complex analytic methods to the geometric questions of quantum computing.

The author would also like to express his gratitude to Prof. A. Agrachev as host of the SISSA. His expert conversations on geometric control theory helped the author to clarify some questions concerning control theory for quantum systems.

Approximately half of the contents of the book is available for university students specializing in mathematics, physics, and computer science. In this book there are questions for those highly qualified specialists, new ideas, and possible ways to determine a solution.

Chapter 1

Basics of Quantum Computation

1.1. The Space of Quantum Bits

In classical computers the elementary information unit is known as a bit. The quantum analog of a bit is called a *qubit*. A two state quantum system is described by a unit vector in the Hilbert space $\mathbb{C}^2$. Quantum mechanics is a mathematical theory that is governed by a set of axioms. The consequences of these axioms describe the behavior of quantum systems.

The state space of a quantum system is modeled by a Hilbert space of wave functions. For quantum computing, we deal only with finite quantum systems when it suffices to consider finite dimensional complex vector spaces with an inner product such as for spanned abstract wave functions, for example $|\rightarrow\rangle$.

Quantum state spaces and the transformations acting on them can be described in terms of vectors and matrices or in the more compact bra/ket notation by Dirac. Kets like $|x\rangle$ denote column vectors and are used to describe quantum states. The matching bra, $\langle x|$, denotes the conjugate transpose of $|x\rangle$. For example, the orthonormal basis $\{|0\rangle, |1\rangle\}$ can be expressed as $\{(1,0)^T, (0,1)^T\}$. Any complex linear combination of $|0\rangle$ and $|1\rangle$, $a|0\rangle + b|1\rangle$ can be written as $(a,b)^T$. Note that the choice of the order for these basis vectors is arbitrary.

Combining $\langle x|$ and $|y\rangle$ as in $\langle x||y\rangle$, also written as $\langle x|y\rangle$, denotes the inner product of the two vectors. For instance, since $|0\rangle$ is a unit vector we have $\langle 0|0\rangle = 1$ and since $|0\rangle$ and $|1\rangle$ are orthogonal we have $\langle 0|1\rangle = 0$.

The notation $|x\rangle\langle y|$ is the outer product of $|x\rangle$ and $\langle y|$. For example, $|0\rangle\langle 1|$ is the transformation that maps $|1\rangle$ to $|0\rangle$ and $|0\rangle$ to $(0,0)^T$ since

$$|0\rangle\langle 1||1\rangle = |0\rangle\langle 1|1\rangle = |0\rangle,$$
$$|0\rangle\langle 1||0\rangle = |0\rangle\langle 1|0\rangle = 0|0\rangle = \begin{pmatrix} 0 \\ 0 \end{pmatrix}.$$

Equivalently, $|0\rangle\langle 1|$ can be written in matrix form where $|0\rangle = (1,0)^T$, $\langle 0| = (1,0)$, $|1\rangle = (0,1)^T$, and $\langle 1| = (0,1)$. Then

$$|0\rangle\langle 1| = \begin{pmatrix} 1 \\ 0 \end{pmatrix}(0,1) = \begin{pmatrix} 0 & 1 \\ 0 & 0 \end{pmatrix}.$$

This notation gives us a convenient way of specifying transformations on quantum states in terms of what happens to these basis vectors. For example, in the transformation that exchanges $|0\rangle$ and $|1\rangle$, i.e.,

$$X : |0\rangle \rightarrow |1\rangle \text{ and } |1\rangle \rightarrow |0\rangle$$

is given by the matrix

$$X = |0\rangle\langle 1| + |1\rangle\langle 0|.$$

Definition 1.1.1. *[108]A qubit or quantum bit is a quantum system whose state can be fully described by a superposition of two orthogonal eigenstates labeled $|0\rangle$ and $|1\rangle$.*

The machine state $|\Psi\rangle$ of an n-qubit quantum computer is given

$$|\Psi\rangle = \sum_{(d_0,\ldots,d_{n-1})} c_{d_0,\ldots,d_{n-1}} |d_0 \ldots d_{n-1}\rangle$$

with $\sum |c_{d_0,\ldots,d_{n-1}}|^2 = 1$.

The combined Hilbert space $\mathcal{H}$ is thus a composition of $n-1$-qubit Hilbert spaces $H_j \cong \mathbb{C}^2$, i. e.,

$$\mathcal{H} \cong \mathcal{H}_1 \otimes \mathcal{H}_2 \ldots \otimes \mathcal{H}_n \cong \mathbb{C}^{2^n}$$

The state of an multiple quantum system can be described completely by relating the state of each of its component pieces separately. A unintuitive aspect when considering a state space of an n particle quantum system is that the state of the system cannot always be described in terms of the state of its component pieces. It is when examining the systems of more than one qubit that one can first get a glimpse of where the computational power of quantum computers could come from. As we saw, the state of a qubit can be represented by a vector in the two-dimensional complex vector space spanned by $|0\rangle$ and $|1\rangle$. In classical physics, the possible states for a system of n particles, whose individual states can be described by a vector in a two-dimensional vector space, form a vector space of $2n$ dimensions. A system of n qubits has a state space of 2^n dimensions. The state space is the set of normalized vectors in this 2^n dimensional space, just as the state $a|0\rangle + b|1\rangle$ of a qubit is normalized to $|a|^2 + |b|^2 = 1$. It is this exponential growth of the state space with the number of particles that suggests a possible exponential speed-up of computation on quantum computers over classical computers.

Individual state spaces of n particles combine classically through the Cartesian product. Quantum states however, combine through the tensor product.

Let V and W both be the two-dimensional complex vector spaces with bases $\{v_1, v_2\}$ and $\{w_1, w_2\}$, respectively. The Cartesian product of these two spaces can take as its basis, the union of these bases for its component spaces $\{v_1, v_2, w_1, w_2\}$. Note that, the order of this basis was chosen arbitrarily. In particular, the dimension for the state space of multiple classical particles grows linearly with the number of particles, since $\dim(X \times Y) = \dim(X) + \dim(Y)$. The tensor product of V and W has the basis $\{v_1 \otimes w_1, v_1 \otimes w_2, v_2 \otimes w_1, v_2 \otimes w_2\}$. Note also, that the order of this basis, again, is arbitrary. So the state space for two qubits, each with the basis $\{|0\rangle, |1\rangle\}$, has a basis $\{|0\rangle \otimes |0\rangle, |0\rangle \otimes |1\rangle, |1\rangle \otimes |0\rangle, |1\rangle \otimes |1\rangle\}$ which then can be written more compactly as $\{|00\rangle, |01\rangle, |10\rangle, |11\rangle\}$. Generally, we write $|x\rangle$ to mean $|b_n b_{n-1} \ldots b_0\rangle$ where b_i are the binary digits of the number x.

A basis for a three qubit system is

$$\{|000\rangle, |001\rangle, |010\rangle, |011\rangle, |100\rangle, |101\rangle, |110\rangle, |111\rangle\}$$

In general, an n qubit system has 2^n basis vectors. The tensor product $X \otimes Y$ has dimension $dim(X) \times dim(Y)$.

The state $|00\rangle + |11\rangle$ is an example of a quantum state that cannot be described in terms of a state for each of its components (qubits) separately. In other words, we cannot find a_1, a_2, b_1, b_2 such that $(a_1|0\rangle + b_1|1\rangle) \otimes (a_2|0\rangle + b_2|1\rangle) = |00\rangle + |11\rangle$ since

$$(a_1|0\rangle + b_1|1\rangle) \otimes (a_2|0\rangle + b_2|1\rangle) = a_1a_2|00\rangle + a_1b_2|01\rangle + b_1a_2|10\rangle + b_1b_2|11\rangle$$

and $a_1b_2 = 0$ implies that either $a_1a_2 = 0$ or $b_1b_2 = 0$. States which cannot be decomposed in this way are called *entangled* states. These entangled states represent situations that have no classical counterpart and these are also the states that provide the *exponential growth* of quantum state spaces with a number of particles.

The *measurement* postulate of quantum mechanics states that any device measuring a 2-dimensional system has an associated orthonormal basis with respect to which the quantum measurement takes place. Measurement of a state transforms the state into one of the measuring device's associated with basis vectors. The probability that the state is measured as a basis vector $|u\rangle$ is the square for the amplitude norm of the component in the original state for the direction of the basis vector $|u\rangle$. The result of this measurement is *probabilistic* and the process of measurement changes the state to that which is measured. For example, the measurement of a single qubit projects the quantum state on to one of the basis states associated with the measuring device.

Consider the example of measurement in a two qubit system. Any two qubit state can be expressed as $a|00\rangle + b|01\rangle + c|10\rangle + d|11\rangle$, where a, b, c and d are complex numbers such that $|a|^2 + |b|^2 + |c|^2 + |d|^2 = 1$. Suppose we wish to measure the first qubit with respect to the standard basis $\{|0\rangle, |1\rangle\}$. For convenience, we will rewrite the state as follows:

$$a|00\rangle + b|01\rangle + c|10\rangle + d|11\rangle = |0\rangle \otimes (a|0\rangle + b|1\rangle) + |1\rangle \otimes (c|0\rangle + d|1\rangle) =$$

$$= u|0\rangle \otimes (a/u|0\rangle + b/u|1\rangle) + v|1\rangle \otimes (c/v|0\rangle + d/v|1\rangle).$$

For $u = \sqrt{|a|^2 + |b|^2}$ and $v = \sqrt{|c|^2 + |d|^2}$, the vectors $a/u|0\rangle + b/u|1\rangle$ and $c/v|0\rangle + d/v|1\rangle$ are of unit length. Once the state has been rewritten as described earlier, as a tensor product of the bit being measured and a second vector of unit length; the probabilistic result of a measurement is easy to read off. Measurement of the first bit will with probability $u^2 = |a|^2 + |b|^2$ return $|0\rangle$ projecting the state to $|0\rangle \otimes (a/u|0\rangle + b/u|1\rangle)$ or with probability $v = |c|^2 + |d|^2$ yield $|1\rangle$ projecting the state to $|1\rangle \otimes (c/v|0\rangle + d/v|1\rangle)$. As $|0\rangle \otimes (a/u|0\rangle + b/u|1\rangle)$ and $|1\rangle \otimes (c/v|0\rangle + d/v|1\rangle)$ are both unit vectors, no scaling is necessary. Measuring the second bit works similarly. In this example, the state space is a Cartesian product of the subspace consisting of all states whose first qubit is in state $|0\rangle$ and the orthogonal subspace of states whose first qubit is in state $|1\rangle$. Any quantum state can be written as the sum of two vectors, one for each of the subspaces.

Particles are not entangled if the measurement of one has no effect on the other. For instance, the state $\frac{1}{\sqrt{2}}(|00\rangle + |11\rangle)$ is *entangled* since the probability that the first bit is measured to be $|0\rangle$ is $1/2$ if the second bit has not been measured. However, if the second bit had been measured, the probability that the first bit is measured as $|0\rangle$ is either 1 or 0, depending on whether the second bit was measured as $|0\rangle$ or $|1\rangle$, respectively. Thus, the probable result of measuring the first bit is changed by a measurement of the second bit. On the other hand, the state $\frac{1}{\sqrt{2}}(|00\rangle + |01\rangle)$ is not entangled: since $\frac{1}{\sqrt{2}}(|00\rangle + |01\rangle) = |0\rangle \otimes \frac{1}{\sqrt{2}}(|0\rangle + |1\rangle)$, any measurement of the first bit will yield $|0\rangle$ regardless of whether the second bit was measured. *Note that entanglement, in the sense that measurement of one particle has an effect on measurements for another particle, is equivalent to our previous definition of entangled states as states that cannot be written as a tensor product for individual states.*

1.2. Logical Gates for Quantum Computer

In classical theory of computing, it is known that any Boolean function $f : \{0,1\}^n \to \{0,1\}^m$ can be expressed in disjunctive normal form using binary operators OR, AND, and unary operator NOT. These operators are sufficient, i.e., they form a *basis* and are called *fundamental* or *universal gates* in classical computation.

Construction of a basis for quantum computing is one of the fundamental problems whose investigation is found in many papers; let us refer to the seminal works [31], [82], [19], [24]. By practical considerations, it is convenient to construct a basis whose elements are of local nature, i.e., operators act on a small amount of neighborly qubits.

Most quantum algorithms involve a sequence of quantum state transformations followed by a measurement. For classical computers, there are sets of gates that are universal in the sense that any classical computation can be performed using a sequence of these gates. Similarly, there are sets of primitive quantum state transformations, called *quantum gates*, that are universal for quantum computation. Given enough quantum bits, it is possible to construct a universal quantum Turing machine.

Quantum physics puts restrictions on the types of transformations that can be done. In particular, all quantum state transformations, and therefore all quantum gates and all quantum computations, must be reversible. Yet, all classical algorithms can be made reversible and can be computed on a quantum computer in comparable time.

The following examples are useful for *single-qubit* (or *one - qubit*) quantum state *transformations*. Because of linearity, the transformations are fully specified by their effect on the basis vectors. The associated matrix, with $\{|0\rangle, |1\rangle\}$ as the preferred ordered basis is also shown:

$$\begin{array}{rlcl} X: & |0\rangle & \to & |1\rangle \\ & |1\rangle & \to & |0\rangle \end{array} \quad \begin{pmatrix} 0 & 1 \\ 1 & 0 \end{pmatrix},$$

$$\begin{array}{rlcl} Y: & |0\rangle & \to & -|1\rangle \\ & |1\rangle & \to & |0\rangle \end{array} \quad \begin{pmatrix} 0 & 1 \\ -1 & 0 \end{pmatrix},$$

where X is *negation*, Z is a *phase shift* operation.

Another important single-bit transformation is the Hadamard transformation defined by

$$\begin{array}{rlcl} H: & |0\rangle & \to & \frac{1}{\sqrt{2}}(|0\rangle + |1\rangle) \\ & |1\rangle & \to & \frac{1}{\sqrt{2}}(|0\rangle - |1\rangle). \end{array}$$

The transformation H has a number of important applications. When applied to $|0\rangle$, H creates a superposition state $\frac{1}{\sqrt{2}}(|0\rangle + |1\rangle)$. Applied to n bits individually, H generates a superposition of all 2^n possible states, which can be viewed as the binary representation of the numbers from 0 to $2^n - 1$.

$$(H \otimes H \otimes \cdots \otimes H)|00\ldots0\rangle =$$

$$= \frac{1}{\sqrt{2^n}}((|0\rangle + |1\rangle) \otimes (|0\rangle + |1\rangle) \otimes \cdots \otimes (|0\rangle + |1\rangle)) = \frac{1}{\sqrt{2^n}} \sum_{x=0}^{2^n-1} |x\rangle.$$

The transformation that applies H to n bits is called the Walsh, or Walsh-Hadamard, transformation W. It can be defined as a recursive decomposition of the form

$$W_1 = H, W_{n+1} = H \otimes W_n.$$

The unitary property implies that quantum states cannot be *copied* or *cloned.*

It means, that there is no unitary operation that can reliably clone unknown quantum states.

Indeed, assume that U is a unitary transformation that clones, in that $U(|a0\rangle) = |aa\rangle$ for all quantum states $|a\rangle$. Let $|a\rangle$ and $|b\rangle$ be two orthogonal quantum states. Let us say $U(|a0\rangle) = |aa\rangle$ and $U(|b0\rangle) = |bb\rangle$. Consider $|c\rangle = (1/\sqrt{2})(|a\rangle + |b\rangle)$. By linearity,

$$U(|c0\rangle) = \frac{1}{\sqrt{2}}(U(|a0\rangle) + U(|b0\rangle)) = \frac{1}{\sqrt{2}}(|aa\rangle + |bb\rangle).$$

But, if U is a cloning transformation then

$$U(|c0\rangle) = |cc\rangle = 1/2(|aa\rangle + |ab\rangle + |ba\rangle + |bb\rangle),$$

which is not equal to $(1/\sqrt{2})(|aa\rangle + |bb\rangle)$. It is clear that cloning is not possible by using measurement since measurement is both probabilistic and destructive for states not in the measuring device's associated subspaces.

It is important to understand what sort of cloning is and isn't allowed. It is possible to clone a known quantum state. What the no cloning principle tells us is that it is impossible to reliably clone an unknown quantum state. Also, it is possible to obtain n particles in an entangled state $a|00\ldots0\rangle + b|11\ldots1\rangle$ from an unknown state $a|0\rangle + b|1\rangle$. Each of these particles will behave in exactly the same way when measured with respect to the standard basis for quantum computation $\{|00\ldots0\rangle, |00\ldots01\rangle, \ldots, |11\ldots1\rangle\}$, but not when measured with respect to other bases. It is not possible to create the n particle state $(a|0\rangle + b|1\rangle) \otimes \cdots \otimes (a|0\rangle + b|1\rangle)$ from an unknown state $a|0\rangle + b|1\rangle$.

For example consider the so-called controlled U-operator, which is defined for every unitary operator $U \in \mathrm{U}(2)$ as

$$\Lambda_k(U)|x_1,\ldots,x_k\rangle \otimes |\xi\rangle = \begin{cases} |x_1,\ldots,x_k\rangle \otimes U|\xi\rangle & \text{if } \bigwedge_{j=1}^k x_j = 1 \\ |x_1,\ldots,x_k\rangle \otimes |\xi\rangle & \text{if } \bigwedge_{j=1}^k x_j = 0 \end{cases}$$

for all $x_1,\ldots,x_k \in \mathbf{B}$. Here, $\bigwedge_{j=1}^k x_j$ denotes action of the operator AND on the Boolean variables $\{0,1\}$. For example, if $k=0$, then $\Lambda_0(\begin{pmatrix} 0 & 1 \\ 1 & 0 \end{pmatrix}) = \mathrm{NOT}_q : \mathcal{H} \to \mathcal{H}$, which is called the *quantum NOT* operator and by definition one has $\sigma_x = \mathrm{CNOT}$, which acts on the state $\alpha|0\rangle + \beta|1\rangle \in \mathcal{H}$ via $\alpha|0\rangle + \beta|1\rangle \mapsto \beta|0\rangle + \alpha|1\rangle$. If the quantum state $\alpha|0\rangle + \beta|1\rangle$ is written in vector form as $(\alpha,\beta)^\top$, then $\mathrm{NOT}_q(\alpha,\beta) = (\beta,\alpha)$. Second important example is the so-called *controlled-not* operator — cNOT: $\mathcal{H} \otimes \mathcal{H} \to \mathcal{H} \otimes \mathcal{H}$, which acts on the pairs of qubits and carries out bitwise summation $\mathrm{cNOT}|u,v\rangle = |u, v \oplus u\rangle$, where $\oplus$ is addition mod

2. Let us consider one more operator — the *controlled controlled NOT* operator ccNOT: $\mathcal{H}\otimes\mathcal{H}\otimes\mathcal{H}\to\mathcal{H}\otimes\mathcal{H}\otimes\mathcal{H}$. The classical one and two-bit operators OR and AND can be expressed via ccNOT as follows:

$$\text{ccNOT}|1,1,a\rangle = |1,1,(\text{NOT } a)\rangle,$$
$$\text{ccNOT}|a,b,0\rangle = |a,b,(a \text{ AND } b)\rangle,$$

for $a,b\in\mathbf{B}$.

Knowing that an arbitrary classical function f with m input and k output bits can be implemented on quantum computer, we assume the existence of a *quantum gate array* U_f that implements f. U_f is a $m+k$ bit transformation of the form $U_f : |x,y\rangle\to|x,y\oplus f(x)\rangle$ where $\oplus$ denotes the bitwise exclusive-OR. Quantum gate arrays U_f, defined in this way, are unitary for any function f. To compute $f(x)$, we apply U_f to $|x\rangle$ tensored with k zeros $|x,0\rangle$. Since $f(x)\oplus f(x)=0$ we have $U_fU_f=I$.

On the other hand, every Boolean function f can be expressed in the basis {NOT, OR, AND} and therefore there exists as the unitary operator

$$U_f : \mathcal{H}\otimes\ldots\otimes\mathcal{H}\to\mathcal{H}\otimes\ldots\otimes\mathcal{H}$$

such that $U_f|x,0\rangle = |x,f(x)\rangle$.

Indeed, since U_f is a linear transformation, it is applied to all basis vectors in the superposition simultaneously and will generate a superposition as the results. In this way, it is possible to compute $f(x)$ for n values of x in a single application of U_f. This effect is called *quantum parallelism*.

Start with an n-qubit state $|00\ldots0\rangle$. Apply the Walsh-Hadamard transformation W to get a superposition

$$\frac{1}{\sqrt{2^n}}(|00\ldots0\rangle+|00\ldots1\rangle+\cdots+|11\ldots1\rangle)=\frac{1}{\sqrt{2^n}}\sum_{x=0}^{2^n-1}|x\rangle$$

which should be viewed as the superposition of all integers $0\le x<2^n$. Add a k-bit register $|0\rangle$ then by linearity

$$U_f(\frac{1}{\sqrt{2^n}}\sum_{x=0}^{2^n-1}|x,0\rangle)=\frac{1}{\sqrt{2^n}}\sum_{x=0}^{2^n-1}U_f(|x,0\rangle)=\frac{1}{\sqrt{2^n}}\sum_{x=0}^{2^n-1}|x,f(x)\rangle$$

where $f(x)$ is the function of interest.

The power of quantum algorithms comes from taking advantage of quantum *parallelism* and *entanglement*. So most quantum algorithms begin by computing a function of interest on a superposition of all values.

Two and three qubit operations, respectively, cNOT and ccNOT, are represented by one qubit operations as follows:

$$\text{cNOT} = |0\rangle\langle 1|\otimes\mathbf{1}+|1\rangle\langle 1|\otimes\text{NOT}$$

and

$$\text{ccNOT} = |0\rangle\langle 0|\otimes\mathbf{1}\otimes\mathbf{1}+|1\rangle\langle 1|\otimes\text{cNOT}.$$

This means that if it is possible to find a finite subset of $U(2)$ which generates a dense subset of $U(2)$, then we obtain all operators for quantum computing.

Theorem 1.2.1. *The gate* cNOT *together with all 1-bit quantum gates is a universal gate set.*

Therefore, it suffices to include the following one-bit transformations

$$\begin{pmatrix} \cos\alpha & \sin\alpha \\ -\sin\alpha & \cos\alpha \end{pmatrix}, \begin{pmatrix} e^{i\alpha} & 0 \\ 0 & e^{-i\alpha} \end{pmatrix}$$

for all $0 \leq \alpha \leq 2\pi$ together with the C_{not} to obtain a universal set of gates.

Note that a quantum model of a physical system, as opposed to the classical one, naturally admits existence of d-level systems, and assertions made below for 2-level systems can be extended with inessential change to d-level systems. In this case, the term "qubit" is replaced by "qudit".

A qudit is a state in the Hilbert space $\mathbb{C}^d$. A n-qudit is a state in the tensor product Hilbert space $(\mathbb{C}^d)^{\otimes n} \cong \mathbb{C}^d \otimes ... \otimes \mathbb{C}^d$. A quantum gate on n-qudits is a unitary operator $U : (\mathbb{C}^d)^{\otimes n} \to (\mathbb{C}^d)^{\otimes n}$, which is an element in the unitary group $\mathbf{U}(d^n)$. A sequence $U_1, U_2, ..., U_k \in \mathbf{U}(d^n)$ of gates constitutes a quantum circuit on n-qudits.

1.3. Construction of Single Qubit Gates

The solutions of the Schrödinger equation represents the evolution of the initial state (input). Here, before treating the Schrödinger equation, the time evolution for the state (the unitary operators) is known as a discrete evolution and is determined by the unitary matrices. Our problem is to find out which type of the Schrödinger equation has evolutionary solutions that are the same as the evolutions given by the quantum gates.

For example, we consider a simply model of construction for the computation gates by coherent oscillations in two-level systems (see [42]).

Let us consider an atom with 2 energy levels E_0 and E_1 ($E_1 > E_0$) which interacts with external periodic field with $g\cos(\omega t + \phi)$. The Hamiltonian in the dipole approximation is given by

$$H = H_0 + g\cos(\omega t + \phi)\sigma_1, \tag{1.3.1}$$

where

$$H_0 = E_0 \mathbf{1}_2 + \begin{pmatrix} 0 & 0 \\ 0 & \Delta \end{pmatrix},$$

where $\Delta = E_1 - E_0$ is an energy difference, ω is the frequency of the external field, g the coupling constant between the external field and the atom.

We use the rotating wave approximation

$$cos(\omega t + \phi) = \frac{1}{2} e^{i(\omega t+\phi)} (1 + e^{-2i(\omega t+\phi)}) \approx \frac{1}{2} e^{i(\omega t+\phi)},$$

from this

$$cos(\omega t + \phi)\sigma_1 \approx \frac{1}{2} \begin{pmatrix} 0 & e^{i(\omega t+\phi)} \\ e^{-i(\omega t+\phi)} & 0 \end{pmatrix},$$

Therefore, the Hamiltonian is given by the redefinition of $g/2 \to g$ by

$$H \;=\; E_0 \mathbf{1}_2 + \frac{\Delta}{2}(\mathbf{1}_2 - \sigma_3) + g\left(e^{i(\omega t+\phi)}\sigma_+ + e^{-i(\omega t+\phi)}\sigma_-\right),$$

or

$$H = E_0 \mathbf{1}_2 + \begin{pmatrix} 0 & ge^{i(\omega t+\phi)} \\ ge^{-i(\omega t+\phi)} & \Delta \end{pmatrix}. \tag{1.3.2}$$

Consider the Schrödinger equation

$$i\frac{d}{dt}\Psi = H\Psi, \tag{1.3.3}$$

We would like to solve this equation. For that, let us decompose H in (1.3.2) into

$$\begin{pmatrix} 0 & ge^{i(\omega t+\phi)} \\ ge^{-i(\omega t+\phi)} & \Delta \end{pmatrix} = \begin{pmatrix} 1 & 0 \\ 0 & e^{-i(\omega t+\phi)} \end{pmatrix}\begin{pmatrix} 0 & g \\ g & \Delta \end{pmatrix}\begin{pmatrix} 1 & 0 \\ 0 & e^{i(\omega t+\phi)} \end{pmatrix}, \tag{1.3.4}$$

so if we set

$$\Phi = e^{itE_0}\begin{pmatrix} 1 & \\ & e^{i(\omega t+\phi)} \end{pmatrix}\Psi \iff \Psi = e^{-itE_0}\begin{pmatrix} 1 & 0 \\ 0 & e^{-i(\omega t+\phi)} \end{pmatrix}\Phi \tag{1.3.5}$$

then it is not difficult to see

$$i\frac{d}{dt}\Phi = \begin{pmatrix} 0 & g \\ g & \Delta-\omega \end{pmatrix}\Phi, \tag{1.3.6}$$

which is easily solved. For simplicity, we set the resonance condition

$$\Delta = \omega, \tag{1.3.7}$$

then the solution of (1.3.6) is

$$\Phi(t) = \exp\left\{-igt\begin{pmatrix} 0 & 1 \\ 1 & 0 \end{pmatrix}\right\}\Phi(0) = \begin{pmatrix} \cos(gt) & -i\sin(gt) \\ -i\sin(gt) & \cos(gt) \end{pmatrix}\Phi(0).$$

As a result, the solution of the equation (1.3.3) is given as

$$\begin{aligned} \Psi(t) &= e^{-itE_0}\begin{pmatrix} 1 & 0 \\ 0 & e^{-i(\omega t+\phi)} \end{pmatrix}\Phi(t) \\ &= e^{-itE_0}\begin{pmatrix} 1 & 0 \\ 0 & e^{-i(\omega t+\phi)} \end{pmatrix}\begin{pmatrix} \cos(gt) & -i\sin(gt) \\ -i\sin(gt) & \cos(gt) \end{pmatrix}\Phi(0) \end{aligned} \tag{1.3.8}$$

by (1.3.5). If we choose $\Phi(0) = (1,\ 0)^T$ as an initial condition, then

$$\Psi(t) = e^{-itE_0}\begin{pmatrix} \cos(gt) \\ -ie^{-i(\omega t+\phi)}\sin(gt) \end{pmatrix}. \tag{1.3.9}$$

This is a well known model for the coherent oscillation.

For the use later, we set

$$U(t,0) = e^{-itE_0}\begin{pmatrix} 1 & 0 \\ 0 & e^{-i(\omega t+\phi)} \end{pmatrix}\begin{pmatrix} \cos(gt) & -i\sin(gt) \\ -i\sin(gt) & \cos(gt) \end{pmatrix} \tag{1.3.10}$$

and

$$U(t_f,t_i) = U(t_f - t_i, 0) \tag{1.3.11}$$

because $\Delta = \omega$ due to the resonance condition (1.3.7). Moreover, for $g = 0$ (no interaction with external field), then we have

$$V(t_f,t_i) = \mathrm{e}^{-i(t_f-t_i)E_0} \begin{pmatrix} 1 & 0 \\ 0 & \mathrm{e}^{-i\Delta(t_f-t_i)} \end{pmatrix}. \tag{1.3.12}$$

We construct single qubit quantum logic gates useful by combining $U(t_f,t_i)$ and $V(t_f,t_i)$ for appropriate $t_i < t_f$ by change a time t and a phase ϕ as free parameters.

First, we choose t_1 as $t_1 = \frac{3\pi}{2\Delta}$ satisfying

$$V(t_1,0) = \mathrm{e}^{-iE_0t_1} \begin{pmatrix} 1 & 0 \\ 0 & \mathrm{e}^{-i\Delta t_1} \end{pmatrix} = \mathrm{e}^{-iE_0t_1} \begin{pmatrix} 1 & 0 \\ 0 & i \end{pmatrix},$$

and next choose $t_2 - t_1$ as $t_2 - t_1 = \frac{\pi}{2g}$ satisfying

$$U(t_2,t_1) = \mathrm{e}^{-iE_0(t_2-t_1)} \times$$

$$\times \begin{pmatrix} 1 & 0 \\ 0 & \mathrm{e}^{-i\{\Delta(t_2-t_1)+\phi\}} \end{pmatrix} \begin{pmatrix} \cos(g(t_2-t_1)) & -i\sin(g(t_2-t_1)) \\ -i\sin(g(t_2-t_1)) & \cos(g(t_2-t_1)) \end{pmatrix} = \tag{1.3.13}$$

$$= \mathrm{e}^{-iE_0(t_2-t_1)} \begin{pmatrix} 1 & 0 \\ 0 & \mathrm{e}^{-i\{\Delta(t_2-t_1)+\phi\}} \end{pmatrix} \begin{pmatrix} 0 & -i \\ -i & 0 \end{pmatrix}.$$

Then

$$U(t_2,t_1)V(t_1,0) = \mathrm{e}^{-iE_0t_2} \begin{pmatrix} 1 & 0 \\ 0 & \mathrm{e}^{-i\{\Delta(t_2-t_1)+\phi\}} \end{pmatrix} \begin{pmatrix} 0 & 1 \\ -i & 0 \end{pmatrix}.$$

Moreover, multiplying $V(t_3,t_2)$ from the left

$$V(t_3,t_2)U(t_2,t_1)V(t_1,0) = \mathrm{e}^{-iE_0t_3} \begin{pmatrix} 1 & \\ & \mathrm{e}^{-i\{\Delta(t_3-t_1)+\phi\}} \end{pmatrix} \begin{pmatrix} 0 & 1 \\ -i & 0 \end{pmatrix}$$

Here, we choose $t_3 = 2k\pi/E_0$ $(> t_2 > t_1)$ as $\mathrm{e}^{-iE_0t_3} = 1$ and the phase ϕ as $\mathrm{e}^{-i\{\Delta(t_3-t_1)+\phi\}} = i$, so that

$$V(t_3,t_2)U(t_2,t_1)V(t_1,0) = \begin{pmatrix} 1 & 0 \\ 0 & i \end{pmatrix} \begin{pmatrix} 0 & 1 \\ -i & 0 \end{pmatrix} = \begin{pmatrix} 0 & 1 \\ 1 & 0 \end{pmatrix} = \sigma_1.$$

Next, we construct σ_θ. We choose t_1 as $t_1 = 2\pi/g$

$$U(t_1,0) = \mathrm{e}^{-iE_0t_1} \begin{pmatrix} 1 & 0 \\ 0 & \mathrm{e}^{-i(\Delta t_1+\phi)} \end{pmatrix} \begin{pmatrix} \cos(gt_1) & -i\sin(gt_1) \\ -i\sin(gt_1) & \cos(gt_1) \end{pmatrix} =$$

$$= \mathrm{e}^{-iE_0t_1} \begin{pmatrix} 1 & 0 \\ 0 & \mathrm{e}^{-i(\Delta t_1+\phi)} \end{pmatrix}.$$

Then multiplying $V(t_2,t_1)$ from the left

$$V(t_2,t_1)U(t_1,0) = \mathrm{e}^{-iE_0t_2}\begin{pmatrix} 1 & 0 \\ 0 & \mathrm{e}^{-i(\Delta t_2+\phi)} \end{pmatrix}.$$

Here, we choose $t_2 = 2k\pi/E_0$ ($> t_1$) for some k and the phase ϕ as $e^{-i(\Delta t_2+\phi)} = e^{i\theta}$ for any θ, so that

$$V(t_2,t_1)U(t_1,0) = \begin{pmatrix} 1 & 0 \\ 0 & \mathrm{e}^{i\theta} \end{pmatrix} = \sigma_\theta.$$

In particular, we obtain

$$\begin{pmatrix} 1 & 0 \\ 0 & i \end{pmatrix}, \quad \sigma_3 = \begin{pmatrix} 1 & 0 \\ 0 & -1 \end{pmatrix}, \quad \begin{pmatrix} 1 & 0 \\ 0 & -i \end{pmatrix}$$

for $\theta = \pi/2,\ \pi,\ 3\pi/2$, respectively.

We construct the Walsh–Hadamard matrix. For $V(t_1,0)$, $V(t_3,t_2)$ with $t_1 = 3\pi/2\Delta$ and $t_3 - t_2 = 3\pi/2\Delta$

$$V(t_1,0) = \mathrm{e}^{-iE_0t_1}\begin{pmatrix} 1 & 0 \\ 0 & i \end{pmatrix}, \quad V(t_3,t_2) = \mathrm{e}^{-iE_0(t_3-t_2)}\begin{pmatrix} 1 & 0 \\ 0 & i \end{pmatrix}$$

and $U(t_2,t_1)$ with $t_2 - t_1 = \pi/4g$

$$U(t_2,t_1) = \mathrm{e}^{-iE_0(t_2-t_1)}\begin{pmatrix} 1 & 0 \\ 0 & \mathrm{e}^{-i\{\Delta(t_2-t_1)+\phi\}} \end{pmatrix}\begin{pmatrix} \frac{1}{\sqrt{2}} & \frac{-i}{\sqrt{2}} \\ \frac{-i}{\sqrt{2}} & \frac{1}{\sqrt{2}} \end{pmatrix},$$

we have

$$V(t_3,t_2)U(t_2,t_1)V(t_1,0) = \mathrm{e}^{-iE_0t_3}\begin{pmatrix} 1 & 0 \\ 0 & \mathrm{e}^{-i\{\Delta(t_2-t_1)+\phi\}} \end{pmatrix}\begin{pmatrix} \frac{1}{\sqrt{2}} & \frac{1}{\sqrt{2}} \\ \frac{1}{\sqrt{2}} & -\frac{1}{\sqrt{2}} \end{pmatrix}.$$

We cannot remove the phase $\mathrm{e}^{-iE_0t_3}$, so we multiply this by $V(t_4,t_3)$ to obtain

$$V(t_4,t_3)V(t_3,t_2)U(t_2,t_1)V(t_1,0) = \mathrm{e}^{-iE_0t_4}\begin{pmatrix} 1 & 0 \\ 0 & \mathrm{e}^{-i\{\Delta(t_4-t_3+t_2-t_1)+\phi\}} \end{pmatrix}\begin{pmatrix} \frac{1}{\sqrt{2}} & \frac{1}{\sqrt{2}} \\ \frac{1}{\sqrt{2}} & -\frac{1}{\sqrt{2}} \end{pmatrix}.$$

By choosing $t_4 = 2k\pi/E_0$ for some k and the phase ϕ as $e^{-i\{\Delta(t_4-t_3+t_2-t_1)+\phi\}} = 1$, we finally obtain

$$V(t_4,t_3)V(t_3,t_2)U(t_2,t_1)V(t_1,0) = \frac{1}{\sqrt{2}}\begin{pmatrix} 1 & 1 \\ 1 & -1 \end{pmatrix} = H.$$

Other construction examples for the universal set of quantum gates will be considered in subsequent sections.

1.4. The Entangled Operator

Construction by single qubit gates for a two-bit gate is not difficult. If it is known that for two quantum systems with two-dimensional state space Hamiltonians $\mathcal{H}_1$, H_1, and $\mathcal{H}_2$, H_2, then $H_1 \otimes 1_2 + 1_1 \otimes H_2$ is the Hamiltonian of this joint system $\mathcal{H}_1 \otimes \mathcal{H}_2$ and the corresponding evolution operators acts on the states in the form $u \otimes v$, where $u \in \mathcal{H}_1$ and $v \in \mathcal{H}_2$ as $e^{itH_1} \otimes e^{itH_2}$.

Consider the two qubit state

$$|\phi>= \frac{|00> + |11>}{\sqrt{2}}.$$

This state has the property that these are not single qubit states $|a>$ and $|b>$, such that $|\phi>= |a> |b>$.

Definition 1.4.1. *A state with a composite system cannot be written as a product of states for its component system called entangled state.*

Let $|00>, |01>, |10>, |11>$ the standard basis of $\mathbb{C} \otimes \mathbb{C}$. Consider the following entangled states of $\mathbb{C} \otimes \mathbb{C}$:

$$|00> \rightarrow \frac{|00> + |11>}{\sqrt{2}}, \; |01> \rightarrow \frac{|00> - |11>}{\sqrt{2}},$$

$$|10> \rightarrow \frac{|10> + |01>}{\sqrt{2}}, \; |11> \rightarrow \frac{|01> - |10>}{\sqrt{2}}.$$

These states are known as the *Bell basis* (or Bell states or EPR pairs).

Proposition 1.4.1. *Bell basis is a orthonormal basis for the two qubit state space.*

One qubit gate has elements of $SU(2)$ and act on $\mathbb{C}^2$. For realization of quantum algorithms one needs all operators from $SU(2^n)$. This problem can be reduced to finding such elements M_i of $SU(4)$ which together with $SU(2) \otimes SU(2)$ generate the whole $SU(4)$.

Definition 1.4.2. *[24] A two qubit gate A is a universal for the quantum computation, if A, together with local unitary transformations (i.e., transformation of one qubit states) generates all unitary transformations for the complex vector space of dimension 2^n to itself.*

For example, this is known as a controlled NOT

$$CNOT = \begin{pmatrix} 1 & 0 & 0 & 0 \\ 0 & 1 & 0 & 0 \\ 0 & 0 & 0 & 1 \\ 0 & 0 & 1 & 0 \end{pmatrix}$$

and is the universal gate.

Definition 1.4.3. *A gate A is said to be entangling, if there is a vector $|ab>= |a> \otimes |b> \in \mathbb{C}^2 \otimes \mathbb{C}^2$ such that $A|ab>$ is not decomposable as a tensor product of two qubits.*

Theorem 1.4.1. *[24] Two qubit gate A is a universal, if and only if it is entangling.*

Next, we give an example following Kauffman [74], a topological description of the entangling operator.

Let

$$R : V \otimes V \to V \otimes V$$

linear map. If R satisfies braided Yang-Baxter equation

$$(R \otimes \mathbb{I})(\mathbb{I} \otimes R)(R \otimes \mathbb{I}) = (\mathbb{I} \otimes R)(R \otimes \mathbb{I})(\mathbb{I} \otimes R),$$

then R called *braiding*.

Theorem 1.4.2. *[34] All 4×4 unitary solutions to the Yang-Baxter equation are similar to one of the following matrix types:*

$$R1 = \begin{pmatrix} \frac{1}{\sqrt{2}} & 0 & 0 & \frac{1}{\sqrt{2}} \\ 0 & \frac{1}{\sqrt{2}} & -\frac{1}{\sqrt{2}} & 0 \\ 0 & \frac{1}{\sqrt{2}} & \frac{1}{\sqrt{2}} & 0 \\ -\frac{1}{\sqrt{2}} & 0 & 0 & \frac{1}{\sqrt{2}} \end{pmatrix}, R2 = \begin{pmatrix} a & 0 & 0 & 0 \\ 0 & 0 & b & 0 \\ 0 & c & 0 & 0 \\ 0 & 0 & 0 & d \end{pmatrix}$$

and

$$R3 = \begin{pmatrix} 0 & 0 & 0 & a \\ 0 & b & 0 & 0 \\ 0 & 0 & c & 0 \\ d & 0 & 0 & 0 \end{pmatrix},$$

where a, b, c, d are unit complex numbers.

Operator $R1$ acts on the standard basis of $\mathbb{C}^2 \otimes \mathbb{C}^2$ and transforms it into the Bell basis. As for the operators $R2$ and $R3$, these are described in the following proposition.

Proposition 1.4.2. *[74] The braidings $R2$ and $R3$ are universal gates when $ad - bc \neq 0$ for certain internal parameters a, b, c and d.*

From this proposition, it follows that not every braiding is universal. For example, if $a = b = c = d = 1$, we obtain a so-called *SWAP* operator, which is not universal. By the *SWAP* operator, when by the definition act on the tensor product $u \otimes v \in V \otimes V$; for the following manner $SWAP(u \otimes v) = v \otimes u$, we define the operator $R_{13} : V \otimes V \otimes V \to V \otimes V \otimes V$ as $R_{13} = (\mathbb{I} \otimes SWAP)(R \otimes \mathbb{I})(\mathbb{I} \otimes SWAP)$. Denoted by $R_{12} = R \otimes \mathbb{I}$ and $R_{23} = \mathbb{I} \otimes R$.

Definition 1.4.4. *The equation (respect to R)*

$$R_{12}R_{13}R_{23} = R_{23}R_{13}R_{12}$$

called algebraic Yang-Baxter equation.

Proposition 1.4.3. *R is a solution for the algebraic Yang-Baxter equation iff $R \circ SWAP$ is solution for the braided Yang-Baxter equation.*

From these results, it follows a simple proof of universality called *phase gate*

$$D = \begin{pmatrix} 1 & 0 & 0 & 0 \\ 0 & 1 & 0 & 0 \\ 0 & 0 & 1 & 0 \\ 0 & 0 & 0 & -1 \end{pmatrix}.$$

Indeed, if D is a solution for the algebraic Yang-Baxter equation, then $D \circ SWAP$ is the solution for the braided Yang-Baxter equation and $ad - bc = -2 \neq 0$; and, is therefore a universal gate.

Kauffman [74] investigated the relation between the topological and quantum entangling to prove that the *unitary solution of Yang-Baxter equation detect topological linking iff the gates corresponding to these solutions can entangle quantum states.*

Let the braid group representation b-matrix and the quantum Yang-Baxter equation solution $\check{R}$-matrix are $n^2 \times n^2$ matrices acting on $V \otimes V$ where V is an n-dimensional vector space. If b or $\check{R}$ act on the tensor product $V_i \otimes V_{i+1}$, we denote them by b_i and $\check{R}_i$, respectively.

It is known, that the braid group representation b-matrix has to satisfy the braid relation

$$b_i b_{i+1} b_i = b_{i+1} b_i b_{i+1}, \tag{1.4.1}$$

and quantum Yang-Baxter equation has the form

$$\check{R}_i(x) \check{R}_{i+1}(xy) \check{R}_i(y) = \check{R}_{i+1}(y) \check{R}_i(xy) \check{R}_{i+1}(x) \tag{1.4.2}$$

with the asymptotic condition $\check{R}(x = 0) = b$, and x called the *spectral parameter.* From these two equations both b and $\check{R}(x)$ are fixed up to an overall scalar factor.

In terms of the permutation operator P specified by $P(\xi \otimes \eta) = \eta \otimes \xi$ and the $\check{R}(x)$-matrix, the solution of the algebraic quantum Yang-Baxter reading

$$R_{12}(x) R_{13}(xy) R_{23}(y) = R_{23}(y) R_{13}(xy) R_{12}(x)$$

takes the form $R(x) = P\check{R}(x)$, where R_{ij} is an operator acting on the tensor product $V_i \otimes V_j$.

As a solution for the quantum Yang-Baxter equation (1.4.2), the $\check{R}(x)$-matrix usually depends on a deformation parameter q and the spectral parameter x. From this, there exists two approaches to solve this quantum Yang-Baxter equation:

1. The limit as $q \to 1$ led to the classical r-matrix to satisfy the classical Yang–Baxter equation.

2. The limit as $x \to 0$ led to the braid relation (1.4.1) from the quantum Yang-Baxter equation (1.4.2) and the braid group representation b-matrix from the $\check{R}(x)$-matrix.

To construct the $\check{R}(x)$-matrix from a given braid group representation b-matrix is called Yang–Baxterization.

We consider the case when $n = 4$. Let

1. Let braid group representation b-matrix have two distinct non-vanishing eigenvalues λ_1 and λ_2, then b has the form

$$b = \lambda_1 P_1 + \lambda_2 P_2 \tag{1.4.3}$$

where P_1 and P_2 are the projection matrices satisfying

$$P_1 + P_2 = 1\!\!1, \qquad P_1^2 = P_1,\ P_2^2 = P_2,\ P_1 P_2 = 0. \tag{1.4.4}$$

By the help of Yang–Baxterization, the corresponding $\check{R}(x)$-matrix modulo an overall scalar factor has the form

$$\check{R}(x) = (\lambda_1 + \lambda_2 x) P_1 + (\lambda_2 + \lambda_1 x) P_2 = b + \lambda_1 \lambda_2 x b^{-1} \tag{1.4.5}$$

where the inverse matrix b^{-1} is given by

$$b^{-1} = \frac{1}{\lambda_1} P_1 + \frac{1}{\lambda_2} P_2.$$

2. In the case that the braid group representation b-matrix has three distinct non-vanishing eigenvalues λ_1, λ_2, and λ_3, Yang–Baxterization gives to the following formula

$$\check{R}(x) = \lambda_1 \lambda_3 x (x-1) b^{-1} + (\lambda_1 + \lambda_2 + \lambda_3 + \lambda_1 \lambda_3 \lambda_2^{-1}) x 1\!\!1 - (x-1) b. \tag{1.4.6}$$

Changing the ordering of these three different eigenvalues, in principle, gives us different solutions for the quantum Yang-Baxter equation (1.4.2). However, this formula is symmetric with respect to interchanging λ_1 and λ_3, so usually we obtain three types of the $\check{R}(x)$-matrices.

In two formulas (1.4.5) and (1.4.6), the $\check{R}(x)$-matrix is proportional to the unit matrix $1\!\!1$ at $x = 1$, namely,

$$\check{R}(x = 1) \propto 1\!\!1 \tag{1.4.7}$$

except the case of $\lambda_1 + \lambda_2 + \lambda_3 + \lambda_1 \lambda_3 \lambda_2^{-1} = 0$.

Following [74], we present a method of constructing the Hamiltonian from the unitary $\check{R}(x)$-matrix ($\check{R}(\theta)$-matrix) for the six-vertex and eight-vertex model.

The wave function $\psi(x)$ is specified by the unitary $\check{R}(x)$-matrix, with $\psi(x) = \check{R}(x)\psi$, the pure state ψ independent of the time (or the spectral parameter x). Hence, we obtain the Shrödinger equation corresponding to the time evolution of $\psi(x)$ controlled by the unitary $\check{R}(x)$-matrix

$$\frac{\partial \psi(x)}{\partial x} = H(x)\psi(x), \qquad H(x) = i \frac{\partial \check{R}(x)}{\partial x} \check{R}^{-1}(x). \tag{1.4.8}$$

Here, the unitary $\check{R}(x)$-matrix has to be of the form containing the normalization factor, namely $\rho^{-\frac{1}{2}} \check{R}(x)$ satisfying $(\rho^{-\frac{1}{2}} \check{R})^{-1} = (\rho^{-\frac{1}{2}} \check{R})^{\dagger}$. The "time-dependent" Hamiltonian $H(x)$ is obtained to be

$$H(x) = i \frac{\partial (\rho^{-\frac{1}{2}} \check{R})}{\partial x} (\rho^{-\frac{1}{2}} \check{R})^{-1}(x) = i \frac{\partial (\rho^{-\frac{1}{2}} \check{R})}{\partial x} \rho^{-\frac{1}{2}} \check{R}^{\dagger}(x). \tag{1.4.9}$$

It is expanded into the form

$$H(x) = i(\partial_x \ln \rho^{-\frac{1}{2}}(x) + \partial_x \check{R}(x) \check{R}^{-1}(x)) =$$

$$= i\rho^{-1}(x)(-\frac{1}{2}\frac{\partial \rho(x)}{\partial x} + \frac{\partial \check{R}(x)}{\partial x} \check{R}^{\dagger}(x)),$$

where the formula in terms of $\check{R}^{\dagger}(x)$ is applied in the following since the calculation of $\check{R}^{\dagger}(x)$ is easier than that of $\check{R}^{-1}(x)$.

We take the spectral parameter $\|x\| = 1$, namely $x = e^{i\theta}$ so the Hamiltonian $H(\theta)$ has another form

$$H(\theta) = i\rho^{-1}(\theta)(-\frac{1}{2}\frac{\partial\rho(\theta)}{\partial\theta} + \frac{\partial\check{R}(\theta)}{\partial\theta}\check{R}^{\dagger}(\theta)) \tag{1.4.10}$$

which gives the following Schrödinger equation

$$\frac{\partial\psi(\theta)}{\partial\theta} = H(\theta)\psi(\theta). \tag{1.4.11}$$

It is known, that

$$|\phi >= a|00> +b|01> +c|10> +d|11>$$

is an entangled state, if and only if $ad - bc \neq 0$. The following proposition gives way to construct entangled operators.

Proposition 1.4.4. *For any two-qubit gate U, we have the following canonical decomposition*

$$U = (A_1 \otimes B_1)e^{i(\theta_x X\otimes X + \theta_y Y\otimes Y + \theta_z Z\otimes Z)}(A_2 \otimes B_2),$$

where X,Y,Z are the Pauli matrices, $A_j, B_j, j = 1,2$ are one qubit gates and $-\frac{\pi}{4} < \theta_x, \theta_y, \theta_z \leq \frac{\pi}{4}$.

For the proof of this result see [90].

1.5. Geometric Character for Quantum Computing

First, we consider *geometric* or *Berry phase*, as studied by M. Berry in [15]. (For review on geometric phase see the book by [130].)

A phase is, for our purposes, not a state of matter but a complex number of unit modules, an element of the group $U(1)$ or more general elements of matrix group $U(N)$. The phase is often obtained by cyclic evolution in a physical system. Cyclic variation of external parameters often provides a net evolution involving a phase that depends only on the geometry of the path traversed in parameter space. The natural mathematical context for geometric phases in the theory of $U(N)$ fiber bundles. Here, one defines a phase, known as holonomy, that depends on the geometry of a loop, and is independent of any coordinate choice.

Examples of geometric phases abound in many areas of physics: in optics, Pancharatnm's phase led to measurable interference effects; in molecular physics, the molecular Aharonov-Bohm effect was found to have a significant effect on molecular dynamics as to degenerate the universal significance of these phases, to show that whenever an adiabatic approximation applies, we may expect to find a geometric phase [158], [130]. The crucial role of adiabatically is to make sure that the cyclic variable of parameters does not lead to cyclic evolution. In [151], F. Wilczeck and A. Zee considers non-Abelian phases associated with adiabatic evolution of degenerate Hamiltonians. Presence of the geometric phase in quantum systems and its properties have been used by P. Zanardi and M. Rasetti [156] to construct a new model for quantum computing, which they called the holonomic quantum computation. Subsequently, quite a number of papers have been devoted to the holonomic quantum computing (see for example [158], [117], [42], [35] and references there in).

1.5.1. Adiabatic Theorem

Let M be a smooth manifold, which in our case plays the role of parameter space. Let a family of Hamiltonians H_λ be given, depending continuously on parameters $\lambda \in M$, all of which have a set of n degenerate levels. By a simple renormalization of these energies, we can suppose that these levels are at $E = 0$.

We use the adiabatic theorem:

Theorem 1.5.1. *[151] If the parameters are slowly varied from an initial value λ_0 to some final value λ_1 over long time interval T, and the given space of degenerate levels does not cross other levels. Then, these solutions for*

$$H_{\lambda_0}\Psi(t) = 0 \tag{1.5.1}$$

are mapped onto solutions of

$$H_{\lambda_1}\Psi(t) = 0 \tag{1.5.2}$$

by solving the time-dependent Schrödinder equation

$$i\frac{\partial\Psi(t)}{\partial t} = H(\lambda(t))\Psi(t)$$

with the boundary conditions

$$\lambda(0) = \lambda_0, \lambda(T) = \lambda_1.$$

The essence of this theorem is the fact that, if one can find eigenvalues for the parameter-dependent Hamiltonian H_λ as a value λ_0 parameter, then it is also possible to find its eigenvalues for the value λ_1 parameter.

Quantum computation has been considered from a complexity-theoretic viewpoint by E. Bernstein and U. Vazirani in the famous work by [16]. Bernstein and Vazirani were able to show the existence of an efficient universal quantum Turing machine in D. Deutsch's model of a quantum Turing machine. In this paper, the first formal evidence is given as to why quantum Turing machines violate the modern (complexity-theoretic) formulation of the Church-Turing thesis. Bernstein and Vazirani showed the existence of a problem, relative to an oracle, that can be solved in polynomial time on a quantum Turing machine, but requires superpolynomial time on a bounded-error probabilistic Turing machine, and thus is not in the class BPP. The languages in class BQP are efficiently decidable (with small error probability) on a quantum Turing machine satisfies $BPP \subseteq BQP \subseteq P^{\sharp P}$. Therefore, *there is no possibility of giving a mathematical proof that quantum Turing machines are more powerful than classical probabilistic Turing machines* (in the unrelativized setting) *unless there is a major breakthrough in complexity theory.*

Recently, in papers by [77], [78] Kieu, using the adiabatic theorem, argues that quantum Turing machines may be used to solve Hilbert's *tenth problem*, which requires solving *Diophantine equations*.

Hilbert's tenth problem. A Diophantine equation is an equation of the form $p(x_1, \ldots x_n) = 0$ where p is a polynomial with integer coefficients. Hilbert's tenth problem is as follows:

Given a Diophantine equation with any number of unknown quantities and with rational integral numerical coefficients: to devise a process according to which it can be determined in a finite number of operations whether the equation is solvable in rational integers.

Yu. Matijasevich [102] shows that every recursively enumerable relation is diophantine and so completes the solution of Hilbert's tenth problem in the negative.

Since solving Diophantine equations is known to be uncomputable, Kieu's claim would imply that more is computable quantumly than classically. In general, this contradicts the established results of E. Bernstein and U.V. Vazirani [16]. Assuming the correctness of these results, the model assumed by Kieu extends the standard model of quantum Turing machines.

Inspired by the connection with operators acting on some infinite-dimensional Hilbert space, Kieu reformulated Hilbert's tenth problem in the domain of integer arithmetic into a problem involving either a set of infinitely coupled nonlinear differential equations or a Schrödinger propagator with some appropriate kernel. From this, Kieu first derived a set of nonlinearly coupled differential equations, then also a linear Schrödinger equation, each of which separately contains the sought-after decision result. If one could find a universal procedure to derive certain information from these differential equations (not necessarily by solving them explicitly but possibly by other means), then one would have settled Hilbert's tenth problem in the affirmative in the quantum setting.

1.5.2. Holonomy and Monodromy Operators

Suppose that the Hamiltonian of the system depends on an external parameter τ, whose cyclic adiabatic change induces appearance of an additional phase factor, the evolution operator. This phase factor depends on the geometry of the parameter space. After this work by M. Berry [15], it became clear that the operator $e^{-iH_\tau t}$ can be represented as a product of operators $e^{-iH^1_\tau} e^{-iH^2 t}$, where the first factor is exactly the holonomy operator or the geometric phase. This holonomy operator is defined on the total space of a bundle over the parameter space as base. As already known, the adequate mathematical formalism for description of the geometric phase is a bundle with connection. As noted, holonomic quantum computation is based on a smooth bundle $E \to X$ with connection Ω. Quantum gates for computing are obtained from the holonomy operator $\Gamma_\Omega(\gamma)$ acting on the fiber F of the given bundle, which is an n-dimensional vector bundle. When the connection is fixed within one gauge equivalence class, to construct the needed gate one has to choose the integration path γ in an appropriate way.

Next, we give mathematical theory of gauge computation.

Let E be a fiber bundle with base space X and fiber F. Let G be a group and suppose we have left action of G on F and let the group homomorphism. Suppose the transition function $h_{ij} : U_j \cap U_i \to Aut(F)$ belong in $\rho(G)$ for all $x \in U_j \cap U_i$ and all i, j with $U_j \cap U_i \neq \emptyset$, then we say that G is the structure group of the bundle. Consider a fiber bundle $\pi : E \to M$ with fiber F. At each point $x \in E$ there is a canonical subspace $V_x \subset T_x E$ of the tangent space at x to E called the vertical subspace consisting of all vectors v such that $d\pi v$. These are vectors tangent to fiber $F_{\pi(x)}$, which is a submanifold. A connection on E we mean a smooth choice at each $x \in E$ of a subspace complementary to V_xE. One has $T_{(x,y)}(U \times F) \cong T_xU \times T_vF$. Let $(\xi, y) \in T_xU \times T_vF$. Such a vector is vertical if and only if $\xi = 0$. Thus π^{hor}_{ver} must have the form $\pi^{hor}_{ver}(\xi, y) = (0, \Gamma(x, v)\xi)$, where $\Gamma(x, v)$ is linear map $T_xU \to T_vF$. From this follows that $\pi^{hor}_{ver}(\xi, y) = (\xi, -\Gamma(x, v)\xi)$. The differential of this map must interwrite the local representatives of π^{hor}, the projection maps on the vertical tangent subspaces, specifically, $d\Psi_{(x,y)} \circ \pi^{ver}_{(x,y)} = \pi^{ver}_{(x,y)} \circ d\Psi_{(x,y)}$. One has

$$dh_{VU}(x)(v)(x,y) = (\xi, d_1h_{VU}(x)(v)\xi + d_2h_{VU}(x)(v)y)$$

The intertwining relation with π^{hor} gives us

$$d_1h_{VU}(x)(v) + \Gamma_V(x, h_{VU}(x)(v)) = d_2h_{VU}(x)(v)\Gamma_U(x,v). \tag{1.5.3}$$

Theorem 1.5.2. *Let $\pi : E \to X$ be a bundle with fiber F. Suppose we have a set of trivializations $h_U : \pi^{-1}(U) \to U \times F$ such that the open sets U cover X. Suppose that for each trivialization and each we have linear maps $\Gamma_U(x,y) : T_xU \to T_vF$ varying smoothly with respect to (x, v) and such that relation (1.5.3) is satisfied for each pair of trivializations such that $U_i \cap U_j \neq \emptyset$. Then there is a unique connection of the bundle for which the representative with respect to the given trivializations are the given $\Gamma_U(x, v)$.*

In a locally trivialized vector bundle using the mentioned identification $T_{(x,y)}(U \times F) \cong T_xU \times T_vF \cong T_xU \times F$ for a connection to be linear one must have $\Gamma_U(x, v)\xi = \Gamma(x)(\xi)v$ where $\Gamma(x)$ can now be interpreted as a linear map $T_xU \to End(F)$, in other words an $End(F)$-value 1-form on U.

Let P be a principal G-bundle over a base space X. A gauge transformation is a bundle isomorphism $\varphi : P \to G$ that commutes with right action. For this commute with right action it is necessary and sufficient that $(pg)\gamma(pg) = (p\gamma(p))g$ and so

$$\gamma(pg) = g^{-1}\gamma(p)g. \tag{1.5.4}$$

Thus, the set of all gauge transformations is the set of all maps $\gamma : P \to G$ satisfying (1.5.4). The set of all gauge transformations of P is called the *gauge group* of P.

Consider a connection on $E \to X$ with vertical tangent space projectors π^{ver}_x. A smooth curve in E is said horizontal if at each points its tangent lies in the horizontal tangent space of that point. To be horizontal is to be an integral curve of a differential equation. In fact consider a parametrized smooth curve in E and consider its image $u(t) = (x(t), v(t))$ in a trivialization $U \times F$. This then has $\pi^{ver}(u'(t)) = (0, v'(t)) + \Gamma(x(t), v(t)x'(t))$ and so the condition for being horizontal is

$$v'(t) + \Gamma(x(t), v(t))x'(t) = 0. \tag{1.5.5}$$

This is local coordinates in an ordinary differential equation for $u(t)$. Note that only the components $v(t)$ are required to obey a differential equation and that $x(t)$ can be given with arbitrary parametrization. This allows us to determine $v(t)$ from $x(t)$. Given a smooth curve σ in X a horizontal lifting of σ is a horizontal curve $\widetilde{\sigma}$ in E such that $\pi(\widetilde{\sigma})$. In a trivialization this means that once σ is parametrized, then $\widetilde{\sigma}$ satisfies inheriting a parametrization from that of σ. By the existence, uniqueness and regularity theorems for solutions of ordinary differential equations, any smooth curve in X has at least a local unique horizontal lifting passing through any point $v \in F_x$ for any $x\sigma$.

Let σ smooth curve in X with initial point x_0 and end point x_1. Let $v_0 \in F_{x_0}$ and assume that there as a global horizontal lifting of σ with initial point v_0. The endpoint $v_1 \in F_{x_0}$ of σ is called the parallel transport of v_0 along σ. It is obviously unique if it exists. If the parallel transport exists for all $v_0 \in F_{x_0}$ then the map $v_0 \to v_1$ defines a diffeomorphism $F_{x_0} \to F_{x_1}$ is a linear isomorphism.

It is useful to have explicit forms for the parallel transport map. This is given by a construction known as time-ordered exponential integrals. Let F be a finite dimensional vector space and consider the following non-autonomous differential equation in F:

$$\frac{v(t)}{dt} = A(t)v(t) \tag{1.5.6}$$

where $A(t)$ is a linear operator which is a C^∞-function of t. By the existence, uniqueness and regularity theorems for solutions of ordinary differential equations, for any $v \in F$, there is a unique solution $v(t)$ with $v(a)$ given. In differential equation theory, one generally introduces what is known as the fundamental solution of (1.5.6), that is, an $End(F)$ valued function which satisfies

$$\frac{\partial}{\partial t}\Phi(t,a) = A(t)\Phi(t,a),$$
$$\Phi(a,a) = 1.$$

One now has

$$v(t) = \Phi(t,a)v(t).$$

Another representation for the $v(t)$ is:

$$v(t) = Texp(\int_a^t A(s)ds)v(a)$$

where called time-ordered exponential integral.

Let now A be an $End(F)$-valued 1-form on a manifold X and let σ be a smooth oriented curve in X. One can also define the path-ordered exponential integral

$$Pexp(\int_\sigma A) = lim_{N\to\infty} P\prod_{i=1}^{N} exp(\int_{\sigma_i} A)$$

where the curve σ has been partitioned into successive arcs each one inheriting its orientation from σ. Then limit is to be understood in relation to a fixed parametrization of σ with the maximum parameter length of the σ_I tending to zero. Such a path-ordered exponential integral is an element of $End(F)$. We can use the new path-ordered exponential integrals to express the effect of parallel transport. In a trivialization of a vector bundle the parallel transport equation reads

$$\frac{dv(t)}{dt} = \Gamma(x(t))(x'(t))v,$$

where $x(t)$ is a parametrized path σ in U with initial point $x(0)$ and final point $x(1)$. Thus, we have

$$V(1) = Texp(-\int_0^1 \Gamma(x(t)))(x'(t))dt)v(0) = Pexp(\Gamma_\sigma)v(0)$$

and so $Pexp(\Gamma_\sigma \Gamma)v(0)$ is the parallel transport operator for the oriented curve σ.

Let G be a Lie group with Lie algebra $\mathfrak{g}$ and A a $\mathfrak{g}$-value 1-form on an open set U in a manifold M. Such a path exponential integral solves the parallel transport equation for a principal G-bundle with an invariant connection. In fact, consider a trivialization $U \times G$ of such a bundle and a parametrized curve $p(t) = (x(t), g(t))$ in it. The tangent vector at $p(t)$ is $(x'(t), g'(t)) \in T_{x(t)}U \times T_{g(t)}G$. With $T_{g(t)}G$ identified with $\mathfrak{g}$ by right action we represent this tangent vector now by $(x'(t), g'(t)g(t)^{-1}) \in T_{x(t)}U \times \mathfrak{g} \cdot A$ vector is horizontal if and only if $g(t)^{-1}(A(x')) + g'(t)g(t)^{-1}) = 0$ that is, if and only if $A(x'(t)) + g'(t)g(t)^{-1} = 0$. Thus, the parallel transport equation in a principal G-bundle with an invariant connection is

$$\frac{dg(t)}{dt} = A(x(t))(x'(t))g(t).$$

It is now easy to see that

$$g(t) = Texp(-\int_0^t A(x(s))x'(s))dt)g(0).$$

The effect of parallel transporting a group element over an oriented curve σ in U is to multiply it on the left by $Pexp(-\int_\sigma A)$.

This result shows that parallel transport is globally defined as the path-oriented exponential integral exists for any compact oriented curve with end points.

If a physical system or theory is described by the principal bundle $P \longrightarrow M$, with structural group G, then we say that it is a *gauge theory*. The group G is called a gauge group. The choice of a trivialization of the principal bundle is called *gauge*. Analog classical elementary field in gauge theory, which is a connection on the principal bundle called *gauge field* or *vector potential. Gauge transformation (change of gauge)* is a automorphism of the principal bundle which covers the identity of the base manifold. Maxell's electro-magnetism $U(1)$-gauge theory) and Yang-Mills theory are examples correspondingly abelian and nonabelian gauge theories.

Theorem 1.5.3. *[156] Any computation code can be realized by connection* Γ.

Let us now give a simplified quantum-mechanical interpretation for this holonomy operator.

Suppose $\{|\phi_j>\}$ is a set of eigenstates for the Hamiltonian H, i. e.,

$$H|\phi_j>= E_j|\phi_j> .$$

From this, any state has form $|\psi(0)>= \sum_j a_j|\phi_j>$ and the state $|\psi(t)>$ satisfies the Shrödinger equation

$$-iH|\psi(t)>= \frac{\partial}{\partial t}|\psi(t)>,$$

from this $|\phi(t)>= e^{iHt}|\psi(0)$ and therefore

$$|\psi(t)>= \sum_j a_j e^{iE_j t}|\phi_j> .$$

The term $e^{iE_j t}$ is called *dynamical phase*. In case $H = H(t)$, we have

$$|\psi(t)>= \sum_j a_j(t) T e^{-i\int_0^\tau H(\tau)d\tau}|\phi_j(t)>, \tag{1.5.7}$$

where $\{|\phi_j(t)>\}$ are a basis of eigenstates of $H(t)$ and T denotes the time ordering as before.

Suppose that the Hamiltonian H depends on the parameter $\lambda \in M$, where M is an n-dimensional manifold; such Hamiltonian is called parametric Hamiltonian. Using the parametric Hamiltonian identity (1.5.7) becomes:

$$|\phi(t)>= \sum_j a_j(t) T e^{-i\int_0^\tau H(\lambda(\tau))d\tau}|\phi_j(\lambda(t))> .$$

Using the adiabatic theorem 4 with $|\psi(0) = |\phi_j(0)>$,, we have that

$$|\psi(t) = e^{-i\gamma_j(t)} e^{-i\int_0^\tau E_j(\tau)d\tau}|\phi_j(t)>,$$

where $e^{-i\gamma_j}$ is obtained from adiabatic evolution.

If we substitute this into the Schödinger equation, we have

$$i\frac{\partial\gamma_j}{\partial t}|\phi_j(t)>= \frac{\partial}{\partial t}|\phi_j(t)>,$$

from this

$$\frac{\partial\gamma_j}{\partial t} = -i < \phi_j(t)|\frac{\partial}{\partial t}|\phi_j(t)> .$$

If we integrate this identity over a closed path in the parameter space M, we obtain the Berry's phase γ_j as holonomy:

$$\gamma_j = -i\int < \phi_j(t)|\frac{\partial}{\partial t}|\phi_j(t)> dt = -i\int_0^\tau < \phi_j(t)|\frac{\partial}{\partial t}|\phi_j(t)> dt$$

where τ is defined in such a way that $H(\lambda(t)) = H(\lambda(0))$.

In general, let $E \to X$ be a vector bundle with connection ω and let γ be a loop in X at point x_0: $\gamma : [0,1] \to X, \gamma(0) = \gamma(1)$. For this γ, one has a *holonomy* operator $\Gamma_\omega(\gamma) = Pexp(-\int_\gamma \omega)$. It acts on the fiber F_{x_0} at x_0 for the vector bundle $E \to X$ as follows: $E \ni v \mapsto \Gamma_\omega(\gamma)v$. Suppose $\Gamma_\omega(\gamma) \in SU(n)$ and fiber of the bundle is isomorphic to $\mathbb{C}^2 \otimes \ldots \otimes \mathbb{C}^2$. In this case we speak about *gauge quantum computer*.

In gauge quantum computing the *encoding space of information* is the fiber F of the vector bundle $E \to X$ and *processing of information* is the holonomy operator defined by a connection ω of this bundle and acts on the fiber as $v \mapsto \Gamma_\omega(\gamma)v$.

Consider the trivial bundle $\mathbb{C}^n \times X \to X$. A fixed trivialization gives a one-to-one correspondence between connections ∇ on this bundle and matrix valued 1-forms $\omega \in \Lambda X \otimes \mathfrak{gl}(n,\mathbb{C})$ according to the rule

$$\nabla f = df - \omega f,$$

where ΛX are smooth 1-forms on manifold X, $f : X \to \mathbb{C}^n$ is smooth vector function. Connection ∇ is flat if and only if the 1-form is complete integrable, i.e.,

$$d\omega + \omega \wedge \omega = 0.$$

We can define the *iterated integral* [27] on the space of piecewise smooth path on X, which we denote by PX:

$$\int_\gamma \omega_1 \dots \omega_r : PX \to \mathbb{R}.$$

The holonomy (or parallel transport) $\Gamma_\omega : PX \to GL(n,\mathbb{C})$ of the connection ω on $C^n \times X \to X$ is given by

$$\Gamma_\omega(\gamma) = 1 + \int_\gamma \omega + \int_\gamma \omega\omega + \int_\gamma \omega\omega\omega + \dots.$$

When ω is integrable, the $\Gamma_\omega(\gamma)$ depends only on homotopy class γ relative to its endpoints. Therefore holonomy operator Γ_ω induces the *monodromy* representation

$$\rho : \pi_1(X, x_0) \to GL(n,\mathbb{C}),$$

$$\gamma \mapsto \Gamma_\omega(\gamma).$$

This property of integrable connection is based on the monodromic approach to quantum computation considered in chapter 5, which is the main idea of this book.

Chapter 2

Quantum Computers Based on Exactly Solvable Models and Geometric Phases

Presented in this chapter, as shown in [142], the way of obtaining a universal set of gates is by using time-dependent Hamiltonians admitting exact solutions. The quantum state of N independent qubits can be expressed as a vector in a space of dimension 2^N: $|i_1>, |i_2>, \ldots |i_N> \in (\mathrm{C}^2)^N$, where $|i_k>=|0>$ or $|1>$.

We shall say, that there is a quantum register if there is an ensemble of quantum independent 2-state systems and each single state one can transform to another state independently.

Having an explicit form for the evolution matrix of a 2-level system it is not difficult to construct a register from N qubits. An universal 1-qubit logic gate can be constructed from the time evolution matrices which we obtain in a closed analytic form. In our approach, time-dependent periodic Hamiltonian admitting exact solutions are obtained from time-independent soluble Hamiltonian and a set of unitary time-dependent transformations.

If the quantum system undergoes a cyclic evolution, the initial- and final-state vectors are related by a geometric phase factor.

Recently it has been suggested to construct the holonomic quantum computation [156]–[120] by using non-abelian Berry phase [15]. We show how it is possible to explicitly generate logic gates with desired properties by producing an universal set of unitary transformations. In our approach, the geometric phase can play the role of the additional control parameter for implementing the controlled gates.

We give an example of the quantum register and discuss a way of obtaining entanglement operator with mathematical point of view. It is known by [24], the universal set of gates are given 2×2 unitary operators and one unitary entangled operator R: which acts on $C^2 \otimes C^2$ and transforms decomposed tensor into not decomposed one. It means that there is a vector $|uv>=|u>|v> \in C^2 \bigotimes C^2$ such that $R|uv>$ is not decomposed as a tensor product of two qubits. In our case, the entanglement operator will be obtained from two independent systems with the use of unitary gauge time-dependent transformations, which led to time-dependent periodic operator and entanglement of states.

2.1. Construction of a Time-Dependent Hamiltonian

In this section, we describe a method for implementing the controlled gates with the use of time-dependent soluble Hamiltonians.

The time evolution of quantum system is governed by the Schrödinger equation

$$i\frac{d|\Psi(r,t)>}{\partial t} = H(r,t)|\Psi(r,t)> \tag{2.1.1}$$

with $\hbar = 1$ and T periodic time-dependent Hamiltonian, $H(t) = H(t+T)$. The equation (2.1.1) can be solved by algebraic dynamical method [141,150].

To this end, we use the time-independent Hamiltonian $\widetilde{H}(r)$ and a unitary time-dependent transformation $\mathcal{S}(t)$

$$|\Psi(r,t)>= \mathcal{S}(t)|\Phi(r,t)>, \tag{2.1.2}$$

by means of which the known time-independent Hamiltonian is changed to a time-dependent one

$$H(t) = \mathcal{S}(t)\widetilde{H}\mathcal{S}^\dagger(t) + i\dot{\mathcal{S}}(t)\mathcal{S}^\dagger(t). \tag{2.1.3}$$

Here, $|\Phi(r,t)>$ satisfies the equation of motion (2.1.1) with the time-independent Hamiltonian $\widetilde{H}(r)$ and it is of the form

$$|\Phi(r,t)>= \exp(-i\widetilde{H}(r)t)|\Phi(r,0)> . \tag{2.1.4}$$

Clearly, the solutions $|\Psi(r,t)>$ and $|\Phi(r,t)>$ can be properly defined by solutions for the time-independent problem

$$\widetilde{H}|\Phi(\widetilde{\mathcal{E}})>= \widetilde{\mathcal{E}}|\Phi(\widetilde{\mathcal{E}})> . \tag{2.1.5}$$

Note, if the system of equations (2.1.5) with some known time-independent Hamiltonian $\widetilde{H}(r)$ is exactly soluble, the system of equations (2.1.1) with the time-dependent Hamiltonian (2.1.3) admits exact solutions, too. The result depends on transformation operators $\mathcal{S}(t)$ and the choice of initial states.

Now consider reconstruction of the 2×2 periodic time-dependent Hamiltonian taken in the form

$$H(t,r) = \left(p_r^2 + q(r)\right)\hat{I} + 2(\mathbf{s}\cdot\mathbf{B}(r,t)). \tag{2.1.6}$$

We start with the time-independent Hamiltonian

$$\widetilde{H}(r) = \left(p_r^2 + q(r)\right)\hat{I} + 2(\mathbf{s}\cdot\mathbf{B}(r)), \tag{2.1.7}$$

By means of a unitary time-dependent transformation taken in the form

$$\mathcal{S}(t) = \exp(-i\mathbf{s}\cdot\mathbf{h}(t)) = \exp(-i\sum_{i=1}^{3} s_i h_i(t)), \tag{2.1.8}$$

the time-independent Hamiltonian (2.1.7) with regard to equations (2.1.1) and (2.1.3) turns to the time-dependent Hamiltonian

$$H(r,t) = \left(p_r^2 + q(r)\right)\hat{I} + 2\exp(-i\mathbf{s}\cdot\mathbf{h}(t))(\mathbf{s}\cdot\mathbf{B}(r))\exp(i\mathbf{s}\cdot\mathbf{h}(t)) \tag{2.1.9}$$

Here $\mathbf{s} = (1/2)\sigma$ is the spin operator, $\sigma = (\hat{\sigma}_1, \hat{\sigma}_2, \hat{\sigma}_3)$ and $\hat{\sigma}_i$ are the Pauli matrices and a dot means a time-derivative. The solutions of (2.1.1) with Hamiltonian (2.1.9), according to (2.1.2) and (2.1.4), are represented as

$$|\Psi(r,t)> = \exp(-i\mathbf{s}\cdot\mathbf{h}(t)\exp(-i\widetilde{H}(r)t)|\Phi(r,t=0)> . \tag{2.1.10}$$

Obviously, the transformation (2.1.8) does not change the first term of (2.1.7) and transforms the second term. The Hamiltonians in the form of (2.1.6) and (2.1.9) can be used to describe the motion of a spin 1/2-particle in the space-nonuniform and time-dependent magnetic field or can be applied in investigating multi-level atoms and nuclei.

In terms of the evolution operator $\mathcal{U}(t) = \mathcal{U}(t,0)$, the solution $|\Psi(r,t)>$ is

$$|\Psi(r,t)> = \mathcal{U}(t)|\Psi(r,0)>, \quad \mathcal{U}(0) = 1. \tag{2.1.11}$$

It is easy to find that from (2.1.10) and (2.1.11), there is an important relationship between the operators $\mathcal{U}(t)$ and $\mathcal{S}(t)$: $\mathcal{U}(t) = \mathcal{S}(t)\exp(-i\widetilde{H}t)\mathcal{S}(0)$. In this case, when $|\Psi(r,0)> = |\Phi(r,0)>$, then it is

$$\mathcal{U}(t) = \mathcal{S}(t)\exp -i\widetilde{H}t = \exp(-i\mathbf{s}\cdot\mathbf{h}(t))\exp(-i\widetilde{H}t). \tag{2.1.12}$$

Now, let us consider cyclic solutions that after one period T $(T = 2\pi/\omega)$ are recovered up to this phase, i.e., initial states $|\Psi_\nu(0)>$ are eigenvectors of the evolution operator in one period, $\mathcal{U}(T) = \exp(-i\mathbf{s}\cdot\mathbf{h}(T))\exp(-i\widetilde{H}T)$,

$$|\Psi_\nu(r,T)> = \mathcal{U}(T)|\Psi_\nu(r,0)> = \exp(-i\beta_\nu)|\Psi_\nu(r,0)>, \tag{2.1.13}$$

where $\exp(-i\beta_\nu)$ are eigenvalues of $\mathcal{U}(T)$ and β_ν is the total phase. Let us demand that initial states are eigenvectors of the time-independent Hamiltonian $\widetilde{H}$, $|\Psi_\nu(0)> = |\Phi_\nu>$. It is possible if $\mathcal{U}(T)$ and $\widetilde{H}$ commute $[\mathcal{U}(T), \widetilde{H}] = 0$. Then, we immediately obtain that $\mathcal{S}(T)$ and $\widetilde{H}$ commute: $[\mathcal{S}(T), \widetilde{H}] = 0$. It is one of the conditions on the choice of transformations $\mathcal{S}(t)$. Other properties of $\mathcal{S}(t)$ are evident: they have to be unitary and possess the same matrix dimension as $\widetilde{H}$. In allowance of (2.1.10), the recurrent solutions at any time are written as

$$|\Psi_\nu(r,t)> = \exp(-i\mathbf{s}\cdot\mathbf{h}(t))\exp(-i\widetilde{E}_\nu t)|\Phi_\nu(r)> . \tag{2.1.14}$$

It is evident now that in order to determine the cyclic solutions, we need time-independent solutions. Thus, whenever $\widetilde{H}$ is an exactly soluble time-independent Hamiltonian, the properly generated time-dependent system of equations has cyclic exact solutions.

The evolution of an arbitrary initial state $|\Psi(0)> = \sum_\nu \alpha_\nu|\Psi_\nu(0)>$ can be represented as a superposition for a basis set of recurrent linearly independent vector solutions, $|\Psi(t)> = \sum_\nu \alpha_\nu|\Psi_\nu(t)>$.

Now, let us consider some particular examples of constructing time-dependent with corresponding solutions by using $SU(2)$ transformation (2.1.8) in which the components of $\mathbf{h}(t)$ are linear functions of time, $h_i(t) = \omega_i t$. Let $\mathcal{S}(t)$ be chosen as an operator of rotation around z-axis

$$\mathcal{S}_3(t) = \exp(-i\hat{\sigma}_3\omega t/2) = \begin{pmatrix} \exp(-i\omega t/2) & 0 \\ 0 & \exp(i\omega t/2) \end{pmatrix}, \tag{2.1.15}$$

where ω is a constant angular velocity. Then, in accordance with (2.1.9), the time-independent Hamiltonian (2.1.7) is transformed into the time-dependent one as

$$H(r,t) = \left(p_r^2 + q(r)\right)\hat{I} + \begin{pmatrix} B_3(r) + \frac{\omega}{2} & B_1(r)\exp(-i\omega t) \\ B_1(r)\exp(i\omega t) & -B_3 + \frac{\omega}{2} \end{pmatrix}. \tag{2.1.16}$$

It is evident that $H(r,t)$ is T-periodic, $H(r,t=0) = H(r,t=T), T = 2\pi/\omega$. Its Routhian is $\widetilde{H}(r)$ connected with $H(r,t=0)$ by $H(r,0) = \bar{H}(r) + \frac{\omega}{2}\hat{\sigma}_3$. By using (2.1.14), the cyclic solutions can be written as

$$|\Psi_\nu(r,t)> = \begin{pmatrix} \exp(-i(\widetilde{\mathcal{E}}_1^\nu + \omega/2)t) & 0 \\ 0 & \exp(-i(\widetilde{\mathcal{E}}_2^\nu - \omega/2)t) \end{pmatrix} |\Phi_\nu(r)>, \tag{2.1.17}$$

and after one period

$$|\Psi_\nu(r,(T+t))> = \begin{pmatrix} \exp(-i\widetilde{\mathcal{E}}_1^\nu T - i\pi) & 0 \\ 0 & \exp(-i\widetilde{\mathcal{E}}_2^\nu T + i\pi) \end{pmatrix} |\Psi_\nu(r,t)>,$$

where $\widetilde{\mathcal{E}}_\alpha^\nu = (\widetilde{\mathcal{E}}_\nu - \Delta_\alpha)$ and Δ_α can be the energy of the inner structure (threshold). If $\Delta_\alpha = 0$, then the eigenstates $|\Phi_\nu(r)>$ are two-fold degenerated.

Let us present the time-independent Hamiltonian (2.1.7) and the time-dependent one (2.1.16) in forms (2.1.7) and (2.1.6); suitable for describing the motion of a spin $1/2$-particle in the inhomogeneous magnetic field [149]:

$$\widetilde{H}(r) = \left(p_r^2 + q(r)\right)\hat{I} - \widetilde{\Omega}(r) \begin{pmatrix} \cos\widetilde{\theta}(r) & \sin\widetilde{\theta}(r) \\ \sin\widetilde{\theta}(r) & -\cos\widetilde{\theta}(r) \end{pmatrix}, \tag{2.1.18}$$

where $\widetilde{\mathbf{B}}(r)$ is expressed as

$$\widetilde{\mathbf{B}}(r) = \widetilde{\Omega}(r)(\sin\widetilde{\theta}(r), 0, \cos\widetilde{\theta}(r)), \quad \widetilde{\Omega}(r) = \sqrt{B_3^2(r) + B_1^2(r)}, \tag{2.1.19}$$

The time-dependent Hamiltonian (2.1.16) can be rewritten in the form (2.1.6)

$$H(t,r) = (p_r^2 + q(r))\hat{I} + \Omega(r) \begin{pmatrix} \cos\theta(r) & \sin\theta(r)\exp(-i\omega t \\ \sin\theta(r)\exp(i\omega t & -\cos\theta(r) \end{pmatrix} \tag{2.1.20}$$

for a spin $1/2$-particle in nonhomogeneous magnetic field $\mathbf{B}(t,r)$ precessing around $0z$ axis

$$\mathbf{B}(t,r) = \Omega(r)(\sin\theta(r)\cos(\omega t), \sin\theta(r)\sin(\omega t), \cos\theta(r)), \tag{2.1.21}$$

where

$$\Omega(r) = \sqrt{(B_3(r)+\omega/2)^2 + B_1(r)^2},$$

$$\sin\theta(r) = \frac{B_1(r)}{\Omega(r)}, \ \cos\theta(r) = \frac{B_3(r)+\omega/2}{\Omega(r)}.$$

By means of $S(t)$ transformation (2.1.15) the Hamiltonian (2.1.18) turns to the time-dependent Hamiltonian written in terms of the components for the time-independent field

$$H(t,r) = (p_r^2 + q(r))\hat{I} + \widetilde{\Omega}(r)\begin{pmatrix} \cos\widetilde{\theta}(r) + \omega/2\widetilde{\Omega}(r) & \sin\widetilde{\theta}(r)\exp(-i\omega t) \\ \sin\widetilde{\theta}(r)\exp(+i\omega t) & -\cos\widetilde{\theta}(r) - \omega/2\widetilde{\Omega}(r) \end{pmatrix}. \quad (2.1.22)$$

The comparison of (2.1.20) and (2.1.22) gives us

$$\Omega(r) = \widetilde{\Omega}(r)\Big(1 + \frac{\omega^2}{4\widetilde{\Omega}^2(r)} + \frac{\omega}{\widetilde{\Omega}(r)}\cos\widetilde{\theta}(r)\Big)^{1/2}, \quad (2.1.23)$$

$$\sin\theta(r) = \frac{\widetilde{\Omega}(r)\sin\widetilde{\theta}(r)}{\Omega(r)}, \ \cos\theta(r) = \frac{\widetilde{\Omega}(r)\cos\widetilde{\theta}(r)}{\Omega(r)} + \frac{\omega}{\Omega(r)}.$$

As a result of the unitary transformation specified by $S(t)$, the $B_2(t,r)$ – component of the vector $\mathbf{B}(t,r)$ arises.

If $S(t)$ is chosen as an operator of rotation around y-axis

$$S_2(t) = \exp(-i\hat{\sigma}_2\frac{\omega_2 t}{2}) = \begin{pmatrix} \cos(\frac{\omega_2 t}{2}) & -\sin(\frac{\omega_2 t}{2}) \\ \sin(\frac{\omega_2 t}{2}) & \cos(\frac{\omega_2 t}{2}) \end{pmatrix}, \quad (2.1.24)$$

then in accordance with (2.1.9) the time-independent Hamiltonian (2.1.7) turns to the time-dependent Hamiltonian $H(t)$ in the form

$$H(r,t) = (p_r^2 + q(r))\hat{I} + \widetilde{\Omega}(r)\begin{pmatrix} C(r)c - S(r)s & S(r)c + C(r)s - \frac{i\omega_2}{2\widetilde{\Omega}(r)} \\ S(r)c + C(r)s + \frac{i\omega_2}{2\widetilde{\Omega}(r)} & -(C(r)c - S(r)s) \end{pmatrix},$$

where $c = \cos\omega_2 t$, $s = \sin\omega_2 t$ and $C = \cos\widetilde{\theta}(r)$, $S = \sin\widetilde{\theta}(r)$ and can be used in the description of the spin-1/2 particle in a time-dependent space-inhomogeneous magnetic field precessing around 0y axis. In accordance with (2.1.14), the recurrent solutions of equation (2.1.1) are immediately written as

$$|\Psi_\nu(r,t)> = \begin{pmatrix} \cos(\omega_2 t/2)\exp(-i\widetilde{E}_1^\nu t) & -\sin(\omega_2 t/2)\exp(-i\widetilde{E}_2^\nu t) \\ \sin(\omega_2 t/2)\exp(-i\widetilde{E}_1^\nu t) & \cos(\omega_2 t/2)\exp(-i\widetilde{E}_2^\nu t) \end{pmatrix} |\Phi_\nu(r)> \quad (2.1.25)$$

and in one period

$$|\Psi_\nu(r,(t+T))> = -\begin{pmatrix} \exp(-i\widetilde{E}_1^\nu T) & 0 \\ 0 & \exp(-i\widetilde{E}_2^\nu T) \end{pmatrix} |\Psi_\nu(r,t)> .$$

When there is no dependence on the space variable, the time-dependent Hamiltonian (2.1.25) and corresponding *cyclic solutions* (2.1.25) become

$$H(t)=\widetilde{\Omega}\begin{pmatrix} \cos\widetilde{\theta}\cos\omega_2 t-\sin\widetilde{\theta}\sin\omega_2 t & \sin\widetilde{\theta}\cos\omega_2 t+\cos\widetilde{\theta}\sin\omega_2 t-\frac{i\omega_2}{2\widetilde{\Omega}} \\ \sin\widetilde{\theta}\cos\omega_2 t+\cos\widetilde{\theta}\sin\omega_2 t+\frac{i\omega_2}{2\widetilde{\Omega}} & -(\cos\widetilde{\theta}\cos\omega_2 t-\sin\widetilde{\theta}\sin\omega_2 t) \end{pmatrix},$$

$$|\Psi_1(t)>=\cos(\omega_2 t/2)\exp(-i\widetilde{\mathcal{E}}_1 t)\cos\widetilde{\theta}/2-$$
$$\sin(\omega_2 t/2)\exp(-i\widetilde{\mathcal{E}}_2 t)\sin\widetilde{\theta}/2\sin(\omega_2 t/2)\exp(-i\widetilde{\mathcal{E}}_1 t)\cos\widetilde{\theta}/2$$
$$+\cos(\omega_2 t/2)\exp(-i\widetilde{\mathcal{E}}_2 t)\sin\widetilde{\theta}/2),$$

$$|\Psi_2(t)>=-\cos(\omega_2 t/2)\exp(-i\widetilde{\mathcal{E}}_1 t)\sin\widetilde{\theta}/2-$$
$$\sin(\omega_2 t/2)\exp(-i\widetilde{\mathcal{E}}_2 t)\cos\widetilde{\theta}/2-\sin(\omega_2 t/2)\exp(-i\widetilde{\mathcal{E}}_1 t)\sin\widetilde{\theta}/2+$$
$$\cos(\omega_2 t/2)\exp(-i\widetilde{\mathcal{E}}_2 t)\cos\widetilde{\theta}/2.$$

This problem is identical to the motion of a spin-$1/2$ in the uniformly rotating magnetic field in the $x-z$ plane.

Let $\mathcal{S}(t)$ be chosen as an operator of rotation around x-axis

$$\mathcal{S}(t)=\begin{pmatrix} \cos(\omega_1 t/2) & -i\sin(\omega_1 t/2) \\ -i\sin(\omega_1 t/2) & \cos(\omega_1 t/2) \end{pmatrix}. \tag{2.1.26}$$

By this analogy with the previous cases, the time-independent Hamiltonian (2.1.7) is transformed into the time-dependent one and can be represented in the form for a spin-$1/2$ particle in the space-inhomogeneous magnetic field precessing around $0x$ axis:

$$H(r,t)=\left(p_r^2+q(r)\right)\hat{I}+ \tag{2.1.27}$$

$$\widetilde{\Omega}(r)\begin{pmatrix} \cos\widetilde{\theta}(r)\cos(\omega_1 t) & \sin\widetilde{\theta}(r)-\frac{\omega_1}{2\widetilde{\Omega}(r)}+i\cos\widetilde{\theta}(r)\sin(\omega_1 t) \\ \sin\widetilde{\theta}(r)-\frac{\omega_1}{2\widetilde{\Omega}(r)}-i\cos\widetilde{\theta}(r)\sin(\omega_1 t) & -\cos\widetilde{\theta}(r)\cos(\omega_1 t) \end{pmatrix}.$$

The cyclic solutions of (2.1.1) with account of (2.1.14) are written as

$$|\Psi_\nu(r,t)>=\begin{pmatrix} \cos(\omega_1 t/2)\exp(-i\widetilde{\mathcal{E}}_1^\nu t) & -i\sin(\omega_1 t/2)\exp(-i\widetilde{\mathcal{E}}_2^\nu t) \\ -i\sin(\omega_1 t/2)\exp(-i\widetilde{\mathcal{E}}_1^\nu t) & \cos(\omega_1 t/2)\exp(-i\widetilde{\mathcal{E}}_2^\nu t) \end{pmatrix}|\Phi_\nu(r)>. \tag{2.1.28}$$

Thus, by a different choice of transformation operators $\mathcal{S}_i(t), i=1,2,3$, three families of time-dependent potential matrices with the corresponding cyclic solutions are generated in a closed form from one family of time-independent potential matrices. The obtained expressions turn into ones for a spin-$1/2$ particle evolving in a periodic time-dependent space-uniform magnetic field. It is evident, in all considered cases the Hamiltonians (2.1.16), (2.1.25), (2.1.27) are T-periodic, $T=2\pi/\omega$, and the solutions (2.1.17), (2.1.25) and (2.1.28) are $2T=4\pi/\omega$-periodic and change their sign after one period. As one can easily see, if $\mathcal{S}(t)$ is chosen to be T-periodic, e.g., $\mathcal{S}_i(t)=\exp(-i\hat{\sigma}_i\omega_i t)$, then the Hamiltonian becomes

$T/2 = \pi/\omega$-periodic, and the solutions are T-periodic. In this case, the total phase is multiply of 2π, $\beta_\nu = \widetilde{\mathcal{E}}_\nu T + 2\pi$. Clearly, a more general transformations can be taken: i) as a direct product of the $\mathcal{S}_i(t)$, $\mathcal{S}(t) = \prod_{i=1}^{3} \mathcal{S}_i(t)$, ii) with a more complicated dependence of $\mathcal{S}$ on time.

Now, we show that for all considered examples with $h_i(t) = \omega_i t$ the expectation values of $H(t)$ and spin-expectation values for corresponding cyclic solutions are time-independent. Indeed, the relation (2.2.2), with account of $\mathcal{S}(t)$ as in (2.1.15), (2.1.24), and (2.1.26), gives the expectation values of $H(t)$

$$\varepsilon_\nu^i =< \Phi_\nu(r)|\widetilde{H}(r)|\Phi_\nu(r) > +\frac{\omega_i}{2} < \Phi_\nu(r)|\hat{\sigma}_i|\Phi_\nu(r) >= \widetilde{\mathcal{E}}_\nu + \frac{\omega_i}{2}\bar{\tilde{\sigma}}_\nu^i, \tag{2.1.29}$$

where the quantities $\bar{\tilde{\sigma}}_\nu^i$ are the spin-expectation values

$$\bar{\tilde{\sigma}}_\nu^i =< \Psi_\nu(r,t)|\hat{\sigma}_i|\Psi_\nu(r,t) >=< \Phi_\nu(r)|\hat{\sigma}_i|\Phi_\nu(r) > \tag{2.1.30}$$

do not depend on time in these cases. Therefore, ε_ν^i in (2.1.29) do not depend on time, too.

2.2. Geometric Phases

Now, we can proceed to calculations of physical quantities, such as a geometric phase as associated with the evolution of cyclic solutions [15], a dynamical phase and an expectation value of Hamiltonian $H(t)$. To find the dynamical phase δ_ν, we need the expectation value $\varepsilon_\nu(t)$

$$\delta_\nu = \int_0^T \varepsilon_\nu(t)dt, \tag{2.2.1}$$
$$\varepsilon_\nu(t) =< \Psi_\nu(r,t)|H(t)|\Psi_\nu(r,t) >=< \Psi_\nu(r,0)|\mathcal{U}^\dagger(t)H(t)\mathcal{U}(t)|\Psi_\nu(r,0) > .$$

By using (2.1.12) and (2.1.3) we have $\mathcal{U}^\dagger(t)H(t)\mathcal{U}(t) = e^{i\widetilde{H}(r)t}\Big(\widetilde{H}(r) - i\mathcal{S}^\dagger(t)\dot{\mathcal{S}}(t)\Big)e^{-i\widetilde{H}(r)t}$. Since for cyclic solutions $|\Psi_\nu(r,0)>=|\Phi_\nu(r)>$ and $|\Phi_\nu(r)>$ is an eigenstate vector of $\widetilde{H}(r)$ with eigenvalue $\widetilde{\mathcal{E}}_\nu$, we express the expectation value $\varepsilon_\nu(t)$ and the dynamical phase δ_ν as

$$\varepsilon_\nu(t) = \widetilde{\mathcal{E}}_\nu - < \Phi_\nu(r)|i\mathcal{S}^\dagger(t)\dot{\mathcal{S}}(t)|\Phi_\nu(r) >, \tag{2.2.2}$$
$$\delta_\nu = \widetilde{\mathcal{E}}_\nu T - \int_0^T < \Phi_\nu(r)|i\mathcal{S}^\dagger(t)\dot{\mathcal{S}}(t)|\Phi_\nu(r) > dt. \tag{2.2.3}$$

The geometric phase φ_ν, given by removing the dynamical phase from the total phase β_ν, is

$$\varphi_\nu = \beta_\nu - \delta_\nu = \beta_\nu - \widetilde{\mathcal{E}}_\nu T + \int_0^T < \Phi_\nu(r)|i\mathcal{S}^\dagger(t)\dot{\mathcal{S}}(t)|\Phi_\nu(r) > dt. \tag{2.2.4}$$

It is known that the classical periodic system returns to its initial state after a period, while the quantum system multiplies by an additional Berry phase. For the dynamical phase

δ_ν from (2.2.3) with (2.1.29), we get

$$\delta^i_\nu = \int_0^T \varepsilon^i_\nu(t)dt = \widetilde{\mathcal{E}}_\nu T + \pi\bar{\tilde{\sigma}}^i_\nu. \tag{2.2.5}$$

Removing the dynamical phase (2.2.5) from the total phase $\beta_\nu = \pi + \widetilde{\mathcal{E}}_\nu T$, we get the geometric phase φ^i_ν

$$\varphi^i_\nu = (\beta_\nu - \delta^i_\nu) = \pi(1 - \bar{\tilde{\sigma}}^i_\nu). \tag{2.2.6}$$

Obviously, the geometric phase is determined by the spin-expectation value $\bar{\tilde{\sigma}}^i_\nu$ along the rotating axis. The so-called spin alignment occurs. If $\bar{\tilde{\sigma}}^i_\nu = 0$, changing the dynamic phase angle in a period is determined as $\beta_\nu = \widetilde{\mathcal{E}}_\nu T$; respectively, the geometric phase is $\varphi^i_\nu = \pi$. (For example, in the case 2 direct calculation gives $\bar{\tilde{\sigma}}^2_\nu = 0$). For example, changing the geometric phase in n periods is equal to $n\pi$. It means for an even number of periods $n = 2,4,6,...$, φ^i_ν is multiply 2π and the geometric Berry phase has no effect on time-dependent periodic solutions. It is a property of stationary solutions. As evident from (2.2.6) for an odd number of periods, the wave function's sign changes. If within one period $\bar{\tilde{\sigma}}^i_\nu = 1/2$, then for the odd number of periods $n = 1,3,5..$ φ^i_ν is multiple of $\pi/2$, i.e., $\varphi^i_\nu = n\pi/2$; for an even number of periods $n = 2,4,6,...$ $\varphi^i_\nu = n\pi$ is multiple of π and for $n = 4,8,12...$ the geometric phase again has no effect on periodic solutions. So, quantization of Berry phase is related to the spin-alignment quantization and it is important for nuclear high-spin physics (see [125, 150]).

Note also, that the position probability density in a given point of space-time

$$P(r,t) = ||\Psi_\nu(r,t) > |^2 = ||\Phi_\nu(r) > |^2 \tag{2.2.7}$$

does not depend on time. From (2.1.29) – (2.2.7) it follows that Eq. (2.1.1), for the obtained family of the time-dependent potential matrices, possesses solutions and properties like time-independent ones. So, one of the applications of this approach is modeling quantum systems such as wells and wires with the properties of dynamic localization.

2.3. Construction of an Universal Set of Gates

In this section, we present the way to obtain a universal set of gates by using presented time-dependent Hamiltonians. Two-qubit gates described by 2×2 matrices with the special matrices, whose application led to the interchange of the states, form an universal set of quantum gates. All of these matrices can be obtained from the dynamic evolution of quantum system, which will provide quantum program with needed unitary matrices at each fixed moment. It is known that three-parametric family is enough for approximation of all elements from $SU(2)$.

Using the evolution operator determination (2.1.12) and cyclic solutions $\Psi_1(t), \Psi_2(t)$; we obtain evolution matrix

$$U(t) = \begin{pmatrix} u_{11}(t) & u_{12}(t) \\ u_{21}(t) & u_{22}(t) \end{pmatrix}, \tag{2.3.1}$$

where

$$u_{11}(\omega_2,\widetilde{\mathcal{E}}_1^{\nu},\widetilde{\mathcal{E}}_2^{\nu},\widetilde{\theta},t)=\cos(\omega_2 t/2)\exp(-i\widetilde{\mathcal{E}}_1 t)\cos\widetilde{\theta}/2-\sin(\omega_2 t/2)\exp(-i\widetilde{\mathcal{E}}_2 t)\sin\widetilde{\theta}/2,$$

$$u_{12}(\omega_2,\widetilde{\mathcal{E}}_1^{\nu},\widetilde{\mathcal{E}}_2^{\nu},\widetilde{\theta},t)=-\cos(\omega_2 t/2)\exp(-i\widetilde{\mathcal{E}}_1 t)\sin\widetilde{\theta}/2-\sin(\omega_2 t/2)\exp(-i\widetilde{\mathcal{E}}_2 t)\cos\widetilde{\theta}/2,$$

$$u_{21}(\omega_2,\widetilde{\mathcal{E}}_1^{\nu},\widetilde{\mathcal{E}}_2^{\nu},\widetilde{\theta},t)=\sin(\omega_2 t/2)\exp(-i\widetilde{\mathcal{E}}_1 t)\cos\widetilde{\theta}/2+\cos(\omega_2 t/2)\exp(-i\widetilde{\mathcal{E}}_2 t)\sin\widetilde{\theta}/2,$$

$$u_{22}(\omega_2,\widetilde{\mathcal{E}}_1^{\nu},\widetilde{\mathcal{E}}_2^{\nu},\widetilde{\theta},t)=-\sin(\omega_2 t/2)\exp(-i\widetilde{\mathcal{E}}_1 t)\sin\widetilde{\theta}/2+\cos(\omega_2 t/2)\exp(-i\widetilde{\mathcal{E}}_2 t)\cos\widetilde{\theta}/2.$$

Introduce the next notations $\frac{\omega_2}{2}=\alpha$, $\widetilde{\mathcal{E}}_1^{\nu}=-\widetilde{\mathcal{E}}_2^{\nu}=\lambda$, $\widetilde{\theta}/2=\beta$

$$U(\alpha,\beta,\lambda;t)=\begin{pmatrix} u_{11}(\alpha,\beta,\lambda;t) & u_{12}(\alpha,\beta,\lambda;t) \\ u_{21}(\alpha,\beta,\lambda;t) & u_{22}(\alpha,\beta,\lambda;t)\end{pmatrix},$$

where

$$\begin{aligned}
u_{11}(\alpha,\beta,\lambda;t)&=\cos(\alpha t)\exp(-i\lambda t)\cos(\beta)-\sin(\alpha t)\exp(i\lambda t)\sin(\beta),\\
u_{12}(\alpha,\beta,\lambda;t)&=-\cos(\alpha t)\exp(-i\lambda t)\sin(\beta)-\sin(\alpha t)\exp(i\lambda t)\cos(\beta),\\
u_{21}(\alpha,\beta,\lambda;t)&=\sin(\alpha t)\exp(-i\lambda t)\cos(\beta)+\cos(\alpha t)\exp(i\lambda t)\sin(\beta),\\
u_{22}(\alpha,\beta,\lambda;t)&=-\sin(\alpha t)\exp(-i\lambda t)\sin(\beta)+\cos(\alpha t)\exp(i\lambda t)\sin(\beta)
\end{aligned}$$

and rewrite the given evolution operator as a product of 3 unitary matrices

$$U(t)=\begin{pmatrix}\cos(\alpha t) & \sin(\alpha t)\\ \sin(\alpha t) & \cos(\alpha t)\end{pmatrix}\begin{pmatrix}\exp(-i\lambda t) & 0\\ 0 & \exp(i\lambda t)\end{pmatrix}\begin{pmatrix}\cos(\beta) & -\sin(\beta)\\ \sin(\beta) & \cos(\beta)\end{pmatrix} \tag{2.3.2}$$

Evidently, $U(\alpha,\beta,\lambda,t)$ generates all matrices from $SU(2)$. The action of the right matrix on the basis state $|0>$ or $|1>$ yields all coherent states, which by using the diagonal matrix from the given form of $U(t)$ can be rotated around y axis in a controlled way.

For any α,β,γ, at $t=0$ from (2.3.2) we obtain the gate

$$U(\alpha,\beta,\gamma;t)=\begin{pmatrix}1 & -1\\ 1 & 1\end{pmatrix}. \tag{2.3.3}$$

To obtain Hadamard gate

$$H=\frac{1}{\sqrt{2}}\begin{pmatrix}1 & 1\\ 1 & -1\end{pmatrix} \tag{2.3.4}$$

we multiply $U(\alpha,\beta,\gamma;t)$ from the left on the gate

$$NOT=\begin{pmatrix}0 & 1\\ 1 & 0\end{pmatrix}. \tag{2.3.5}$$

At $t=\frac{\pi}{2\alpha}$ and $\lambda=0$, we have $U(\alpha,\beta,0;\frac{\pi}{2\alpha})=NOT$ and therefore Hadamard gate H is a result of sequence of gates

$$H=U(\alpha,\beta,0;\frac{\pi}{2\alpha})U(\alpha,\beta,\gamma;0).$$

The single qubit gate U can be obtain as a product of unitary matrices

$$U = U(\alpha_k, \beta_k, \gamma_k; t_k)U(\alpha_{k-1}, \beta_{k-1}, \gamma_{k-1}; t_{k-1})...U(\alpha_0, \beta_0, \gamma_0; t_0 = 0).$$

Here, $\alpha_j, \beta_j, \gamma_j,\ j = 0, ..., k$ any real parameters and the parameters of discrete time $t_j,\ j = 0, ..., k$ satisfy the conditions $0 = t_0 < t_1 < ... < t_k$.. Note, in our approach, the effect of geometric phase is naturally taking into account the construction of single qubit gates.

Indeed, for the periodic evolution matrix with the period $T = 2\pi m$,

from (2.2.6), it follows that the state changes sign when $\alpha_j t_j = 2\pi m$ or $\gamma_j t_j = 2\pi m$. By this way, we arrive at the binary representation of the numbers from 0 to $2^n - 1$:

$$(H \otimes H \otimes \cdots \otimes H)|00\ldots 0>=$$

$$= U(\alpha, \beta, 0; \frac{\pi}{2\alpha} + 2n\pi)U(\alpha, \beta, \gamma; 2n\pi) \otimes \cdots \otimes U(\alpha, \beta, 0; \frac{\pi}{2\alpha})U(\alpha, \beta, \gamma; 0)|00\ldots 0>=$$

$$= \frac{1}{\sqrt{2^n}}((|0>+|1>) \otimes (|0>+|1>) \otimes \cdots \otimes (|0>+|1>)) = \frac{1}{\sqrt{2^n}} \sum_{x=0}^{2^n-1} |x>.$$

It should be noted, for this presented sequence of gates in obtaining states $\frac{1}{\sqrt{2^n}} \sum_{x=0}^{2^n-1} |x>$ are not unique. All examples given in the section 1.3 can be used analogously for building single qubit gates.

Let us go to construction of two qubit gates. Note, the operator $SWAP : C^2 \otimes C^2 \rightarrow C^2 \otimes C^2$, acting on tensor product of vectors on the rooles $SWAP(u \otimes v) = v \otimes u$ and tacking in the standard basis the form

$$SWAP = \begin{pmatrix} 1 & 0 & 0 & 0 \\ 0 & 0 & 1 & 0 \\ 0 & 1 & 0 & 0 \\ 0 & 0 & 0 & 1 \end{pmatrix}$$

does not the entangled operator, and the operator, obtained from the Burau representation $bur = \begin{pmatrix} 0 & 1 & 0 & 0 \\ 1 & 0 & 0 & 0 \\ 0 & 0 & 1 & 0 \\ 0 & 0 & 0 & 1 \end{pmatrix}$ (see chapter 5) as well as the operator $CNOT$ is the entangled operator despite all these operators have the same standard Jordan's form.

The Hermitian operator of the form $A = \begin{pmatrix} 0 & 0 & 0 & 0 \\ 0 & 0 & 1 & 0 \\ 0 & 1 & 0 & 0 \\ 0 & 0 & 0 & 0 \end{pmatrix}$ has the rang 2 and commute with the operators of the form

$$bur = \begin{pmatrix} x_{11} & 0 & 0 & x_{14} \\ 0 & a & b & 0 \\ 0 & b & a & 0 \\ x_{41} & 0 & 0 & x_{44} \end{pmatrix}$$

and the operator $exp(-iAt)$ is the entangled one. Indeed,

$$exp(-iAt) = \begin{pmatrix} 1 & 0 & 0 & 0 \\ 0 & \cos(t) & -i\sin(t) & 0 \\ 0 & -i\sin(t) & \cos(t) & 0 \\ 0 & 0 & 0 & 1 \end{pmatrix}$$

takes the form at $t = \pi/4$

$$exp(-iA\pi/4) = \begin{pmatrix} 1 & 0 & 0 & 0 \\ 0 & \sqrt{2}/2 & -i\sqrt{2}/2 & 0 \\ 0 & -i\sqrt{2}/2 & \sqrt{2}/2 & 0 \\ 0 & 0 & 0 & 1 \end{pmatrix}.$$

The universal $CNOT$ consisting from $exp(-iA\pi/4)$ and one-qubit gates as follows

$$CNOT = (X_1 \otimes 1)(1 \otimes 1)(X_2 \otimes 1)(1 \otimes Y_2)\times$$

$$\times exp(-iA\pi/4)(X_3 \otimes 1 + 1 \otimes Y_3)exp(-iA\pi/4)(X_4 \otimes 1 + 1 \otimes Y_4)(1 \otimes Y_5)(X_5 \otimes 1),$$

where

$$X_1 = \sqrt{2}/2\begin{pmatrix} 1 & -1 \\ 1 & 1 \end{pmatrix}, X_2 = \sqrt{2}/2\begin{pmatrix} -i & 1 \\ i & 1 \end{pmatrix}, Y_2 = \sqrt{2}/2\begin{pmatrix} 1 & i \\ i & -i \end{pmatrix},$$

$$X_3 = \begin{pmatrix} 0 & -i \\ i & 0 \end{pmatrix}, Y_3 = 1 = \begin{pmatrix} 1 & 0 \\ 0 & 1 \end{pmatrix}, X_4 = 1/2\begin{pmatrix} 1-i & -1-i \\ -1-i & 1-i \end{pmatrix},$$

$$Y_4 = \sqrt{2}/2\begin{pmatrix} 1 & -1 \\ -i & 1 \end{pmatrix}, X_5 = \sqrt{2}/2\begin{pmatrix} 1 & 1 \\ -1 & 1 \end{pmatrix}, Y_5 = \sqrt{2}/2\begin{pmatrix} 1 & 0 \\ 0 & -i \end{pmatrix}.$$

Proposition 2.3.1. *Let H is a two-dimension diagonal time-independent Hamiltonian in the form*

$$h = \begin{pmatrix} a & 0 \\ 0 & b \end{pmatrix}$$

Then the evolution operator of the matrix Schrödinger equation 2.1.1 with the Hamiltonian $H = h \otimes 1 + 1 \otimes h + \varepsilon A$ is expressed as follows

$$U(t) = (e^{-iht} \otimes e^{-iht})e^{-iAt},$$

where $\varepsilon \in \{0, 1\}$.

Indeed, from this given property of operator A, it follows that the matrix

$$h \otimes 1 + 1 \otimes h = \begin{pmatrix} 2a & 0 & 0 & 0 \\ 0 & a+b & 0 & 0 \\ 0 & 0 & a+b & 0 \\ 0 & 0 & 0 & 2b \end{pmatrix}$$

commutes with A and as consequence $U(t) = e^{-i(h\otimes 1+1\otimes h+A)t} = (e^{-iht}\otimes e^{-iht})e^{-iAt}$. The matrix Pauli $\sigma_3 = \begin{pmatrix} 1 & 0 \\ 0 & -1 \end{pmatrix}$ satisfies to the condition of this proposition. Substitute $h = \sigma_3$ into and get the entangled operator

$$U(t) = \begin{pmatrix} e^{it} & 0 & 0 & 0 \\ 0 & \cos(t) & -i\sin(t) & 0 \\ 0 & -i\sin(t) & \cos(t) & 0 \\ 0 & 0 & 0 & e^{-it} \end{pmatrix}.$$

It is interesting to note, even in the case when the initial Hamiltonian does not give the entangled operator, after the suitable choice of the gauge transformation one can get Hamiltonian to show entanglement. In addition, the geometric phase is obtained from the cyclic evolution of states.

Chapter 3

Quantum Processor Based on the Three-Level Quantum System

In theory computation it is known that the three-state element is most efficient for classical computers, but its realization has remained a problem for many years. Progress was achieved during the 90s (see [98], [157], [9]). Optimal number of states in the physical system for the expression mod p, where p is a prime, is 3. Indeed, if we denote by N_p this number, it is known that $N_p = f(p)N_2$, where $f(p) = \frac{p}{2\log_2 p}$, we obtain that the function $\frac{N_p}{N_2}$ has the minimum when $p = 3$. The necessary mathematical theory in classical many-valued logic is given in [67].

In the quantum world the *many-level systems* are ordinary ones. Thus, in this section, we will consider one possible version of a quantum processor based on a many-level quantum system.

In [36], the extension of the universal quantum logic to the multi-valued domain is considered, where the unit of memory is the *qudit*. Thereby, it is assumed that arbitrary unitary operators on any d-level system; where $d > 2$ is any number, can be decomposed into logical gates that operate on only two levels at a time. For example, the linear ion trap quantum computer uses only two levels in each ion for computing although additional levels can be accessed, and are typically needed, for processing and reading out the state of the ion (see [36] and the literature cited there).

In the case when $d = 3$, the set of universal gates is built in the following manner.

Let us introduce the following unitary operators on $C^3 \to C^3$:

$$X_3(\phi) = \begin{pmatrix} 1 & 0 & 0 \\ 0 & 1 & 0 \\ 0 & 0 & e^{i\phi} \end{pmatrix}, Z_3(c_0, c_1, c_2) = \begin{pmatrix} c_0 & 0 & -c_2 \\ 0 & -c_1 & c_2 \\ -\overline{c}_2 & \overline{c}_2 & \overline{c}_2 \end{pmatrix}.$$

It can be easily checked that $X_3(\phi)$ and $Z_3(c_0, c_1, c_2)$ act on the states $|0>$, $|1>$, $|2>$ as

$$X_3(\phi)|0>= |0>, X_3(\phi)|1>= |1>, X_3(\phi)|2>= e^{i\phi}|2>;$$

$$Z_3(\phi)|0>= c_0|0> -\overline{c}_2|2>, Z_3(\phi)|1>= -c_1|0> +\overline{c}_2|2>,$$

$$Z_3|2>= -c_2|0> +c_2|1> +\overline{c}_2|2> .$$

Furthermore, the gate $Z_3(c_0,c_1,c_2)$ satisfies the following condition:

$$Z_3(c_0,c_1,c_2) = \begin{pmatrix} c_0 & 0 & -c_2 \\ 0 & -c_1 & c_2 \\ -\overline{c}_2 & \overline{c}_2 & \overline{c}_2 \end{pmatrix} (c_0|0> +c_1|1> +c_2|2>) =$$

$$= (-c_2c_0 + c_0c_2, -c_2c_1 + c_1c_2, |c_0|^2 + |c_1|^2 + |c_2|^2 = |2>).$$

In [36], it is proved that the set of gates $\{X_3(\phi), Z_3(c_0,c_1,c_2), G_2[X_3(\phi)], G_2[Z_3(c_0,c)1,c_2)]\}$ is universal for a three-level quantum system, where $G_2[Y] = \begin{pmatrix} I & 0 \\ 0 & Y \end{pmatrix}$ and I is the identity matrix. Recently, a new and interesting way to generate entanglement via two identical three-level atoms splitting in the magnetic field interacting with the cavity field is obtained [153] (see also [140]).

Another approach is given next for the description of quantum system which gives the single *qutrit* gates, in particular, we obtain in closed form evolution operators depending on parameters.

3.1. Atom-photon Interactions Hamiltonian

We will build this special Hamiltonian case for a hydrogen-like atom moving and interacting with a quantized electromagnetic field, by using the notations of [139] and [122]. The general form for such a Hamiltonian, includes the center-of-mass motion and all the modes of the interacting field, as presented in [139]. For our purposes, we consider the following conditions on atom-photon interactions: (i) all the atom-photon interactions are the electric dipole; (ii) only three atomic levels are included in the interaction; (iii) two quantized resonator modes are interacting with this three-level system; (iv) these modes interact with only one for each couple of levels; (v) so, the couple from these three possible levels, only two of them interact directly. Levels of the third couple interact only by means of an intermediate level; (vi) in this work, we permit but do not consider the center-of-mass motion and large detuning from resonances, which cause the state entanglement.

Let us consider atom-photon interactions Hamiltonian in the electric dipole approximation

$$\hat{H}_{\mathbf{r.E}} = -\wp\mathbf{E}(R,t) = -e\mathbf{r}\mathbf{E}(\mathbf{R},t), \qquad (3.1.1)$$

and transform it to three-level atom interacting with two modes of quantized fields of radiation. For these purposes previously described, we must discuss the following operators.

Hamiltonian for internal degrees of freedom in a non-interacting atom. We consider internal degrees of freedom in a three-level atom. Due to assumptions (i-vi) only two couples have energy states that are involved in interaction. In all cases, the intermediate state with energy $E_i \equiv \hbar\omega_i$ is labeled by the state vector $|i>$. From other states, the state with higher energy is named excited, denoted $E_e \equiv \hbar\omega_e$ and labeled by the state vector $|e>$. The remaining energy level is named ground level, denoted $E_g \equiv \hbar\omega_g$ and labeled by the state vector $|g>$ (see Fig.1). All of them are eigenstates of the Hamiltonian $\hat{H}_{atom}$ in a non-interacting three-level atom:

$$\hat{H}_{atom}|e> = \hbar\omega_e|e>, \ \hat{H}_{atom}|i> = \hbar\omega_i|i>, \ \hat{H}_{atom}|g> = \hbar\omega_g|g> . \qquad (3.1.2)$$

Using the completeness relation $|e><e|+|i><i|+|g><g|=1$,, we can write the arbitrary atomic operator $\hat{\mathbf{O}}$ as

$$\hat{\mathbf{O}} = 1\hat{\mathbf{O}}1 = \sum_{j,j'=e,g,i} |j><j|\hat{\mathbf{O}}|j'><j'|. \tag{3.1.3}$$

Utilizing (3.1.3) and the orthonormality condition $<j||j'>=\delta_{ij'}$ for the Hamiltonian in a three-level atom, one can easily obtain

$$\hat{H}_{atom} = (|e><e|+|i><i|+|g><g|)\hat{H}_{atom}(|e><e|+|i><i|+|g><g|) = \tag{3.1.4}$$
$$= E_e|e><e|+E_i|i><i|+E_g|g><g|,$$

where we use (3.1.2) and in the earlier notations $E_e=\hbar\omega_e$, $E_i=\hbar\omega_i$, $E_g=\hbar\omega_g$. Then

$$|e>=\begin{pmatrix}1\\0\\0\end{pmatrix}, <e|=(1,0,0), |i>=\begin{pmatrix}0\\1\\0\end{pmatrix}, <i|=(0,1,0),$$

$$|g>=\begin{pmatrix}0\\0\\1\end{pmatrix}, <g|=(0,0,1).$$

Thus, the Hamiltonian (3.1.4) has the following matrix form

$$\hat{H}_{atom} = E_e\begin{pmatrix}1&0&0\\0&0&0\\0&0&0\end{pmatrix}+E_i\begin{pmatrix}0&0&0\\0&1&0\\0&0&0\end{pmatrix}+E_g\begin{pmatrix}0&0&0\\0&0&0\\0&0&1\end{pmatrix}= \tag{3.1.5}$$

$$=\begin{pmatrix}E_g&0&0\\0&E_i&0\\0&0&E_g\end{pmatrix}=\hbar\begin{pmatrix}\omega_e&0&0\\0&\omega_i&0\\0&0&\omega_g\end{pmatrix}.$$

The dipole moment operator. Now, we can use the completeness relation once more to express the position operator $\hat{\mathbf{r}}$ in terms of energy eigenstates. As with each state, wave functions $\Psi_j(\mathbf{r})$ of a free atom have definite parity, then the diagonal matrix elements equal zero

$$<j|\hat{\mathbf{r}}|j>=\int|\Psi_j(\mathbf{r})|^2\mathbf{r}d^3r=0.$$

Indeed, if the function $|\Psi_j(\mathbf{r})|^2$ is symmetric and the operator $\mathbf{r}$ antisymmetric, the expression under the integral sign is antisymmetric. For the nondiagonal matrix elements we have the following expressions

$$e<i|\hat{\mathbf{r}}|g>=e\int\Psi_i^*(\mathbf{r})\mathbf{r}\Psi_g(\mathbf{r})d^3r=\wp_{ig}, \tag{3.1.6}$$

and

$$e<g|\hat{\mathbf{r}}|i>=e\int\Psi_g^*(\mathbf{r})\mathbf{r}\Psi_i(\mathbf{r})d^3r=\wp_{ig}^*=\wp_{gi}.$$

In the same way

$$e < e|\hat{\mathbf{r}}|i >= e\int \Psi_i^*(\mathbf{r})\mathbf{r}\Psi_g(\mathbf{r})d^3r = \wp_{ei},$$

and

$$e < i|\hat{\mathbf{r}}|e >= e\int \Psi_g^*(\mathbf{r})\mathbf{r}\Psi_i(\mathbf{r})d^3r = \wp_{ei}^* = \wp_{ie}.$$

Thus, the dipole moment operator $e\hat{\mathbf{r}}$ is expressed in the following manner

$$e\hat{\mathbf{r}} = \wp_{ig}|i >< g| + \wp_{gi}|g >< i| + \wp_{ei}|e >< i| + \wp_{ie}|i >< e|. \tag{3.1.7}$$

Note, that this operator describes transitions from the ground state $|g>$ to the intermediate state $|i>$, from intermediate to the excited state $|e>$, and the process in reverse order back to $|g>$. For verification let us apply the operator $e\hat{\mathbf{r}}$ (9) to the ground state $|g>$. Then, we get

$$e\hat{\mathbf{r}}|g >= \wp_{ig}|i >< g||g > +\wp_{gi}|g >< i||g > +\wp_{ei}|e >< i||g > +\wp_{ie}|i >< e||g >= \wp_{ig}|i > .$$

Similarly

$$e\hat{\mathbf{r}}|e >= \wp_{ig}|i >< g||e > +\wp_{gi}|g >< i||e > +\wp_{ei}|e >< i||e > +\wp_{ie}|i >< e||e >= \wp_{ie}|i >$$

and finally

$$e\hat{\mathbf{r}}|i >= \wp_{ig}|i >< g||i > +\wp_{gi}|g >< i||i > +\wp_{ei}|e >< i||i > +\wp_{ie}|i >< e||g >=$$

$$= \wp_{gi}|g > +\wp_{ei}|e > .$$

Thus, the operator

$$\hat{\sigma}_{ig} \equiv |i >< g|,\ |i >< g| = \begin{pmatrix} 0 \\ 1 \\ 0 \end{pmatrix}(0,0,1) = \begin{pmatrix} 0 & 0 & 0 \\ 0 & 0 & 1 \\ 0 & 0 & 0 \end{pmatrix}, \tag{3.1.8}$$

causes the atom transition from the ground to the intermediate state and can be interpreted as a creation operator for the atom in the intermediate state $|i>$.

On the contrary, the operator

$$\hat{\sigma}_{gi} \equiv |g >< i|,\ |g >< i| = \begin{pmatrix} 0 \\ 0 \\ 1 \end{pmatrix}(0,1,0) = \begin{pmatrix} 0 & 0 & 0 \\ 0 & 0 & 0 \\ 0 & 1 & 0 \end{pmatrix}, \tag{3.1.9}$$

annihilates the atom in the intermediate state and so represents the annihilation operator for the intermediate state.

The operator

$$\hat{\sigma}_{ei} \equiv |e >< i|,\ |e >< i| = \begin{pmatrix} 1 \\ 0 \\ 0 \end{pmatrix}(0,1,0) = \begin{pmatrix} 0 & 1 & 0 \\ 0 & 0 & 0 \\ 0 & 0 & 0 \end{pmatrix}, \tag{3.1.10}$$

causes transition of the atom from the intermediate to the excited state and can be interpreted as a creation operator for the atom in the excited state $|e>$.

The operator

$$\hat{\sigma}_{ei} \equiv |i><e|,\ |i><e| = \begin{pmatrix} 0 \\ 1 \\ 0 \end{pmatrix} (1,0,0) = \begin{pmatrix} 0 & 0 & 0 \\ 1 & 0 & 0 \\ 0 & 0 & 0 \end{pmatrix}, \qquad (3.1.11)$$

annihilates the atom in the excited state and so represents the annihilation operator for the excited state. The dipole moment operator (3.1.7) is defined in accordance to the expressions (3.1.8)-(3.1.11), for the creation and annihilation operators, to be

$$e\hat{\mathbf{r}} = \wp_{ig}\hat{\sigma}_{ig} + \wp_{gi}\hat{\sigma}_{ig} + \wp_{ei}\hat{\sigma}_{ei} + \wp_{ie}\hat{\sigma}_{ie}. \qquad (3.1.12)$$

The electric field operator. Now, let us discuss two modes of radiation field. We can assume for determinacy, that the radiation field is formed in three-mirror standing wave resonator and will decompose the electric field strength in the resonator as follows

$$\hat{\mathbf{E}}(\hat{\mathbf{R}},t) = \mathcal{F}_{0a}\mathbf{u}_a(\hat{\mathbf{R}})i(\hat{a}-\hat{a}^+) + \mathcal{F}_{0b}\mathbf{u}_b(\hat{\mathbf{R}})i(\hat{b}-\hat{b}^+), \qquad (3.1.13)$$

where $\mathcal{F}_{0j} \equiv \sqrt{\frac{\hbar\Omega_j}{2\varepsilon_0 V_j}}$, $j = a,b$.

Here, $\mathbf{u}_j(\hat{\mathbf{R}})$ represents the mode functions of the resonator at the position $\mathbf{R}$ of the atom. Where Ω_a and Ω_b are oscillation frequencies in the field. In a classical treatment, $\hat{a}^+$, $\hat{b}^+$ and $\hat{a}$, $\hat{b}$ are time dependent amplitudes. In our present treatment they are operators which obey the commutation relations

$$\hat{a}\hat{a}^+ - \hat{a}^+\hat{a} = 1,\ \hat{b}\hat{b}^+ - \hat{b}^+\hat{b} = 1. \qquad (3.1.14)$$

By their aid, the quantized field Hamiltonian can be written in the form

$$\hat{H}_{field} = \hbar\Omega_a\hat{a}^+\hat{a} + \hbar\Omega_b\hat{b}^+\hat{b}. \qquad (3.1.15)$$

Here, we neglect the zero-point energy.

Back to the interaction Hamiltonian. Now, consider the Hamiltonian of electric dipole atom-photon interactions again (3.1.1)

$$\hat{H}_{\mathbf{rE}} = -e\hat{\mathbf{r}}\hat{\mathbf{E}}(\hat{\mathbf{R}},t).$$

Substituting in this expression the expressions of the dipole moment operator $e\hat{\mathbf{r}}$ (3.1.12) and electric field operator $\hat{\mathbf{r}}\hat{\mathbf{E}}(\hat{\mathbf{R}},t)$, we obtain

$$\hat{H}_{\mathbf{rE}} = -\mathcal{F}_{0a}i[\wp_{ig}\mathbf{u}_a(\hat{\mathbf{R}})\hat{\sigma}_{ig} + \wp_{ei}\mathbf{u}_b(\hat{\mathbf{R}})\hat{\sigma}_{gi}](\hat{a}-\hat{a}^+) - \qquad (3.1.16)$$
$$-\mathcal{F}_{0b}i[\wp_{ei}\mathbf{u}_b(\hat{\mathbf{R}})\hat{\sigma}_{ei} + \wp_{ei}\mathbf{u}_b(\hat{\mathbf{R}})\hat{\sigma}_{ie}](\hat{b}-\hat{b}^+).$$

This expression contains the scalar product $\wp\mathbf{u}$ of the dipole moment and the mode function $\mathbf{u}$. As this product is a complex quantity, we represent it in the following manner

$$\wp\mathbf{u} = |\wp\mathbf{u}|e^{i\varphi}, \qquad (3.1.17)$$

where φ is the phase. Substituting (3.1.17) in (3.1.16) the Hamiltonian of electric dipole atom-photon interactions can be written in the following form

$$\hat{H}_{\mathbf{rE}} = -\hbar\mathcal{F}_{0a}i[\frac{|\wp_{ig}\mathbf{u}_a(\hat{\mathbf{R}})|}{\hbar}\hat{\sigma}_{ig}e^{i\varphi_a} + \frac{|\wp_{ei}\mathbf{u}_b(\hat{\mathbf{R}})|}{\hbar}\hat{\sigma}_{gi}e^{-i\varphi_a}](\hat{a}-\hat{a}^+)- \quad (3.1.18)$$
$$-\hbar\mathcal{F}_{0b}i[\frac{|\wp_{ei}\mathbf{u}_b(\hat{\mathbf{R}})|}{\hbar}\hat{\sigma}_{ei}e^{i\varphi_b} + \frac{|\wp_{ei}\mathbf{u}_b(\hat{\mathbf{R}})|}{\hbar}\hat{\sigma}_{ie}e^{-i\varphi_b}](\hat{b}-\hat{b}^+).$$

Introducing the Rabi frequencies

$$g_a(\mathbf{R}) = \frac{|\wp_{ig}\mathbf{u}_a(\hat{\mathbf{R}})|}{\hbar}\mathcal{F}_{0a},\ g_b(\mathbf{R}) = \frac{|\wp_{ei}\mathbf{u}_b(\hat{\mathbf{R}})|}{\hbar}\mathcal{F}_{0b}$$

the interaction Hamiltonian (3.1.18) now is given by

$$\hat{H}_{\mathbf{rE}} = \hbar g_a(\hat{\mathbf{R}})(-i)(\hat{\sigma}_{ig}e^{i\varphi_a} + \hat{\sigma}_{gi}e^{-i\varphi_a})(\hat{a}-\hat{a}^+)+ \quad (3.1.19)$$
$$\hbar g_b(\hat{\mathbf{R}})(-i)(\hat{\sigma}_{ei}e^{i\varphi_b} + \hat{\sigma}_{ie}e^{-i\varphi_b})(\hat{b}-\hat{b}^+).$$

Introducing the difference of phases $\delta\varphi \equiv \varphi_a - \varphi_b$, in the Hamiltonian (3.1.19), we obtain

$$\hat{H}_{\mathbf{rE}} = \hbar g_a(\hat{\mathbf{R}})(-i)(\hat{\sigma}_{ig}e^{i\varphi_a} + \hat{\sigma}_{gi}e^{-i\varphi_a})(\hat{a}-\hat{a}^+)+ \quad (3.1.20)$$
$$\hbar g_b(\hat{\mathbf{R}})(-i)(\hat{\sigma}_{ei}e^{i(\delta\varphi-\varphi_a)} + \hat{\sigma}_{ie}e^{-i(\delta\varphi-\varphi_a)})(\hat{b}-\hat{b}^+).$$

In quantum mechanics, only the difference in phases is important, so in (3.1.20), we can put $\varphi_a = \pi/2$; and, finally obtain

$$\hat{H}_{\mathbf{rE}} = \hbar g_a(\hat{\mathbf{R}})(\hat{\sigma}_{ig} - \hat{\sigma}_{gi})(\hat{a}-\hat{a}^+) + \hbar g_b(\hat{\mathbf{R}})(\hat{\sigma}_{ei}e^{-i\delta\varphi} - \hat{\sigma}_{ie}e^{i\delta\varphi})(\hat{b}-\hat{b}^+). \quad (3.1.21)$$

Expression (3.1.21) is the final form of the exact Hamiltonian responsible for all atom-photon interactions between the three-level atom and two modes of electromagnetic field. Just this Hamiltonian will be used hereafter.

Total Hamiltonian. Joining up with these main results (3.1.4), (3.1.15), (3.1.21), and by taking this into account

$$\hat{\sigma}_{ee} \equiv |e><e|,\ \hat{\sigma}_{ii} \equiv |i><i|,\ \hat{\sigma}_{gg} \equiv |g><g|,$$

one can build the total Hamiltonian of the whole atom-field system

$$\hat{H} = \frac{\hat{\mathbf{P}}^2}{2M} + \hat{H}_{field} + \hat{H}_{atom} + \hat{H}_{\mathbf{rE}} = \quad (3.1.22)$$
$$\frac{\hat{\mathbf{P}}^2}{2M} + \hbar\Omega_a\hat{a}^+\hat{a} + \hbar\Omega_b\hat{b}^+\hat{b} + E_e\hat{\sigma}_{ee} + E_i\hat{\sigma}_{ii} + E_g\hat{\sigma}_{gg}+$$
$$+\hbar g_a(\hat{\mathbf{R}})(\hat{\sigma}_{ig} - \hat{\sigma}_{gi})(\hat{a}-\hat{a}^+) + \hbar g_b(\hat{\mathbf{R}})(\hat{\sigma}_{ei}e^{-i\delta\varphi} - \hat{\sigma}_{ie}e^{i\delta\varphi})(\hat{b}-\hat{b}^+).$$

This Hamiltonian includes: the operator of kinetic energy $\frac{\hat{\mathbf{P}}^2}{2M}$ at the center of mass movement; free-field Hamiltonian $\hat{H}_{field} = \hbar\Omega_a\hat{a}^+\hat{a} + \hbar\Omega_b\hat{b}^+\hat{b}$; the Hamiltonian responsible for internal states of the atom $\hat{H}_{atom} = E_e\hat{\sigma}_{ee} + E_i\hat{\sigma}_{ii} + E_g\hat{\sigma}_{gg}$ and interaction operator $\hat{H}_{\mathbf{rE}} = \hbar g_a(\hat{\mathbf{R}})(\hat{\sigma}_{ig} - \hat{\sigma}_{gi})(\hat{a}-\hat{a}^+) + \hbar g_b(\hat{\mathbf{R}})(\hat{\sigma}_{ei}e^{-i\delta\varphi} - \hat{\sigma}_{ie}e^{i\delta\varphi})(\hat{b}-\hat{b}^+)$. Connecting the internal states with field operators $\hat{a}-\hat{a}^+$, $\hat{b}-\hat{b}^+$ by means of atom-field coupling constants $g_a(\hat{\mathbf{R}})$, $g_a(\hat{\mathbf{R}})$.

3.2. Simplification of Atom-Photon Interactions Hamiltonian

3.2.1. Selection the Real Transitions and Rotating Wave Approximation

There are two reasons for the simplification of the atom-photon interactions Hamiltonian (3.1.21)

$$\hat{H}_{\mathbf{rE}} = \hbar g_a(\hat{\mathbf{R}})(\hat{\sigma}_{ig}\hat{a}^+ + \hat{\sigma}_{gi}\hat{a} - \hat{\sigma}_{ig}\hat{a} - \hat{\sigma}_{ig}\hat{a}^+) + \tag{3.2.1}$$

$$+\hbar g_b(\hat{\mathbf{R}})(\hat{\sigma}_{ei}\hat{b}^+ e^{i\delta\varphi} - \hat{\sigma}_{ie}\hat{b}e^{-i\delta\varphi} - \hat{\sigma}_{ie}\hat{b}e^{i\delta\varphi} - \hat{\sigma}_{ie}\hat{b}^+ e^{-i\delta\varphi}).$$

Namely, it amounts to confining our analysis to real transitions in which a photon is emitted while the atom goes from its upper state to its lower state, or a photon is absorbed while the atom goes from its lower state to its upper state. The second reason is mathematical and is connected with the averaging procedure – neglecting the fast oscillating terms. Hereafter, we will use the interaction representation (Dirac representation) which is defined by the internal atomic states and the free-field states and under these assumptions we may reduce the Hamiltonian (3.2.1) to the form what is called the *Rotating Wave Approximation* (RWA).

In a three-level system the simplification procedure of interaction Hamiltonian should be carried out separately for each transition scheme. Thus, first let us consider these transition schemes. In the three-level atom due to (i)-(v) shown in section 3.1, there are three different configurations possible for transitions between the three levels depending upon the position of the intermediate level $|i>$. Those are shown in Fig. 3.1 along with the allowed transitions. In the configuration of the first diagrams in Fig. 3.1, the energy of $|i>$ is intermediate between the energies of the other two levels. It is called the *ladder configuration*. Notation 'L' stands for this configuration. The energy of $|i>$ in the Λ or Raman configuration is higher than that of the other two levels (second diagram in Fig. 3.1) whereas the level $|i>$ in the V-system is below the other two levels (third diagram in Fig. 3.1).

Simplification for L-configuration. The terms $\hat{\sigma}_{gi}\hat{a}$, $\hat{\sigma}_{ig}\hat{a}^+$, $\hat{\sigma}_{ie}\hat{b}$ and $\hat{\sigma}_{ei}\hat{b}^+$, in the interaction Hamiltonian (3.2.1) led to the violation of the energy conservation law. Indeed, note that the operators $\hat{\sigma}_{gi}\hat{a}$ and $\hat{\sigma}_{ie}\hat{b}$, annihilate the excited atom (in $|i>$ and $|e>$) and simultaneously annihilate the field excitation (Fig. 3.2c). In the same manner, operators $\hat{\sigma}_{ig}\hat{a}^+$ and $\hat{\sigma}_{ei}\hat{b}^+$ create the field excitation while the atom is annihilated in $|g>$ ($|i>$) and created in $|i>$ ($|e>$) states (Fig. 3.2d).

On the contrary, the operators $\hat{\sigma}_{gi}\hat{a}^+$, $\hat{\sigma}_{ie}\hat{b}^+$ and $\hat{\sigma}_{ig}\hat{a}^+$, $\hat{\sigma}_{ei}\hat{b}$ either annihilate the excitation ($|i>$ or $|e>$) while creating the field excitation (Fig. 3.2b), or create the excited atom (in $|i>$ or $|e>$) while annihilating the field excitation (Fig. 3.2a).

Under the assumptions of real transitions, neglecting the operators $\hat{\sigma}_{gi}\hat{a}$, $\hat{\sigma}_{ie}\hat{b}$, $\hat{\sigma}_{ig}\hat{a}^+$, $\hat{\sigma}_{ie}\hat{b}^+$,, we may reduce the Hamiltonian (3.2.1) to the form

$$\hat{H}^L_{\mathbf{rE}} \cong \hat{H}^L_{int} = \hbar g_a(\hat{\mathbf{R}})(\hat{\sigma}_{gi}\hat{a}^+ + \hat{\sigma}_{ig}\hat{a}) + \hbar g_b(\hat{\mathbf{R}})(\hat{\sigma}_{ie}\hat{b}^+ e^{i\delta\varphi} + \hat{\sigma}_{ei}\hat{b}e^{-i\delta\varphi}). \tag{3.2.2}$$

Simplification for Λ-configuration. For this configuration the operators $\hat{\sigma}_{gi}\hat{a}$, $\hat{\sigma}_{ig}\hat{a}^+$, $\hat{\sigma}_{ie}\hat{b}^+$, $\hat{\sigma}_{ei}\hat{b}$ in the interaction Hamiltonian led to the violation of the energy conservation law (see Fig.1 and Fig. 3.3c,d). Thus, neglecting these terms in the interaction Hamiltonian (3.2.1) and keeping $\hat{\sigma}_{gi}\hat{a}^+$, $\hat{\sigma}_{ig}\hat{a}$, $\hat{\sigma}_{ie}\hat{b}$ and $\hat{\sigma}_{ei}\hat{b}^+$, and (Fig. 3.3a,b), we may reduce the Hamiltonian (3.2.1) to the form

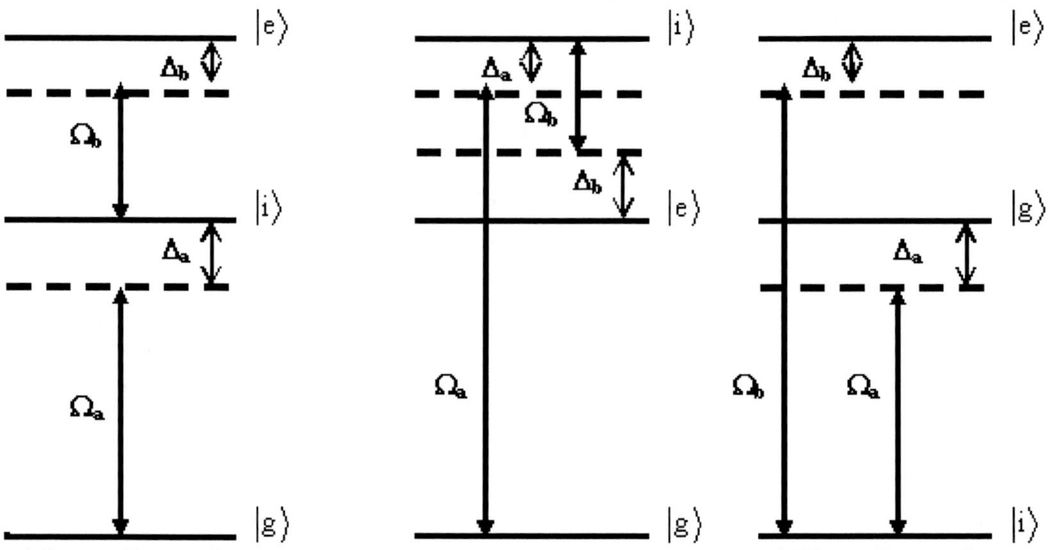

Figure 3.2.1. Three configurations for the three-level atom in a two-mode field where each act on only one transition.

$$\hat{H}^{\Lambda}_{\mathbf{rE}} \cong \hat{H}^{L}_{int} = \hbar g_a(\hat{\mathbf{R}})(\hat{\sigma}_{gi}\hat{a}^{+} + \hat{\sigma}_{ig}\hat{a}) - \hbar g_b(\hat{\mathbf{R}})(\hat{\sigma}_{ie}\hat{b}^{+}e^{i\delta\varphi} + \hat{\sigma}_{ei}\hat{b}e^{-i\delta\varphi}). \tag{3.2.3}$$

Simplification for V-configuration. For this configuration the operators $\hat{\sigma}_{gi}\hat{a}^{+}$, $\hat{\sigma}_{ig}\hat{a}$, $\hat{\sigma}_{ie}\hat{b}$, and $\hat{\sigma}_{ei}\hat{b}^{+}$, of the interaction Hamiltonian led to the violation of the energy conservation law (see Fig.1 and Fig. 3.4c,d). Thus, neglecting these terms in the interaction Hamiltonian (3.2.1) and keeping $\hat{\sigma}_{gi}\hat{a}$, $\hat{\sigma}_{ig}\hat{a}$, $\hat{\sigma}_{ie}\hat{b}$, and $\hat{\sigma}_{ei}\hat{b}$, and (Fig. 3.4a,b), we may reduce the Hamiltonian (3.2.1) to the form

$$\hat{H}^{\Lambda}_{\mathbf{rE}} \cong \hat{H}^{V}_{int} = -\hbar g_a(\hat{\mathbf{R}})(\hat{\sigma}_{gi}\hat{a}^{+} + \hat{\sigma}_{ig}\hat{a}) + \hbar g_b(\hat{\mathbf{R}})(\hat{\sigma}_{ie}\hat{b}^{+}e^{i\delta\varphi} + \hat{\sigma}_{ei}\hat{b}e^{-i\delta\varphi}). \tag{3.2.4}$$

3.2.2. Interaction Representation

Further simplification of the atom-photon interaction Hamiltonians (3.1.21)-(3.2.4) takes place in the interaction representation, where we can use the averaging procedure by fast oscillations. To accomplish this representation, we represent the total Hamiltonian for the whole atom-field system (see (3.1.22))

$$\hat{H} = \frac{\hat{\mathbf{P}}^2}{2M} + \hat{H}_{field} + \hat{H}_{atom} + \hat{H}_{\mathbf{rE}} =$$

$$\frac{\hat{\mathbf{P}}^2}{2M} + \hat{H}_0 + \hbar g_a(\hat{\mathbf{R}})(\hat{\sigma}_{ig} - \hat{\sigma}_{gi})(\hat{a} - \hat{a}^{+}) + \hbar g_b(\hat{\mathbf{R}})(\hat{\sigma}_{ei}e^{i\delta\varphi} - \hat{\sigma}_{ie}e^{-i\delta\varphi}(\hat{b} - \hat{b}^{+}).$$

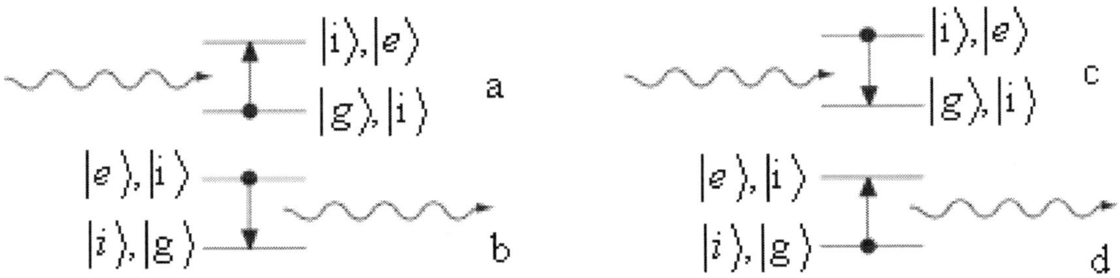

Figure 3.2.2.

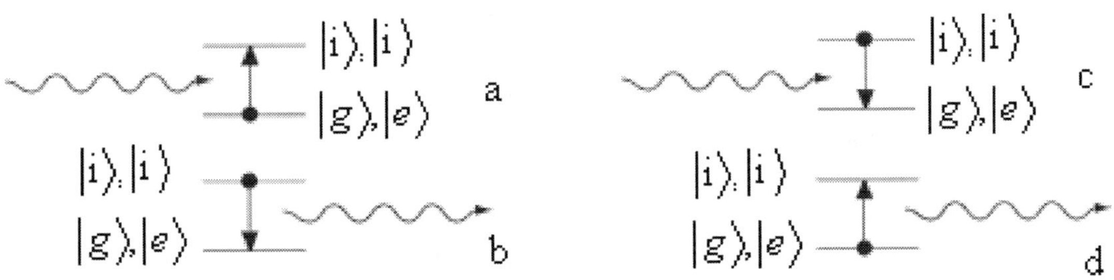

Figure 3.2.3.

by the internal states of the atom and free-field states. Here,

$$\hat{H}_0 = \hat{H}_{fild} + \hat{H}_{atom} = \hbar\Omega_a \hat{a}^+ \hat{a} + \hbar\Omega_b \hat{b}^+ \hat{b} + E_e \hat{\sigma}_{ee} + E_i \hat{\sigma}_{ii} + E_g \hat{\sigma}_{gg},$$

or

$$\hat{H}_0 = \hbar\Omega_a \hat{a}^+ \hat{a} + \hbar\Omega_b \hat{b}^+ \hat{b} + \hbar \begin{pmatrix} \omega_e & 0 & 0 \\ 0 & \omega_i & 0 \\ 0 & 0 & \omega_g \end{pmatrix} \qquad (3.2.5)$$

denotes the sum of free-field $\hat{H}_{field}$ and free atom $\hat{H}_{atom}$ Hamiltonians.

Now, defining the state vector in the interaction representation $|\Phi^{(I)}(t)>$ via

$$|\Phi(t)> \equiv exp[-\frac{i}{\hbar}\hat{H}_0 t]|\Phi^{(I)}(t)> \qquad (3.2.6)$$

and substituting (5.4.5), containing the state vector $|\Phi^{(I)}(t)>$ in the interaction representation, in the Schrödinger equation

$$i\hbar\frac{d|\Phi^{(I)}(t)>}{dt} = (\frac{\hat{\mathbf{P}}^2}{2M} + \hat{H}_0 + \hat{H}_{\mathbf{rE}})|\Phi(t)> .$$

We receive

$$i\hbar\frac{d|\Phi^{(I)}(t)>}{dt} = (\frac{\hat{\mathbf{P}}^2}{2M} + \hat{H}^I_{\mathbf{rE}})|\Phi(t)>,$$

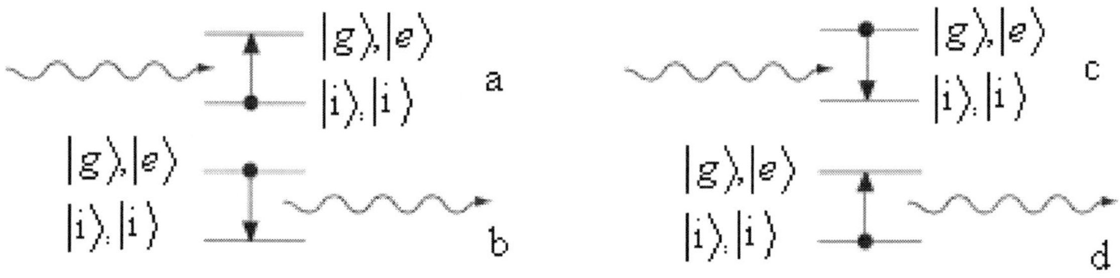

Figure 3.2.4.

where

$$\hat{H}^{(I)}_{\mathbf{rE}} \equiv \exp[\frac{i}{\hbar}\hat{H}_0 t]\hat{H}_{\mathbf{rE}}\exp[-\frac{i}{\hbar}\hat{H}_0 t] \qquad (3.2.7)$$

is the interaction Hamiltonian in the interaction picture. Note, that by reason of (vi) in section 3.2, we do not include the kinetic energy operator (corresponding to the center of mass movement) in these transformations. Now, let us obtain the exact expression for the interaction Hamiltonian $\hat{H}^{(I)}_{\mathbf{rE}}$. Substituting $\hat{H}_0$ (3.2.5) into the expression (3.2.7), under conditions, that the field and atom operators commute with each other; we obtain

$$\hat{H}^{(I)}_{\mathbf{rE}} = \hbar g_a(\hat{\mathbf{R}})e^{\frac{i}{\hbar}\hat{H}_{atom}t}(\hat{\sigma}_{ig} - \hat{\sigma}_{gi})e^{-\frac{i}{\hbar}\hat{H}_{atom}t}e^{i\Omega_a\hat{a}^+\hat{a}t}(\hat{a}-\hat{a}^+)e^{-i\Omega_a\hat{a}^+\hat{a}t} \qquad (3.2.8)$$

$$+\hbar g_a(\hat{\mathbf{R}})e^{\frac{i}{\hbar}\hat{H}_{atom}t}(\hat{\sigma}_{ei}e^{-i\delta\varphi} - \hat{\sigma}_{ie}e^{i\delta\varphi})e^{-\frac{i}{\hbar}\hat{H}_{atom}t}e^{i\Omega_b\hat{b}^+\hat{b}t}(\hat{b}-\hat{b}^+)e^{-i\Omega_b\hat{b}^+\hat{b}t}.$$

At first, let us consider the part containing the atom operators. As the matrix corresponding to $\hat{H}_{atom}$ is diagonal (see (3.1.5), (3.2.5)), then we obtain the following expression

$$exp(\frac{i}{\hbar}\hat{H}_{atom}t) = \begin{pmatrix} e^{i\omega_e t} & 0 & 0 \\ 0 & e^{i\omega_i t} & 0 \\ 0 & 0 & e^{i\omega_g t} \end{pmatrix}. \qquad (3.2.9)$$

Utilizing (3.2.8) in (3.2.9) with the aid of (3.1.8) one can easily obtain that

$$exp(\frac{i}{\hbar}\hat{H}_{atom}t)\hat{\sigma}_{ig}exp(-\frac{i}{\hbar}\hat{H}_{atom}t)\hat{\sigma}_{ig} =$$

$$= \hat{\sigma}_{ig}e^{i(\omega_i-\omega_g)t}$$

In the same way, let us take into account (3.1.9), (3.1.10), and (3.1.11)

$$exp(\frac{i}{\hbar}\hat{H}_{atom}t)\hat{\sigma}_{gi}exp(-\frac{i}{\hbar}\hat{H}_{atom}t)\hat{\sigma}_{ig} = \hat{\sigma}_{gi}e^{-i(\omega_i-\omega_g)t},$$

$$exp(\frac{i}{\hbar}\hat{H}_{atom}t)\hat{\sigma}_{ei}e^{-i\delta\varphi}exp(-\frac{i}{\hbar}\hat{H}_{atom}t) = \hat{\sigma}_{ei}e^{i(\omega_e-\omega_i)t}e^{-i\delta\varphi},$$

$$exp(\frac{i}{\hbar}\hat{H}_{atom}t)\hat{\sigma}_{ie}e^{i\delta\varphi}exp(-\frac{i}{\hbar}\hat{H}_{atom}t)\hat{\sigma}_{ig} = \hat{\sigma}_{ig}e^{-i(\omega_e-\omega_i)t}e^{i\delta\varphi}.$$

Thus, the atomic part of the interaction Hamiltonian (3.2.8) is expressed in the following form

$$e^{\frac{i}{\hbar}\hat{H}_{atom}t}(\hat{\sigma}_{ig}-\hat{\sigma}_{gi})e^{-\frac{i}{\hbar}\hat{H}_{atom}t}+e^{\frac{i}{\hbar}\hat{H}_{atom}t}(\hat{\sigma}_{ei}e^{-i\delta\varphi}-\hat{\sigma}_{ie}e^{i\delta\varphi})e^{-\frac{i}{\hbar}\hat{H}_{atom}t}= \tag{3.2.10}$$

$$=\hat{\sigma}_{ig}e^{i(\omega_i-\omega_g)}t-\hat{\sigma}_{gi}e^{-i(\omega_i-\omega_g)t}+\hat{\sigma}_{ei}e^{i(\omega_e-\omega_i)t}e^{-i\delta\varphi}-\hat{\sigma}_{ei}e^{-i(\omega_e-\omega_i)t}e^{i\delta\varphi}.$$

Now, let us calculate the field operators. The time dependence of the operators in the interaction representation is determined by the Heisenberg equations of motion

$$\frac{d}{dt}\hat{f}=\frac{i}{\hbar}[\hat{H}_{field},\hat{f}]\equiv\frac{i}{\hbar}(\hat{H}_{filed}\hat{f}-\hat{f}\hat{H}_{filed}), \tag{3.2.11}$$

where $\hat{f}$ stands for $\hat{a}$, $\hat{a}^+$, $\hat{b}$ and $\hat{b}$. Inserting $\hat{H}_{field}$ (3.1.15) in (3.2.11), we obtain

$$\frac{d}{dt}\hat{f}=\frac{i}{\hbar}\Omega(\hat{f}^+\hat{f}\hat{f}-\hat{f}\hat{f}^+\hat{f}). \tag{3.2.12}$$

Because of the commutation relation (3.1.14), the equation (3.2.12) is transformed into

$$\frac{d}{dt}\hat{f}=-i\Omega\hat{f}^+. \tag{3.2.13}$$

In the same manner

$$\frac{d}{dt}\hat{f}^+=i\Omega\hat{f}^+. \tag{3.2.14}$$

So the time dependance of field operators is defined from (3.2.13) and (3.2.14) and are equal

$$\hat{a}(t)=\hat{a}(0)e^{-i\Omega_a t},\ \hat{a}(t)^+=\hat{a}^+(0)e^{i\Omega_a t},\ \hat{b}(t)=\hat{b}(0)e^{-i\Omega_b t},\ \hat{b}(t)^+=\hat{b}^+(0)e^{i\Omega_b t}, \tag{3.2.15}$$

Thus, by substituting (3.2.10) and (3.2.15) into the expression (3.2.8), we obtain the atom-photon interactions Hamiltonian in the interaction representation

$$\hat{H}^{(I)}_{\mathbf{rE}}=\hbar g_a(\hat{\mathbf{R}})[\hat{\sigma}_{ig}e^{i(\omega_i-\omega_g)t}-\hat{\sigma}_{gi}e^{-i(\omega_i-\omega_g)t}](\hat{a}e^{-i\Omega_a t}-\hat{a}^+e^{\Omega_a t})+$$

$$+\hbar g_a(\hat{\mathbf{R}})[\hat{\sigma}_{ei}e^{i(\omega_e-\omega_i)t}e^{-i\delta\varphi}-\hat{\sigma}_{ie}e^{-i(\omega_e-\omega_i)t}e^{i\delta\varphi}](\hat{b}e^{-i\Omega_b t}-\hat{b}^+e^{\Omega_b t}).$$

Now we shall introduce the following notation

$$\omega_i-\omega_j=\omega_{ij}. \tag{3.2.16}$$

By taking into account (3.2.16), the Hamiltonian is given by the following expression

$$\hat{H}^{(I)}_{\mathbf{rE}}=\hbar g_a(\hat{\mathbf{R}})[\hat{\sigma}_{ig}e^{i\omega_{ig}t}-\hat{\sigma}_{gi}e^{-i\omega_{ig}t}](\hat{a}e^{-i\Omega_a t}-\hat{a}^+e^{i\Omega_a t})+$$

$$+\hbar g_b(\hat{\mathbf{R}})[\hat{\sigma}_{ei}e^{i\omega_{ei}t}e^{i\delta\varphi}-\hat{\sigma}_{ie}e^{-i\omega_{ei}t}](\hat{b}e^{-i\Omega_b t}-\hat{b}^+e^{i\Omega_b t}),$$

or

$$\hat{H}^{(I)}_{\mathbf{rE}}=\hbar g_a(\hat{\mathbf{R}})(\hat{\sigma}_{gi}\hat{a}^+e^{i(\Omega_a-\omega_{ig})t}+\hat{\sigma}_{ig}\hat{a}e^{-i(\Omega_a-\omega_{ig})t}-\hat{\sigma}_{gi}\hat{a}e^{-i(\Omega_a+\omega_{ig})t}- \tag{3.2.17}$$

$$-e^{-i\delta\varphi}\hat{\sigma}_{ig}\hat{a}^+e^{i(\Omega_a+\omega_{ig})t}+\hbar g_b(\hat{\mathbf{R}})(\hat{\sigma}_{ie}\hat{b}^+e^{i(\Omega_b-\omega_{ei})t+i\delta\varphi}+\hat{\sigma}_{ei}\hat{b}e^{-i(\Omega_b-\omega_{ei})t-i\delta\varphi}-$$

$$-\hat{\sigma}_{ie}\hat{b}e^{-i(\Omega_b+\omega_{ei})t+i\delta\varphi} - \hat{\sigma}_{ei}\hat{b}^+e^{i(\Omega_b+\omega_{ei})t-i\delta\varphi}.$$

Further transformations require choosing a particular transition scheme (Fig. 3.1).

L-configuration. For this configuration (Fig. 3.1)

$$\omega_g < \omega_i < \omega_e \Rightarrow \omega_{ig} > 0, \omega_{ei} > 0. \tag{3.2.18}$$

After introducing the notations

$$\Delta_a \equiv \Omega_a - \omega_{ig}$$

and

$$\Delta_b \equiv \Omega_b - \omega_{ei}.$$

from (3.2.17) and (3.2.18), we obtain the interaction Hamiltonian for L-configuration $\hat{H}_{\mathbf{rE}}^{(I,L)}$

$$\hat{H}_{\mathbf{rE}}^{(I,L)} = \hbar g_a(\hat{\mathbf{R}})(\hat{\sigma}_{gi}\hat{a}^+e^{i\Delta_a t} + \hat{\sigma}_{ig}\hat{a}e^{-i\Delta_a t} - \hat{\sigma}_{gi}\hat{a}e^{-i(\Omega_a+\omega_{ig})t} - \hat{\sigma}_{ig}\hat{a}^+e^{i(\Omega_a+\omega_{ig})t})+$$

$$+\hbar g_b(\hat{\mathbf{R}})(\hat{\sigma}_{ie}\hat{b}^+e^{i\Delta_b t+i\delta\varphi} + \hat{\sigma}_{ei}\hat{b}e^{-i\Delta_b t-i\delta\varphi} - \hat{\sigma}_{ie}\hat{b}e^{-i(\Omega_b+\omega_{ei})t+i\delta\varphi} - \hat{\sigma}_{ei}\hat{b}^+e^{i(\Omega_b+\omega_{ei})t-i\delta\varphi}).$$

It is easy to see that operators $\hat{\sigma}_{gi}\hat{a}$, $\hat{\sigma}_{ig}\hat{a}^+$, $\hat{\sigma}_{ie}\hat{b}$ and $\hat{\sigma}_{ei}^+\hat{b}^+$ which led to the violation of the energy conservation law are multiplied by factors containing the sums of field oscillating and atom transitions frequencies. Contrary, operators $\hat{\sigma}_{gi}\hat{a}^+$, $\hat{\sigma}_{ig}\hat{a}$, $\hat{\sigma}_{ie}\hat{b}^+$ and $\hat{\sigma}_{ei}\hat{b}$ are multiplied by factors containing the differences (detuning) of aforementioned frequencies $\Delta \equiv \Omega - \omega$. So, in the interaction Hamiltonian we have two groups of terms. The terms of the first group oscillate approximately with double optical frequencies and their contribution in the Schrödinger equation is vanishingly small. The terms of another group oscillate with detuning frequencies Δ, and vary slowly. So, their contribution in the Schrödinger equation is significant. These terms are $\hat{\sigma}_{gi}\hat{a}^+$, $\hat{\sigma}_{ig}\hat{a}$, $\hat{\sigma}_{ie}\hat{b}^+$ and $\hat{\sigma}_{ei}\hat{b}$. So, utilize the averaging procedure - neglecting the fast oscillating terms such as $\hat{\sigma}_{gi}\hat{a}$, $\hat{\sigma}_{ig}\hat{a}^+$, $\hat{\sigma}_{ie}\hat{b}$ and $\hat{\sigma}_{ei}\hat{b}^+$ the interaction Hamiltonian reduces to the following form

$$\hat{H}_{\mathbf{rE}}^{(I,L)} \cong \hat{H}_{int}^{L} = \hbar g_a(\hat{\sigma}_{gi}\hat{a}^+e^{i\Delta_a t} + \hat{\sigma}_{ig}\hat{a}e^{-i\Delta_a t}) + \hbar g_b(\hat{\sigma}_{ei}\hat{b}^+e^{i\Delta_b t+i\delta\varphi} + \hat{\sigma}_{ei}\hat{b}e^{-i\Delta_b t-i\delta\varphi}).$$

It is also possible to strongly validate this given approach by the mathematical averaging, contrary to heuristic averaging used here. Also, the high order contributions can be calculated in this case.

Λ-configuration. For this configuration (Fig. 3.1)

$$\omega_g < \omega_i < \omega_e \Rightarrow \omega_{ig} > 0, \omega_{ei} < 0.$$

After introducing notations

$$\Delta_a \equiv \Omega_a - \omega_{ig}$$

and

$$\Delta_b \equiv \Omega_b - \omega_{ie},$$

we obtain the interaction Hamiltonian for Λ-configuration $\hat{H}_{\mathbf{rE}}^{(I,\Lambda)}$ in the interaction representation

$$\hat{H}_{\mathbf{rE}}^{(I,\Lambda)} = \hbar g_a(\hat{\mathbf{R}})(\hat{\sigma}_{gi}\hat{a}e^{i\Delta_a t} + \hat{\sigma}_{ig}\hat{a}e^{-i\Delta_a t} - \hat{\sigma}_{gi}\hat{a}e^{-i(\Omega_a+\omega_{ig})t} - \hat{\sigma}_{ig}\hat{a}^+e^{i(\Omega_a+\omega_{ig})t})+ \tag{3.2.19}$$

$$+\hbar g_b(\hat{\mathbf{R}})(\hat{\sigma}_{ie}\hat{b}^+ e^{i(\Omega_b+\omega_{ie})t+i\delta\varphi} + \hat{\sigma}_{ei}\hat{b}e^{-i\Delta_b t+i\delta\varphi} - \hat{\sigma}_{ei}\hat{b}e^{i\Delta_b t+i\delta\varphi} - \hat{\sigma}_{ei}\hat{b}^+ e^{i\Delta_b t-i\delta\varphi}).$$

Averaging (3.2.19) by fast oscillations led to

$$\hat{H}^{(I,\Lambda)}_{\mathbf{rE}} \cong \hat{H}^{\Lambda}_{int} = \hbar g_a(\hat{\sigma}^-_{gi}\hat{a}^+ e^{i\Delta_a t} + \hat{\sigma}_{gi}\hat{a}e^{-i\Delta_a t}) + \hbar g_b(\hat{\sigma}^-_{ie}\hat{b}^+ e^{-i\Delta_b t+i\delta\varphi} + \hat{\sigma}_{ei}\hat{b}e^{i\Delta_b t-i\delta\varphi}).$$

V-**configuration.** For this configuration (Fig. 3.1)

$$\omega_g < \omega_i < \omega_e \Rightarrow \omega_{ig} < 0, \omega_{ei} > 0.$$

After introducing notations

$$\Delta_a \equiv \Omega_a - \omega_{gi}$$

and

$$\Delta_b \equiv \Omega_b - \omega_{ei},$$

we obtain the interaction Hamiltonian for V-configuration $\hat{H}^{(I,V)}_{\mathbf{rE}}$ in the interaction representation

$$\hat{H}^{(I,\Lambda)}_{\mathbf{rE}} = \hbar g_a(\hat{\mathbf{R}})(\hat{\sigma}_{gi}\hat{a}^+ e^{i(\Omega_a+\omega_{gi})t} + \hat{\sigma}_{ig}\hat{a}e^{-i(\Omega_a+\omega_{gi})t} - \hat{\sigma}_{gi}\hat{a}e^{-i\Delta_a t} - \hat{\sigma}_{ig}\hat{a}^+ e^{i\Delta_a t} + \quad (3.2.20)$$

$$+\hbar g_b(\hat{\mathbf{R}})(\hat{\sigma}_{ie}\hat{b}^+ e^{i\Delta_b t+i\delta\varphi} + \hat{\sigma}_{ei}\hat{b}e^{-i\Delta_b t-i\delta\varphi} - \hat{\sigma}_{ie}\hat{b}e^{-i(\Omega_b+\omega_{ei})t+i\delta\varphi} - \hat{\sigma}_{ei}\hat{b}^+ e^{-i(\Omega_b+\omega_{ei})t-i\delta\varphi}).$$

Neglecting in (3.2.20) fast oscillating terms, we have

$$\hat{H}^{(I,V)}_{\mathbf{rE}} \cong \hat{H}^{V}_{int} = -\hbar g_a(\hat{\sigma}_{gi}\hat{a}e^{-i\Delta_a t} + \hat{\sigma}_{gi}\hat{a}^+ e^{i\Delta_a t}) + \hbar g_b(\hat{\sigma}_{ie}\hat{b}^+ e^{i\Delta_b t+i\delta\varphi} + \hat{\sigma}_{ei}\hat{b}e^{-i\Delta_b t-i\delta\varphi}).$$

Therefore, in this section, we have shown the RWA three-level atom interacting with two modes of standing wave resonator. The interaction representation is described by the following Hamiltonians

L-configuration

$$\hat{H}^{(I,L)}_{\mathbf{rE}} \cong \hat{H}^{L}_{int} = \hbar g_a(\hat{\sigma}_{gi}\hat{a}^+ e^{-i\Delta_a t} + \hat{\sigma}_{ig}\hat{a}e^{-i\Delta_a t}) + \hbar g_b(\hat{\sigma}_{ie}\hat{b}^+ e^{i\Delta_b t+i\delta\varphi} + \hat{\sigma}_{ei}\hat{b}e^{-i\Delta_b t-i\delta\varphi}). \quad (3.2.21)$$

Λ-configuration

$$\hat{H}^{(I,\Lambda)}_{\mathbf{rE}} \cong \hat{H}^{\Lambda}_{int} = \hbar g_a(\hat{\sigma}_{gi}\hat{a}^+ e^{i\Delta_a t} + \hat{\sigma}_{ig}\hat{a}e^{-i\Delta_a t}) + \hbar g_b(\hat{\sigma}_{ie}\hat{b}e^{-i\Delta_b t+i\delta\varphi} + \hat{\sigma}_{ei}\hat{b}e^{i\Delta^+_b t-i\delta\varphi}). \quad (3.2.22)$$

V-configuration

$$\hat{H}^{(I,V)}_{\mathbf{rE}} \cong \hat{H}^{V}_{int} = -\hbar g_a(\hat{\sigma}_{gi}\hat{a}e^{-i\Delta_a t} + \hat{\sigma}_{ig}\hat{a}^+ e^{i\Delta_a t}) + \hbar g_b(\hat{\sigma}_{ie}\hat{b}^+ e^{i\Delta_b t+i\delta\varphi} + \hat{\sigma}_{ei}\hat{b}e^{-i\Delta_b t-i\delta\varphi}). \quad (3.2.23)$$

Further, we will investigate three-level model concentrated on the internal degrees of freedom and neglecting center of mass movement.

3.3. The Quantum Dynamics of a Three-Level Atom

In this section, by following [100], we consider the internal quantum dynamics of a three-level atom in the fields for two modes of standing wave resonator. This model is more complicated and less studied than the well-known two-level system (see [139], [5], [140] and references cited there). So, consider the Schrödinger equation

$$i\hbar\frac{d|\Psi(t)>}{dt} = \hat{H}_{int}|\Psi(t)> \tag{3.3.1}$$

for the state vector $|\Psi>$ in the interaction representation, where the interaction Hamiltonian $\hat{H}_{int}$ due to (3.2.21)-(3.2.23) contains the following terms

$$\hat{H}_{int} = \hat{H}^L_{int} = \hbar g_a(\hat{\sigma}^-_{gi}\hat{a}^+e^{i\Delta_at} + \hat{\sigma}_{gi}\hat{a}e^{i\Delta_at}) + \hbar g_b(\hat{\sigma}^-_{ie}\hat{b}^+e^{i\Delta_bt+i\delta\varphi} + \hat{\sigma}_{ei}\hat{b}e^{-i\Delta_bt-i\delta\varphi}), \tag{3.3.2}$$

$$\hat{H}_{int} = \hat{H}^\Lambda_{int} = \hbar g_a(\hat{\sigma}^-_{gi}\hat{a}^+e^{i\Delta_at} + \hat{\sigma}_{gi}\hat{a}e^{-i\Delta_at}) + \hbar g_b(\hat{\sigma}^-_{ie}\hat{b}^+e^{-i\Delta_bt+i\delta\varphi} + \hat{\sigma}_{ei}\hat{b}e^{i\Delta_bt-i\delta\varphi}), \tag{3.3.3}$$

$$\hat{H}_{int} = \hat{H}^V_{int} = -\hbar g_a(\hat{\sigma}^-_{gi}\hat{a}^+e^{-i\Delta_at} + \hat{\sigma}_{gi}\hat{a}e^{i\Delta_at}) + \hbar g_b(\hat{\sigma}^-_{ie}\hat{b}^+e^{i\Delta_bt+i\delta\varphi} + \hat{\sigma}_{ei}\hat{b}e^{-i\Delta_bt-i\delta\varphi}). \tag{3.3.4}$$

We will study the internal dynamics of the three-level system when the interaction Hamiltonian becomes time independent. It is the most convenient way to obtain the time evolution operator.

Exact resonance in three-level atom. In the case of exact resonance $\Delta = 0$ the interaction Hamiltonians (3.3.2)-(3.3.4) transform into the following

$$\hat{H}_{int} = \hat{H}^L_{int} = \hbar g_a(\hat{\sigma}_{gi}\hat{a}^+ + \hat{\sigma}_{ig}\hat{a}) + \hbar g_b(\hat{\sigma}_{ie}\hat{b}^+e^{i\delta\varphi} + \hat{\sigma}_{ei}\hat{b}e^{-i\delta\varphi}), \tag{3.3.5}$$

$$\hat{H}_{int} = \hat{H}^\Lambda_{int} = \hbar g_a(\hat{\sigma}_{gi}\hat{a}^+ + \hat{\sigma}_{gi}\hat{a}) - \hbar g_b(\hat{\sigma}_{ie}\hat{b}e^{i\delta\varphi} + \hat{\sigma}_{ei}\hat{b}^+e^{-i\delta\varphi}), \tag{3.3.6}$$

$$\hat{H}_{int} = \hat{H}^V_{int} = -\hbar g_a(\hat{\sigma}_{gi}\hat{a}^+ + \hat{\sigma}_{ig}\hat{a}^+) + \hbar g_b(\hat{\sigma}_{ie}\hat{b}^+e^{i\delta\varphi} + \hat{\sigma}_{ei}\hat{b}e^{-i\delta\varphi}). \tag{3.3.7}$$

Now, they do not contain the explicit time dependence and the Schrödinger equation (3.3.1) for (3.3.5)-(3.3.7) can be integrated formally to give

$$|\Psi(t)> = exp(-\frac{i}{\hbar}\hat{H}_{int}t)|\Psi(0)>,$$

where the unitary time evolution operator is defined as

$$\hat{U}(t,t_0=0) \equiv exp(-\frac{i}{\hbar}\hat{H}_{int}t) \tag{3.3.8}$$

Substituting in (3.3.8) the Hamiltonians (3.3.5)-(3.3.7), we obtain the time evolution operators for L-, Λ- and V-configurations of a three-level atom

$$\hat{U}^L(t,t_0=0) = \exp[-ig_at(\hat{\sigma}_{gi}\hat{a}^+ + \hat{\sigma}_{ig}\hat{a}) - i\hbar g_bt(\hat{\sigma}_{ie}\hat{b}^+e^{i\delta\varphi} + \hat{\sigma}_{ei}\hat{b}e^{-i\delta\varphi})], \tag{3.3.9}$$

$$\hat{U}^\Lambda(t,t_0=0) = \exp[-ig_at(\hat{\sigma}_{gi}\hat{a}^+ + \hat{\sigma}_{ig}\hat{a}) - ig_bt(\hat{\sigma}_{ie}\hat{b}^+e^{i\delta\varphi} + \hat{\sigma}_{ei}\hat{b}e^{-i\delta\varphi})], \tag{3.3.10}$$

$$\hat{U}^V(t,t_0=0) = \exp[ig_at(\hat{\sigma}_{gi}\hat{a}^+ + \hat{\sigma}_{gi}\hat{a}) - i\hbar g_bt(\hat{\sigma}_{ie}\hat{b}^+e^{i\delta\varphi} + \hat{\sigma}_{ei}\hat{b}e^{-i\delta\varphi})]. \tag{3.3.11}$$

The initial state vector

$$|\Psi(0) \equiv |\Psi_{field} > \otimes|\Psi_{atom} >= \tag{3.3.12}$$

$$\sum_{n_a=0}^{\infty}\sum_{n_b=0}^{\infty} w_{n_a}w_{n_b}[c_e|e> +c_i|i> +c_g|g>]|n>= \sum_{n_a=0}^{\infty}\sum_{n_b=0}^{\infty} w_{n_a}w_{n_b}\begin{pmatrix} c_e \\ c_i \\ c_g \end{pmatrix}|n_a>|n_b>,$$

represents the direct product of the field state vector

$$|\Psi_{field}>\equiv \sum_{n_a=0}^{\infty}\sum_{n_b=0}^{\infty} w_{n_a}w_{n_b}|n_a>|n_b> \tag{3.3.13}$$

and the atom state vector

$$|\Phi_{atom}>\equiv [c_e|e> +c_i|i> +c_g|g>]. \tag{3.3.14}$$

This is a convenient form for utilizing the unitary transformations of initial atom-field state vector. Where the dynamics of a three-level atom, in the case of exact resonance, is defined by the (3.3.9)-(3.3.11) time evolution operators. In the next section, we calculate these operators and utilize the unitary transformations important for applications of quantum computation.

3.4. Calculation of the Time Evolution Operator and Unitary Transformations

Calculation of the time evolution operator in terms of the exponential matrix. Based on relations (3.1.8)-(3.1.10), let us rewrite the time evolution operators (3.3.9)-(3.3.11) in the matrix form

$$\hat{U}^{(L)}(t,t_0=0)=e^{-i\begin{pmatrix} 0 & e^{-i\delta\varphi}g_bt\hat{b} & 0 \\ e^{i\delta\varphi}g_bt\hat{b}^+ & 0 & g_at\hat{a} \\ 0 & g_at\hat{a}^+ & 0 \end{pmatrix}} \tag{3.4.1}$$

$$\hat{U}^{(\Lambda)}(t,t_0=0)=e^{-i\begin{pmatrix} 0 & -e^{-i\delta\varphi}g_bt\hat{b}^+ & 0 \\ -e^{i\delta\varphi}g_bt\hat{b} & 0 & g_at\hat{a} \\ 0 & g_at\hat{a}^+ & 0 \end{pmatrix}}, \tag{3.4.2}$$

$$\hat{U}^{(V)}(t,t_0=0)=e^{-i\begin{pmatrix} 0 & e^{-i\delta\varphi}g_bt\hat{b} & 0 \\ e^{i\delta\varphi}g_bt\hat{b}^+ & 0 & -g_at\hat{a}^+ \\ 0 & -g_at\hat{a} & 0 \end{pmatrix}}, \tag{3.4.3}$$

or in the general form as

$$\hat{U}(t,t_0=0)=e^{-i\begin{pmatrix} 0 & \hat{B} & 0 \\ \hat{B}^+ & 0 & \hat{A} \\ 0 & \hat{A}^+ & 0 \end{pmatrix}},$$

where

		configurations:	L	Λ	V
$\hat{A}$	$=$		$g_a t\hat{a}$	$g_a t\hat{a}$	$-g_a t\hat{a}^+$
$\hat{A}^+$	$=$		$g_a t\hat{a}^+$	$g_a t\hat{a}^+$	$-g_a t\hat{a}$
$\hat{B}$	$=$		$e^{-i\delta\varphi} g_b t\hat{b}$	$-e^{i\delta\varphi} g_b t\hat{b}^+$	$e^{-i\delta\varphi} g_b t\hat{b}$
$\hat{B}^+$	$=$		$e^{i\delta\varphi} g_b t\hat{b}^+$	$-e^{-i\delta\varphi} g_b t\hat{b}^+$	$e^{i\delta\varphi} g_b t\hat{b}^+$

corresponding operators from (3.4.1)-(3.4.3), that also includes the corresponding scalar factors for convenience.

Semiclassical approach. At this given stage, we substantially simplify the situation by means of simplifying the field operators. Right-hand sides of these equations (3.2.15) include only $\hat{f} \exp i\Omega t$ type terms, which describe contribution due to free evolution of the field. On the other hand, it means that the applied fields are sufficiently intense compared with what can be contributed by the atomic transitions. The field operators at time t may then be replaced by their averages in the initial state [122]. We assume the fields to be in the coherent states (the field of laser radiation) $|\{\alpha_n\}>$ and $|\{\beta_n\}>$ $(|\alpha|, |\beta| >> 1)$, replace the operators $\hat{a}$ $(t_0 = 0)$ and b $(t_0 = 0)$ by α and β in (3.4.1)-(3.4.3) and receive the following

L-configuration

$$\hat{U}^{(L)}(t, t_0 = 0) = e^{-i\begin{pmatrix} 0 & e^{-i\delta\varphi} g_\beta t\beta & 0 \\ e^{i\delta\varphi} g_\beta t\beta^* & 0 & g_a t\alpha \\ 0 & g_a t\alpha^* & 0 \end{pmatrix}},$$

Λ-configuration

$$\hat{U}^{(\Lambda)}(t, t_0 = 0) = e^{-i\begin{pmatrix} 0 & -e^{-i\delta\varphi} g_\beta t\beta^* & 0 \\ -e^{i\delta\varphi} g_\beta t\beta & 0 & g_\alpha t\alpha \\ 0 & g_\alpha t\alpha^* & 0 \end{pmatrix}},$$

V-configuration

$$\hat{U}^{(V)}(t, t_0 = 0) = e^{-i\begin{pmatrix} 0 & e^{-i\delta\varphi} g_\beta t\beta & 0 \\ e^{i\delta\varphi} g_\beta t\beta^* & 0 & -g_\alpha t\alpha^* \\ 0 & -g_\alpha t\alpha & 0 \end{pmatrix}}.$$

In general, we have

$$\hat{U}(t, t_0 = 0) = e^{-i\begin{pmatrix} 0 & B & 0 \\ B^* & 0 & A \\ 0 & A^* & 0 \end{pmatrix}}, \tag{3.4.4}$$

where

		configurations:	L	Λ	V	
A	$=$		$g_\alpha t\alpha$	$g_\alpha t\alpha$	$-g_\alpha t\alpha^*$	(3.4.5)
A^*	$=$		$g_\alpha t a^*$	$g_\alpha t a^*$	$-g_\alpha t\alpha$	
B	$=$		$e^{-i\delta\varphi} g_\beta t\beta$	$-e^{-i\delta\varphi} g_\beta t\beta^*$	$e^{-i\delta\varphi} g_\beta t\beta$	
B^*	$=$		$e^{i\delta\varphi} g_\beta t\beta^*$	$-e^{-i\delta\varphi} g_\beta t\beta^*$	$e^{i\delta\varphi} g_\beta t\beta^*$	

The exponential matrix (3.4.4) can be calculated to give

$$\hat{U}(t,t_0=0)=\begin{pmatrix} \frac{AA^*+BB^*cos(\sqrt{BB^*+AA^*})}{BB^*+AA^*} & -\frac{iBsin(\sqrt{BB^*+AA^*})}{\sqrt{BB^*+AA^*}} & -\frac{BA(cos(\sqrt{BB^*+AA^*})-1)}{BB^*+AA^*} \\ -\frac{iB^*sin(\sqrt{BB^*+AA^*})}{\sqrt{BB^*+AA^*}} & cos(\sqrt{BB^*+AA^*}) & -\frac{iAsin(\sqrt{BB^*+AA^*})}{\sqrt{BB^*+AA^*}} \\ \frac{A^*B^*(cos(\sqrt{BB^*+AA^*})-1)}{BB^*+AA^*} & -\frac{iA^*sin(\sqrt{BB^*+AA^*})}{\sqrt{BB^*+AA^*}} & \frac{BB^*+AA^*cos(\sqrt{BB^*+AA^*})}{BB^*+AA^*} \end{pmatrix}, \tag{3.4.6}$$

Unitary transformations. So, we have found the time evolution operator for the three-level atom interacting with two modes of field in the coherent states. As the initial state vector is represented for the field in the arbitrary state $|\Psi_{field}>\equiv\sum_{n_a=0}^{\infty}\sum_{n_b=0}^{\infty}w_{n_a}w_{n_b}|n_a>|n_b>$, than it is essential to rewrite this state vector for two modes in coherent state. The probability that there are n_a (n_b) photons at the initial coherent state $|\alpha>$ ($|\beta>$) is given by the Poisson distribution [129], i. e.,

$$w_{n_a}=\frac{<n_a>^{n_a}e^{-<n_a>}}{n_a!},w_{n_b}=\frac{<n_b>^{n_b}e^{-<n_b>}}{n_b!}, \tag{3.4.7}$$

where $<n>=|\alpha|^2$ ($|\beta|^2$). Due to (3.4.7), we can now write the initial state of field (3.3.13) and atom-field system (3.3.12) in a more suitable form

$$|\Psi_{field}>\equiv\sum_{n_a=0}^{\infty}\sum_{n_b=0}^{\infty}\frac{<n_a>^{n_a}e^{-<n_a>}}{n_a!}\frac{<n_b>^{n_b}e^{-<n_b>}}{n_b!}|n_a>|n_b>=|\alpha>|\beta>, \tag{3.4.8}$$

$$|\Psi(0)>=|\Psi_{field}>\otimes|\Psi_{atom}>= \tag{3.4.9}$$

$$=\sum_{n_a=0}^{\infty}\sum_{n_b=0}^{\infty}\frac{<n_a>^{n_a}e^{-<n_a>}}{n_a!}\frac{<n_b>^{n_b}e^{-<n_b>}}{n_b!}[c_e|e>+c_i|i>+c_g|g>]|n_a>|n_b>=$$

$$=\sum_{n_a=0}^{\infty}\sum_{n_b=0}^{\infty}\frac{<n_a>^{n_a}e^{-<n_a>}}{n_a!}\frac{<n_b>^{n_b}e^{-<n_b>}}{n_b!}\begin{pmatrix}c_e\\c_i\\c_g\end{pmatrix}|a_a>|n_b>=$$

$$=[c_e|e>+c_i|i>+c_g|g>]|\alpha>|\beta>=\begin{pmatrix}c_e\\c_i\\c_g\end{pmatrix}|\alpha>|\beta>.$$

Now, when applying (3.4.6) to (3.4.9), we obtain the total state vector $|\Psi(t)>$ of a three-level atom in the coherent field of two modes for arbitrary time $t>t_0=0$

$$|\Psi(t)>=\hat{U}(t,t_0=0)[c_e|e>+c_i|i>+c_g|g>]|\alpha>|\beta>=\hat{U}(t,t_0=0)\begin{pmatrix}c_e\\c_i\\c_g\end{pmatrix}|\alpha>|\beta>. \tag{3.4.10}$$

Unitary transformations for *L*-configuration. For L-configuration (3.3.14), (3.4.10) gives

$$|\Psi(t)>=\hat{U}^{(L)}(t,t_0=0)[c_e|e>+c_i|i>+c_g|g>]|\alpha>|\beta>=$$

$$=\hat{U}^{(L)}(t,t_0=0)\begin{pmatrix}c_e\\c_i\\c_g\end{pmatrix}|\alpha>|\beta>.$$

and due to (3.4.6) and (3.4.5), we obtain

$$\hat{U}^{(L)}(t,t_0=0)=\frac{1}{g_a^2|\alpha|^2+g_b^2|\beta|^2}\times\left\{\begin{pmatrix} g_a^2|\alpha|^2 & 0 & -e^{-i\delta\varphi}g_ag_b\beta\alpha \\ 0 & 0 & 0 \\ -e^{i\delta\varphi}g_ag_b\beta^*\alpha^* & 0 & g_b^2|\beta|^2 \end{pmatrix}+\right.$$

$$+cos(t\sqrt{g_a^2|\alpha|^2+g_b^2|\beta|^2})\begin{pmatrix} g_b^2|\beta|^2 & 0 & e^{-i\delta\varphi}g_ag_b\beta\alpha \\ 0 & g_a^2\alpha^2+g_b^2\beta^2 & 0 \\ e^{i\delta\varphi}g_ag_b\beta^*\alpha^* & 0 & g_a^2|\alpha|^2 \end{pmatrix}-$$

$$\left.-sin(t\sqrt{g_a^2|\alpha|^2+g_b^2|\beta|^2})\begin{pmatrix} 0 & e^{-i\delta\varphi}g_b\beta & 0 \\ e^{i\delta\varphi}g_b\beta^* & 0 & g_a\alpha \\ 0 & g_a\alpha^* & 0 \end{pmatrix}\right\}.$$

Thus, from the time evolution operator one can extract three formal parts: time independent, time dependent as $\cos(Gt)$ and time dependent as $\sin(Gt)$. For the unitary operator $\hat{U}^{(L)}(t,t_0=0)$ an initial moment $t=0$, we obtain

$$\hat{U^{(L)}}(t=0)=\begin{pmatrix} 1 & 0 & 0 \\ 0 & 1 & 0 \\ 0 & 0 & 1 \end{pmatrix} \tag{3.4.11}$$

and

$$\Psi(t=0)=\begin{pmatrix} c_e(0)|\alpha>|\beta> \\ c_i(0)|\alpha>|\beta> \\ c_g(0)|\alpha>|\beta> \end{pmatrix}, \tag{3.4.12}$$

or

$$\Psi(t=0)=c_e(0)|e,\alpha,\beta>+c_i(0)|i,\alpha,\beta>+c_g|g,\alpha,\beta>, \tag{3.4.13}$$

where the following notations are introduced $|e>|\alpha>|\beta>\equiv|e,\alpha,\beta>$, $|i>|\alpha>|\beta>\equiv|i,\alpha,\beta>$ and $|g>|\alpha>|\beta>\equiv|g,\alpha,\beta>$. (3.4.11)-(3.4.13) confirm the main feature of the interaction representation, that in the absence of interaction the state vectors do not change [84].

3.5. Probability Amplitude Method

In this section, we present a different but equivalent method to solve for the evolution of the atom-field system described in interaction picture

$$i\hbar\frac{d|\Psi(t)>}{dt}=\hat{H}_{int}|\Psi(t)> \tag{3.5.1}$$

and based on the solutions of the probability amplitudes. The atom-field system is described by the interaction Hamiltonian (3.2.17)

$$\hat{H}_{\mathbf{rE}}^{(I)}=\hbar g_a(\hat{\sigma}_{gi}\hat{a}^+e^{i(\Omega_a-\omega_{ig})t}+\hat{\sigma}_{ig}\hat{a}e^{-i(\Omega_a-\omega_{ig})t}-\hat{\sigma}_{gi}\hat{a}e^{-i(\Omega_a+\omega_{ig})t}+\hat{\sigma}_{ig}\hat{a}^+e^{i(\Omega_a+\omega_{ig})t}+ \tag{3.5.2}$$
$$+\hbar g_b(\hat{\sigma}_{ie}\hat{b}^+e^{i(\Omega_b-\omega_{ei})t+i\delta\varphi}+\hat{\sigma}_{ei}\hat{b}e^{-i(\Omega_b-\omega_{ei})t-i\delta\varphi}-\hat{\sigma}_{ie}\hat{b}e^{-i(\Omega_b+\omega_{ei})t+i\delta\varphi}-\hat{\sigma}_{ei}\hat{b}^+e^{i(\Omega_b+\omega_{ei})t-i\delta\varphi}.$$

where the RWA should be utilized. Further calculations are based on amplitude method of QED [129] and are very similar for different configurations. So, we perform the calculations only for L-configuration and then present the final result for all of them.

In the RWA, the Hamiltonian (3.5.2), due to configuration (see Fig. 3.1), transforms into the ((3.3.2)-(3.3.4)) interaction operators. Especially for L-configuration

$$\hat{H}_{int} = \hat{H}^L_{int} = \hbar g_a(\hat{\sigma}_{gi}\hat{a}^+ e^{i\Delta_a t} + \hat{\sigma}_{ig}\hat{a}e^{-i\Delta_a t}) + \hbar g_b(\hat{\sigma}_{ie}\hat{b}^+ e^{i\Delta_b t + i\delta\varphi} + \hat{\sigma}_{ei}\hat{b}e^{-i\Delta_b t - i\delta\varphi}). \quad (3.5.3)$$

In the exact resonances case ($\Delta = 0$) (3.5.3) transforms into

$$\hat{H}^L_{int} = \hbar g_a(\hat{\sigma}_{gi}\hat{a}^+ + \hat{\sigma}_{ig}\hat{a}) + \hbar g_b(\hat{\sigma}_{ie}\hat{b}^+ e^{i\delta\varphi} + \hat{\sigma}_{ei}\hat{b}e^{-i\delta\varphi}). \quad (3.5.4)$$

Now, acting on the wave vector $|e, n_a, n_b>$ by the operator (3.5.4); step-by-step we obtain

$$\hat{H}^L_{int}|e, n_a, n_b >= \hbar g_b e^{i\delta\varphi}\sqrt{n_b+1}|e, n_a, n_b >, \quad (3.5.5)$$

$$\hat{H}^L_{int}|e, n_a, n_b+1 >= \hbar g_a\sqrt{n_a+1}|g, n_a+1, n_b+1 > +\hbar g_b e^{-i\delta\varphi}\sqrt{n_b+1}|e, n_a, n_b >, \quad (3.5.6)$$

$$\hat{H}^L_{int}|g, n_a+1, n_b+1 >= \hbar g_a\sqrt{n_a+1}|i, n_a, n_b+1 >, \quad (3.5.7)$$

where n_a and n_b represent the number of $\hbar\Omega_a$ and $\hbar\Omega_b$ photons. From (3.5.7) it is also easy to see that for $n_a = 0$ and for arbitrary n_b (let us $n_b + 1 = m$)

$$\hat{H}^L_{int}|g, 0, m >= 0. \quad (3.5.8)$$

Now with the aid of these obtained amplitudes (3.5.5)-(3.5.8), one can proceed to solve the equation of motion (3.5.1) for a state vector of a total system $|\Psi(t)>$, i. e.,

$$|\Psi(t)>= \sum_{n_a=0,n_b=0}^{\infty} \Psi_{e,n_a,n_b}(t)|e, n_a, n_b > +\Psi_{i,n_a,n_b+1}(t)|i, n_a, n_b+1 > + \quad (3.5.9)$$

$$+\Psi_{g,n_a+1,n_b+1}(t)|g, n_a+1, n_b+1 > +\Psi_{g,0,m}(t)|g, 0, m>$$

Now, substitute (3.5.9) in (3.5.1) and then project the resulting equations onto $< e, n_a, n_b|$, $< i, n_a, n_b+1|, < g, n_a+1, n_b+1|$, $< g, 0, m|$

wecanobtainthefollowingsystemofequations : $\dot{\Psi}_{e,n_a,n_b}(t) = -ig_b e^{-i\delta\varphi}\sqrt{n_b+1}\Psi_{i,n_a,n_b+1}(t)$, (3.5.10)

$$\dot{\Psi}_{i,n_a,n_b+1}(t) = -ig_b e^{i\delta\varphi}\sqrt{n_b+1}\Psi_{e,n_a,n_b}(t) - ig_a\sqrt{n_a+1}\Psi_{g,n_a+1,n_b+1}(t), \quad (3.5.11)$$

$$\dot{\Psi}_{g,n_a+1,n_b+1}(t) = -ig_a e^{i\delta\varphi}\sqrt{n_a+1}\Psi_{i,n_a,n_b+1}(t), \quad (3.5.12)$$

$$\dot{\Psi}_{g,0,m}(t) = 0. \quad (3.5.13)$$

To solve this system of equations (3.5.10)-(3.5.12), we should define the initial conditions first. Under initial conditions the state vector has the following form (see (3.4.8))

$$|\Psi(t=0)>= |\Psi_{atom} \otimes |\Psi_{field} >\equiv [c_e|e> +c_i|i> +c_g|g>] \otimes \sum_{n_a=0}^{\infty}\sum_{n_b=0}^{\infty} w_{n_a}w_{n_b}|n_a>|n_b>,$$

or in notations of (3.5.9)

$$|\Psi(t=0)>=\sum_{n_a,n_b=0}^{\infty}[w_{n_a,n_b}c_e|e,n_a,n_b>+w_{n_a,n_b+1}c_i|i,n_a,n_b+1>+ \qquad (3.5.14)$$

$$+w_{n_a+1,n_b+1}c_g|g,n_a+1,n_b+1>]+w_{0,m}c_g|g,0,m>$$

where $w_{n_a,n_b}=w_{n_a}w_{n_b}$.

Comparing (3.5.9) for $t=0$ and (3.5.14), we obtain the initial conditions

$$\Psi_{e,n_a,n_b}(t=0)=w_{n_a,n_b}c_e, \qquad (3.5.15)$$

$$\Psi_{i,n_a,n_b+1}(t=0)=w_{n_a,n_b+1}c_i, \qquad (3.5.16)$$

$$\Psi_{g,n_a+1,n_b+1}(t=0)=w_{n_a+1,n_b+1}c_g, \qquad (3.5.17)$$

$$\Psi_{g,0,m}(t=0)=\Psi_{g,0,m}(t=0)=w_{0,m}c_g. \qquad (3.5.18)$$

Now, let us solve this system of equations (3.5.10)-(3.5.12) under initial conditions defined by (3.5.15)-(3.5.18). Multiplying (3.5.10)-(3.5.12) by e^{-st} and using the Laplace transformation for the probability amplitudes the system of equations transforms to the following:

$$e^{-st}\dot{\Psi}_{e,n_a,n_b}(t)=-ig_b\sqrt{n_b+1}e^{-st}e^{-i\delta\varphi}\Psi_{i,n_a,n_b+1}(t), \qquad (3.5.19)$$

$$e^{-st}\dot{\Psi}_{i,n_a,n_b+1}(t)=-ig_b\sqrt{n_b+1}e^{-st}e^{i\delta\varphi}-ig_ae^{-st}\sqrt{n_a+1}\Psi_{g,n_a+1,n_b+1}(t), \qquad (3.5.20)$$

$$e^{-st}\dot{\Psi}_{e,n_a,n_b}(t)=-ig_b\sqrt{n_a+1}e^{-st}\Psi_{i,n_a,n_b+1}(t). \qquad (3.5.21)$$

Now, integrating (3.5.19)-(3.5.21) by t, we obtain

$$s\overline{\Psi}_{e,n_a,n_b}(s)+ig_be^{-i\delta\varphi}\sqrt{n_b+1}\overline{\Psi}_{e,n_a,n_b+1}(s)=\Psi_{e,n_a,n_b}(t=0),$$

$$ig_be^{i\delta\varphi}\sqrt{n_b+1}\overline{\Psi}_{e,n_a,n_b+1}(s)+s\overline{\Psi}_{e,n_a,n_b+1}(s)+ig_b\sqrt{n_a+1}\overline{\Psi}_{g,n_a+1,n_b+1}(s)=$$
$$=\Psi_{i,n_a,n_b+1}(t=0),$$

$$ig_a\sqrt{n_a+1}\overline{\Psi}_{i,n_a,n_b+1}(s)+s\overline{\Psi}_{g,n_a+1,n_b+1}(s)=\Psi_{g,n_a+1,n_b+1}(t=0).$$

Here $\overline{\Psi}(s)=\int_0^{\infty}\Psi(t)e^{-st}dt$ is the Laplace transform of $\Psi(t)$. For this system of equations, we can also write the following matrix form:

$$U\begin{pmatrix}\overline{\Psi}_{e,n_a,n_b}(s)\\ \overline{\Psi}_{i,n_a,n_b+1}(s)\\ \overline{\Psi}_{g,n_a+1,n_b+1}(s)\end{pmatrix}=\begin{pmatrix}w_{n_a,n_b}c_e\\ w_{n_a,n_b+1}c_i\\ w_{n_a+1,n_b+1}c_g\end{pmatrix},$$

where

$$U=\begin{pmatrix}s & ig_ae^{-i\delta\varphi}\sqrt{n_b+1} & 0\\ ig_be^{i\delta\varphi}\sqrt{n_b+1} & s & ig_a\sqrt{n_a+1}\\ 0 & ig_a\sqrt{n_a+1} & s\end{pmatrix}.$$

Now, if we invoke the inverse Laplace transform we obtain the following expression for probability amplitudes

$$\begin{pmatrix} \Psi_{e,n_a,n_b}(t) \\ \Psi_{i,n_a,n_b+1}(t) \\ \Psi_{g,n_a+1,n_b+1}(t) \end{pmatrix} = \frac{1}{2\pi i}\int_C e^{st}U^{-1}(s)ds \begin{pmatrix} w_{n_a,n_b}c_e \\ w_{n_a,n_b+1}c_i \\ w_{n_a+1,n_b+1}c_g \end{pmatrix}, \quad (3.5.22)$$

where

$$U^{-1} = \frac{1}{s(s+i\lambda)(1-i\lambda)}\times$$

$$\times \begin{pmatrix} s^2+g_a^2(n_a+1) & -isg_a e^{-i\delta\varphi}\sqrt{n_b+1} & -g_b e^{-i\delta\varphi}\sqrt{n_b+1}g_a\sqrt{n_a+1} \\ -isg_a e^{-i\delta\varphi}\sqrt{n_b+1} & s^2 & -isg_a\sqrt{n_a+1} \\ -g_b e^{-i\delta\varphi}\sqrt{n_b+1}g_a\sqrt{n_a+1} & i-isg_a\sqrt{n_a+1} & s^2+g_a^2(n_a+1) \end{pmatrix},$$

and $\lambda = \sqrt{g_a^2(n_a+1)+g_b^2(n_b+1)}$. The poles of (3.5.22) coincide with roots of the determinant U and equal $0, i\lambda, -i\lambda$. From this, due to the Cauchy second integral theorem it follows

$$\Psi_{e,n_a,n_b}(t) = \left[cos(\lambda t)+\frac{g_a^2}{\lambda^2}(n_a+1)(1-cos(\lambda t))\right]\Psi_{i,n_a,n_b+1}(0)-$$

$$\left[i\frac{g_b}{\lambda}e^{-i\delta\varphi}\sqrt{n_b+1}sin(\lambda t)\right]\Psi_{i,n_a,n_b+1}(0)-$$

$$-\left[\frac{g_b g_a}{\lambda^2}e^{-i\delta\varphi}\sqrt{n_b+1}\sqrt{n_a+1}(1-cos(\lambda t))\right]\Psi_{g,n_a+1,n_b+1}(0),$$

$$\Psi_{e,n_a,n_b+1}(t) = \left[\frac{g_a}{\lambda}e^{i\delta\varphi}\sqrt{n_b+1}sin(\lambda t)\Psi_{e,n_a,n_b}(0)+cos(\lambda t)\right]\Psi_{i,n_a,n_b+1}(0)-$$

$$-\left[i\frac{g_b}{\lambda}\sqrt{n_a+1}sin(\lambda t)\right]\Psi_{g,n_a+1,n_b+1}(0),$$

$$\Psi_{e,n_a,n_b}(t) = \left[\frac{g_b g_a}{\lambda_n^2}e^{-i\delta\varphi}\sqrt{n_b+1}\sqrt{n_a+1}(1-cos(\lambda t))\right]\Psi_{i,n_a,n_b}(0)-$$

$$-\left[i\frac{g_a}{\lambda}\sqrt{n_a+1}sin(\lambda t)\right]\Psi_{i,n_a,n_b+1}(0)-$$

$$-\left[cos(\lambda t)+\frac{g_a^2}{\lambda_n^2}(n_b+1)(1-cos(\lambda t))\right]\Psi_{g,n_a+1,n_b+1}(0).$$

Amplitudes for the configurations Λ and V are obtained in similar manner.
The evolution operator

$$\begin{pmatrix} u_{11}(t) & u_{12}(t) & u_{13}(t) \\ u_{21}(t) & u_{22}(t) & u_{23}(t) \\ u_{31}(t) & u_{32}(t) & u_{33}(t) \end{pmatrix}$$

for the L, Λ, V configurations has the following form:

L-configuration, $\lambda = \sqrt{g_a^2(n_a+1)+g_b^2(n_b+1)}$.

$$u_{11}^L(t) = cos(\lambda t)+\frac{g_a^2}{\lambda^2}(n_a+1)(1-cos(\lambda t)),$$

$$u^L_{12}(t) = -i\frac{g_b}{\lambda}e^{-i\delta\varphi}\sqrt{n_b+1}sin(\lambda t),$$

$$u^L_{13}(t) = \frac{g_b g_a}{\lambda^2}e^{-i\delta\varphi}\sqrt{n_b+1}\sqrt{n_a+1}(1-cos(\lambda t)),$$

$$u^L_{21}(t) = -i\frac{g_b}{\lambda}e^{i\delta\varphi}\sqrt{n_b+1}sin(\lambda t),$$

$$u^L_{22}(t) = cos(\lambda t),$$

$$u^L_{23}(t) = -i\frac{g_b}{\lambda}e^{-i\delta\varphi}\sqrt{n_a+1}sin(\lambda t),$$

$$u^L_{31}(t) = \frac{g_b g_a}{\lambda^2}e^{-i\delta\varphi}\sqrt{n_b+1}\sqrt{n_a+1}(1-cos(\lambda t)),$$

$$u^L_{32}(t) = -i\frac{g_b}{\lambda_n}e^{-i\delta\varphi}\sqrt{n_a+1}sin(\lambda t),$$

$$u^L_{33}(t) = cos(\lambda t) + \frac{g_a^2}{\lambda^2}(n_a+1)(1-cos(\lambda t)).$$

Λ-configuration, $\lambda = \sqrt{g_a^2(n_a+1)+g_b^2 n_b^2}$.

$$u^\Lambda_{11}(t) = cos(\lambda t) + \frac{g_a^2}{\lambda^2}(n_a+1)(1-cos(\lambda t)),$$

$$u^\Lambda_{12}(t) = i\frac{g_b}{\lambda}e^{-i\delta\varphi}\sqrt{n_b}sin(\lambda t),$$

$$u^\Lambda_{13}(t) = \frac{g_b g_a}{\lambda^2}e^{-i\delta\varphi}\sqrt{n_b}\sqrt{n_a+1}(1-cos(\lambda t)),$$

$$u^\Lambda_{21}(t) = i\frac{g_b}{\lambda}e^{i\delta\varphi}\sqrt{n_b}sin(\lambda t),$$

$$u^\Lambda_{22}(t) = cos(\lambda t),$$

$$u^\Lambda_{23}(t) = -i\frac{g_b}{\lambda}\sqrt{n_a+1}sin(\lambda t),$$

$$u^\Lambda_{31}(t) = \frac{g_b g_a}{\lambda^2}e^{i\delta\varphi}\sqrt{n_b}\sqrt{n_a+1}(1-cos(\lambda t)),$$

$$u^\Lambda_{32}(t) = -i\frac{g_a}{\lambda}\sqrt{n_a+1}sin(\lambda t),$$

$$u^\Lambda_{33}(t) = cos(\lambda t) + \frac{g_b^2}{\lambda^2}n_b(1-cos(\lambda t)).$$

V-configuration, $\lambda = \sqrt{g_a^2 n_a + g_b^2(n_b+1)}$.

$$u^V_{11}(t) = cos(\lambda t) + \frac{g_a^2}{\lambda^2}(n_a)(1-cos(\lambda t)),$$

$$u^V_{12}(t) = -i\frac{g_b}{\lambda}e^{-i\delta\varphi}\sqrt{n_b+1}sin(\lambda t),$$

$$u^V_{13}(t) = \frac{g_b g_a}{\lambda^2}e^{-i\delta\varphi}\sqrt{n_b+1}\sqrt{n_a}(1-cos(\lambda t)),$$

$$u_{21}^V(t) = -i\frac{g_b}{\lambda}e^{i\delta\varphi}\sqrt{n_b+1}sin(\lambda t),$$

$$u_{22}^V(t) = cos(\lambda t),$$

$$u_{23}^V(t) = i\frac{g_a}{\lambda}\sqrt{n_a}sin(\lambda t),$$

$$u_{31}^V(t) = \frac{g_b g_a}{\lambda^2}e^{i\delta\varphi}\sqrt{n_b+1}\sqrt{n_a}(1-cos(\lambda t)),$$

$$u_{32}^V(t) = i\frac{g_a}{\lambda}\sqrt{n_a}sin(\lambda t),$$

$$u_{33}^V(t) = cos(\lambda t) + \frac{g_b^2}{\lambda^2}(n_b+1)(1-cos(\lambda t)).$$

Under these previously described assumptions, we have constructed evolution operators in the explicit form. Performed analysis has shown that unlike the case methods of semiclassical analysis — where the atom is quantized whereas the field is classical, the completely quantum approach shows that in the absence of fields the atom still oscillates and this must be taken into account in quantum calculations. Moreover, quantum processing may be performed on the level of single photons.

3.6. Control of Three-Level Quantum System

The paper [18] is devoted to an optimal control problem for a three-level quantum system. A reduction of a quantum system to the first three eigenstates is described by a three-dimensional bilinear control system. The problem of optimal transfer of the system between eigenstates corresponding to the first and the third eigenvalues is stated as an optimal control problem with a quadratic cost. The problem is transformed to a sub-Riemannian problem on the spheres S^2 and S^5 (the real and complex cases). In these cases, the problem is lifted to a right-invariant sub-Riemannian problem on the Lie groups $SO(3)$ and $SU(3)$. Application of the *Pontryagin Maximum Principle* on Lie groups and other techniques of sub-Riemannian geometry yields a complete description of optimal controls and trajectories.

Description of physical picture. Assume that dynamics is governed by the time dependent Schrödinger equation (in a system of units such that $\hbar = 1$):

$$i\frac{d\psi(t)}{dt} = H\psi(t), \tag{3.6.1}$$

where $\psi(.) : \mathbb{R} \to \mathbb{C}^3$ and:

$$H = \begin{pmatrix} E_1 & \Omega_1 & 0 \\ \Omega_1^* & E_2 & \Omega_2 \\ 0 & \Omega_2^* & E_3 \end{pmatrix}. \tag{3.6.2}$$

Let the controls $\Omega_1(.), \Omega_2(.)$, different from zero only in a fixed interval $[0,T]$, be connected to the physical parameters by $\Omega_j(t) = \mu_j F_j(t)/2$, $j = 1,2$, with F_j the external pulsed field and μ_j are the couplings (intrinsic to the quantum system), restricted to couple only levels j and $j+1$ by pairs.

This Hamiltonian is the sum of a "drift-term" H_0, plus a time dependent potential $V(t)$ (the control term, i.e., the lasers). The drift term is assumed to be diagonal with eigenvalues (energy levels) $\ldots > E_3 > E_2 > E_1$. Then in this spectral resolution of H_0, the control term $V(t)$ is assumed to couple only the energy levels E_1, E_2 and E_2, E_3. The projected problem in the eigenspaces corresponding to E_1, E_2, E_3 is completely decoupled and is described by the Hamiltonian (3.6.2).

The problem is as follows:

Problem. Assume that for time $t \leq 0$ the state of the system lies in the eigenspace corresponding to the ground eigenvalue E_1. We want to determine suitable controls $\Omega_i(.)$, $i = 1, 2$, such that for time $t \geq T$, the system reaches the eigenspace corresponding to E_3, requiring that these controls minimize the cost (energy in the following):

$$J = \int_0^T \left(|\Omega_1(t)|^2 + |\Omega_2(t)|^2\right) dt. \tag{3.6.3}$$

The solution of this problem is the following theorem:

Theorem 3.6.1. *[18] For the three-level problem with complex controls, optimality implies resonance. More precisely, controls $\Omega_1(.), \Omega_2(.)$ are optimal, if and only if they have the following form:*

$$\begin{cases} \Omega_1(t) = \cos(t/\sqrt{3})e^{i[(E_2-E_1)t+\varphi_1]}, \\ \Omega_2(t) = \sin(t/\sqrt{3})e^{i[(E_3-E_2)t+\varphi_2]}. \end{cases} \tag{3.6.4}$$

where φ_1, φ_2 are two arbitrary phases. Here, the final time T is fixed in such a way that sub-Riemannian geodesics are parameterized by arclength, and it is given by $T = \frac{\sqrt{3}}{2}\pi$.

For the Hamiltonians of type (3.2.21), (3.2.22), and (3.2.23) the analogical result to this theorem is an open question.

Chapter 4

Methods of Geometric Control Theory for Quantum Computations

In this chapter, we consider geometric approach to control theory [1]. We follow the research papers of [59], [60] and the geometric control theory methods used for construction of the general theory in geometric quantum computation.

There are three main directions in the theory of quantum computations: (a)*Algebraic:* the problem of constructing a universal set of gates; (b) *practical:* the problem of constructing a physical device implementing a universal set of gates; (c) *Algorithmic:* inventing algorithms for quantum devices to solve various computational problems effectively, provided that a universal set of gates is already constructed (at least theoretically).

The first direction we shall describe is the pure algebraic (perhaps linear algebraic). A universal set of elementary gates, this subset is a group of unitary transformations of some finite-dimensional Hilbert space which generates the entire group. In the case of quantum computer there is a correspondence between computational tasks and unitary transformations, and the quantum computer program solving the problem is a composition of some elementary gates which equals the unitary transformation corresponding to the given problem. Therefore, any conventional (as classical so quantum) computational process is discrete while the evolution of any physical system (as classical so quantum) is continuous and is described with the corresponding Hamiltonian or Schrödinger equation. Hence, it can be stated that during a conventional computational process, when we want to represent a given unitary transformation as a composition of some elementary gates, we have a mechanical system with Hamiltonian controlled by some external means (inputs) and we are able to implement the elementary gates by these inputs. The computational process is considered as discrete because we are interested not with the continuous evolution of the system but with the transformations at some finite moments of time corresponding to the elementary gates. Nowadays, there are several works where the "discrete" approach to the computational process is replaced by "continuous" when the aim is to construct a given transformation for a system not by means of discrete inputs generating the preset gates, but by continuous steering with some external parameter-dependent Hamiltonian (very interesting approach is developed in [79], [80], [81]). The geometric control theory is a tool which unifies these two approaches and the aim of this chapter is to demonstrate the methods that can be used to build a computational process (program) by parameter-dependent Hamiltonian and to investigate what can be achieved by these methods (see [59], [60]).

4.1. Basic Notation from Geometric Control Theory

4.1.1. Dynamical Polysystems and Control Group

Definition 4.1.1 (Control Group, see [97]). *Let I be any nonempty set. Consider the set of all finite sequences of type* $((t_1,i_1)\cdots(t_m,i_m))$ *where* $(t_k,i_k)\in\mathbb{R}\times I,\ k=1,\ldots,m$. *We shall introduce the following reduction rules:*

1. *any term of the type* $(0,i)$ *is removed from the sequence*
2. *if* $i_k=i_{k+1}\equiv i$ *then the pair of consecutive terms* $(t_k,i_k)(t_{k+1},i_{k+1})$ *is replaced by the single term* (t_k+t_{k+1},i)

It is clear that after a finite number of steps from any sequence can be obtained an irreducible sequence (which can be empty and we denote that there is one by 0*). We denote the set of all irreducible sequences by* $C(I)$ *and call the control group over the set I.*

Remark 4.1.1. Another point of view is to consider the control group as a quotient set for the set of all sequences $\{((t_1,i_1)\cdots(t_m,i_m))\}$ under the following equivalence relation: two sequences are equivalent if after reduction they give one and the same irreducible sequence.

The following operations on the set $C(I)$ justify the usage of term *group*. For any two sequences s_1 and s_2 from $C(I)$ let (s_1s_2) be the sequence obtained by their concatenation and let the *product* $s_1\cdot s_2$ be the irreducible sequence obtained by the reduction of (s_1s_2). The operation $(s_1,s_2)\mapsto s_1\cdot s_2$ together with the neutral element 0 (the empty sequence) defines a group structure on $C(I)$.

Remark 4.1.2. The group $C(I)$ is commutative only when I is a one-element set.

The elements of $C(I)$ are called *controls*. For any control $s=((t_1,a_1)\cdots(t_n,a_n))$ its inverse element in the group $C(I)$ is the control $s^{-1}=((-t_n,a_n)\cdots(-t_1,a_1))$. A control $((t_1,a_1)\cdots(t_n,a_n))$ is said to be *positive* (*negative*) if $t_i>0$ ($t_i<0$), $\forall\, i\in\{1,\ldots,n\}$. Let us denote the set of all positive (negative) controls by $C^+(I)$ ($C^-(I)$). In general $C(I)\neq C^+(I)\cup C^-(I)$ because there are many "mixed" control types in $C(I)$. It is clear that $C^+(I)$ and $C^-(I)$ are semigroups.

For any $\lambda\in\mathbb{R}$ and $s=((t_1,a_1)\cdots(t_n,a_n))\in C(I)$ define a control $\lambda\cdot s$ as $\lambda\cdot s=((\lambda t_1,a_1)\cdots(\lambda t_n,a_n))$. Any subset $\{s_1,\ldots,s_p\}\subset C(I)$ defines a mapping

$$\Phi_{s_1\cdots s_p}:\mathbb{R}^p\longrightarrow C(I),\quad \Phi_{s_1\cdots s_p}(\lambda_1,\ldots,\lambda_p)=\lambda_1s_1\cdot\lambda_2s_2\cdots\lambda_ps_p$$

After this, we can introduce a topology on $C(I)$ as the strongest topology for which all the maps of type $\Phi_{s_1\cdots s_p}$ are continuous. Analogically can be introduced a differential structure on $C(I)$: a mapping $f:C(I)\longrightarrow\mathbb{R}$ is said to be smooth if the mapping $f\circ\Phi_{s_1\cdots s_p}:\mathbb{R}^p\longrightarrow\mathbb{R}$ is smooth for any $\Phi_{s_1\cdots s_p}$. The group operations in $C(I)$ are continuous and smooth with respect to this defined topology and smooth structure.

Definition 4.1.2. *For a smooth manifold M a dynamical polysystem on M controlled by $C(I)$ is a smooth action of the group $C(I)$ on M:*

$$C(I) \times M \ni (s,m) \mapsto sm \in M$$

Definition 4.1.3. *The set $C(I)x = \{sx \mid s \in C(I)\}$ is called the orbit of the point $x \in M$. The set $C^+(I)x = \{sx \mid s \in C^+(I)\}$ is called the positive orbit of x and $C^-(I)x = \{sx \mid s \in C^-(I)\}$ – the negative orbit of x.*

In this case, any fixed element $a \in I$ gives a one-parameter subgroup $C_a = \{(t,a) \mid t \in \mathbb{R}\} \subset C(I)$ which induces a one-parameter group of diffeomorphisms of M:

$$\varphi_t^a : M \longrightarrow M, \quad t \in \mathbb{R}, \quad \varphi_t^a(x) = (t,a)x, \quad x \in M$$

The latter, corresponds to some smooth vector field X^a on M. The family of vector fields $\{X^a \mid a \in I\}$ is called the *infinitesimal* transformations of the dynamical polysystem. Conversely, if we have a family of vector fields F on M indexed by the elements of some set I, then we obviously have a dynamical polysystem on M controlled by $C(I)$. In such cases, we denote the control group together with the corresponding dynamical polysystem by $C(F)$. Thus, we can state that there is a one-to-one correspondence between the dynamical polysystems on M and the families of smooth vector fields on M.

4.1.2. The Structure of an Orbit of Dynamical Polysystem

Let F be a family of vector fields on the manifold M.

Definition 4.1.4. *The closure of the family of vector fields F is the family of vector fields defined as*

$$\Delta(F) = \{sX \mid s \in C(F),\ X \in F\}$$

In other words it is the minimal set of vector fields containing F and invariant under the action of the control group $C(F)$.

The *dimension* of the dynamical polysystem $C(F)$ at a point $x \in M$ is the dimension of the subspace of the tangent space $T_x(M)$ generated by the family of vector fields $\Delta(F)$. The main result about the structure of an orbit of dynamical polysystem is based upon the following

Lemma 4.1.1 (see [134]). *For any point $x \in M$ the dimension of the dynamical polysystem $C(F)$ is constant through the orbit $C(F)x$. If m is the dimension of the dynamical polysystem $C(F)$ at the point x then there exists a set of controls $\{s_1,\ldots,s_m\} \subset C(F)$ and a set of vector fields $\{X_1,\ldots,X_m\} \subset F$ such that the mapping*

$$\phi^x : \mathbb{R}^m \longrightarrow C(F)x, \quad \phi^x(t_1,\ldots,t_m) = (s_m\varphi_{t_m}^m s_m^{-1} \cdots s_1\varphi_{t_1}^1 s_1^{-1})$$

where $\{\varphi_t^i \mid t \in \mathbb{R}\}$, $i \in \{1,\ldots,m\}$ is the one-parameter group of diffeomorphisms corresponding to X_i, is a local diffeomorphism at the point $(0,\ldots,0) \in \mathbb{R}^m$.

Let $\exp(F)$ denote the group of diffeomorphisms of M generated by the flows for the elements of F:

$$\exp(F) = \{\exp(t_1X_1)\cdots\exp(t_nX_n) \mid t_i \in \mathbb{R},\ X_i \in F,\ n \in \mathbb{N}\}$$

and $\mathcal{L}(F)$ be the minimal Lie subalgebra of $\mathcal{F}(M)$ containing F and invariant under the action of the group $\exp(F)$.

From this lemma it follows

Theorem 4.1.1 (see [134]). *For any point x in the orbit $C(F)x_0$, $x_0 \in M$, the mapping $\phi^x : \mathbb{R}^m \longrightarrow C(F)x_0$ gives a local coordinate system on $C(F)x_0$. The topology and the differential structure induced on the orbit by these coordinate system coincide with the topology and the differential structure induced by the action of the group $C(F)$.*

and

Theorem 4.1.2 (Orbit Theorem, Nagano-Sussmann). *For any point $x_0 \in M$ the orbit $C(F)x_0 \equiv O_{x_0}$ is a connected immersed submanifold of M and $T_{x_0}(O_{x_0}) = \mathcal{L}(F)_{x_0}$.*

4.1.3. Control System and Accessibility Problem

For a smooth manifold M let $\mathcal{F}(M)$ be the space of smooth vector fields on M. We shall generalize the notion of control group in the following way. Let U be some nonempty set (usually assumed to be a finite-dimensional smooth manifold). We denote by $\mathcal{P}(U)$ the set of all paths in U. At this point, we do not specify what type of paths we will consider

$$\mathcal{P}(U) = \{u : [0,T] \longrightarrow U \mid T \geq 0\}$$

The set $\mathcal{P}(U)$ is a semigroup under the following standard operation: for $u_1 : [0,T_1] \longrightarrow U$ and $u_2 : [0,T_2] \longrightarrow U$ let $u_2u_1 : [0,T_1+T_2] \longrightarrow U$ be

$$(u_2u_1)(t) = \begin{cases} u_1(t), & t \in [0,T_1]; \\ u_2(t), & t \in [T_1, T_1+T_2]. \end{cases}$$

A *control system* on the manifold M has the following data

- a smooth map $X : U \longrightarrow \mathcal{F}(M)$;
- a subspace $C(U)$ of $\mathcal{P}(U)$ (at this stage we do not specify the structure of $C(U)$, but it is assumed to be a *subsemigroup* of $\mathcal{P}(U)$).

The elements of $C(U)$ are called *controls* and any $u \in C(U)$, $u : [0,T] \longrightarrow U$ defines a dynamical system

$$\dot{m} = X^u(t)_{m(t)}, \quad m(t) \in M, \quad t \in [0,T] \tag{4.1.1}$$

where X^u denotes the time-dependent vector field $X^u(t) = (X \circ u)(t)$ and $X^u(t)_{m(t)}$ is the value of this vector field at the point $m(t) \in M$.

Remark 4.1.3. The solution of 4.1.1 is defined as the absolutely continuous map $m : [0,T] \longrightarrow M$ such that the equation is satisfied for almost every $t \in [0,T]$.

For any given control $u : [0,T] \longrightarrow U$ and an initial value $m(0) \equiv m_0$, we denote it by $\gamma(m_0,u,t)$, $t \in [0,T]$ the solution (if exists) of 4.1.1. If for some $t \in [0,T]$ and $x \in M$ we have that $\gamma(m_0,u,t) = x$, then we state that the control u *steers* the state m_0 into the state x in time t, and the state x is *reachable* from m_0 in time t. We denote by $\mathcal{R}(x_0,t)$ the set of points for M is reachable from the point $x_0 \in M$ in time $t \in \mathbb{R}$. Traditionally, the following two sets are the subjects of this investigation

$$\hat{\mathcal{R}}(x_0,t) = \bigcup_{0\le\tau\le t} \mathcal{R}(x_0,\tau) \text{ – the set of points reachable from } x_0 \text{ in time } \le t$$

and

$$\hat{\mathcal{R}}(x_0) = \bigcup_{0\le\tau\le\infty} \mathcal{R}(x_0,\tau) \text{ – the set of points reachable from } x_0$$

The aim of geometric control theory is the investigation of the structures for the sets $\mathcal{R}(x_0,t)$, $\hat{\mathcal{R}}(x_0,t)$, and $\hat{\mathcal{R}}(x_0)$. These problems are considered for various types of space for control parameters U, as well as the map $X : U \longrightarrow \mathcal{F}(M)$ and the class controls for $\mathcal{C}(U)$. We assume that $U = \mathbb{R}^m$ for some integer $m > 0$ and consider the following three types of controls

- $\mathcal{C}_u$ – *unrestricted controls*, which consists of locally bounded and measurable mappings from $[0,T]$ to $\mathbb{R}^m$, $T \ge 0$;
- $\mathcal{C}_r$ – *restricted controls*: the subset of $\mathcal{C}_u$ consisting of the mappings with values in the cube $\{(x_1,\ldots,x_m) \in \mathbb{R}^m \,|||x_i| \le 1,\ i = 1,\ldots,m\}$;
- $\mathcal{C}_b$ – *"bang-bang"* controls, consisting of the all piecewise constant maps from $[0,T]$ to $\mathbb{R}^m$.

It is clear that in the case of bang-bang controls the reachable set is independent of the structure for map $X : U \longrightarrow \mathcal{F}(M)$ and depends only on the image of this map. The set reachable from a point $m \in M$ in time $t \ge 0$ for piecewise constant controls is

$$\mathcal{R}(m,t) = \{\exp(t_1X_1)\cdots\exp(t_kX_k)m \mid$$
$$\sum_{i=1}^{k} t_i = t,\ t_i \ge 0,\ X_i \in \mathrm{Image}(X),\ i = 1,\ldots,k,\ k \in \mathbb{N}\}$$

where $\exp(t\xi)$ denotes an element in the one-parameter group of diffeomorphisms associated with a vector field ξ. In the case when the set $\mathrm{Image}(X)$ is symmetric in the sense that $\xi \in \mathrm{Image}(X) \Longrightarrow -\xi \in \mathrm{Image}(X)$, we have that the set reachable from any point $m \in M$ coincides with the orbit of the dynamical polysystem on M defined by the family of vector fields $\mathrm{Image}(X)$. So, we can apply the classical Orbit Theorem (see Theorem 4.1.1) for the structure of the set $\hat{\mathcal{R}}(m)$, $m \in M$.

Let $\mathcal{L}ie(S)$ be the minimal Lie subalgebra of $\mathcal{F}(M)$ containing the subset $S \subset \mathcal{F}(M)$. It is clear that for any point $x \in M$ the space $\mathcal{L}ie(S)_x \subset T_x(M)$ is a subspace of the tangent space to the orbit of x. In some cases it coincides with the tangent space of the orbit. Particularly, the following theorem is one of the most important corollaries of the Orbit Theorem

Theorem 4.1.3 (Rashevsky-Chow). *If M is a connected smooth manifold and $\mathcal{L}ie(S)_x = T_x(M)$ for every point $x \in M$, then the orbit of the point x under the action of the dynamical polysystem defined by S coincides with the entire manifold M.*

A set $S \subset \mathcal{F}(M)$ which satisfies the condition $\mathcal{L}ie(S)_x = T_x(M), \forall\, x \in M$ is said to be *completely nonholonomic* or *bracket-generating*. As mentioned, if the set S is symmetric then the orbits of the corresponding dynamical polysystem and the reachable sets coincide. Hence, we can conclude that for symmetric ($S = -S$) nonholonomic family of vector fields the corresponding control system is completely controllable.

4.1.4. Two Notions of Controllability

Sometimes it is useful to distinguish the following two notions of controllability (reachability) when regarding a classical method for quantum mechanical systems.

Definition 4.1.5 (Transformation-Controllability). *We call a system transformation-controllable if every transformation of the system can be obtained by applying some finite set of appropriate controls in a finite interval of time.*

Definition 4.1.6 (State-Controllability). *We call a system state-controllable if for every pair of states in the system – s_1 and s_2, there exists a finite set of controls that steers the system from state s_1 into state s_2 in a finite time.*

In the case of classical system, if we assume that the set of all transformations is the entire group of diffeomorphisms for a smooth manifold M – $Diff(M)$, and the control system is the dynamical polysystem defined by some family of vector fields F, then the transformation-controllability means that $\exp(F) = Diff(M)$; and the state-controllability means that for any point $m \in M$: $O_m = M$. These two notions are not equivalent in neither the classical nor in the quantum cases. To demonstrate this let us consider the following situation. Let M be a symplectic manifold and the dynamical polysystem F be the family of all Hamiltonian vector fields on M: $F = ham(M)$. Let $\{\ ,\ \}$ denote the Poisson bracket corresponding to the symplectic structure on M. According to the Darboux theorem for any point $m \in M$ there exists a local coordinate system $\{p_i,\ q_i \mid i = 1,\dots,n\}$ such that $\{p_i, q_j\} = \delta_{ij}$. This implies that locally we have the set of Hamiltonian vector fields $\left\{\frac{\partial}{\partial q_i}, \frac{\partial}{\partial p_i} \mid i = 1,\dots,n\right\}$ which forms a basis for $T_m(M)$. So, then we have for any point $m \in M$, the set of Hamiltonian vector fields forms a basis of $T_m(M)$. Therefore, by the theorem of Rashevsky-Chow the family of all Hamiltonian vector fields is completely nonholonomic and therefore the corresponding system is state-controllable. But, the family $ham(M)$ is not transfomation-controllable because the corresponding family of 1-parameter flows is contained in the group of symplectic diffeomorphisms of the manifold M, and it is clear that this group is just a part of the entire group of diffeomorphisms for M. The classical problem of state-controllability for the family of all Hamiltonian vector fields becomes more complicated if we extend the notion of state and consider not only pure states (i.e., points), but also *mixed states*. Let us recall the formal definition of this notion which is rather well-known from statistical physics. The realization in the algebra of observables as the algebra $C^\infty(M)$ dictates the realization for the space of states as the space of probability distributions, i.e., the linear positive functionals $\rho : C^\infty(M) \longrightarrow \mathbb{R}$ such that $\phi > 0 \Longrightarrow \rho(\phi) > 0$ and $\int_M \rho(p,q)\mathrm{d}p\mathrm{d}q = 1$. Let us denote the space of such functionals by $Stat(M)$. In other words, a state $\rho \in Stat(M)$ is a *positive distribution* (generalized function) such that $\rho(1) = 1$. The pure state corresponding to a point $x \in M$ is the Dirac functional δ_x: $\delta_x(\phi) = \phi(x)$. Sometimes we shall use the scalar product notation $\langle \rho, \phi \rangle$ for $\rho(\phi)$. Any smooth mapping of pure states $f : M \longrightarrow M$ can be extended to $f : Stat(M) \longrightarrow Stat(M)$ as $\langle f(\rho), \phi \rangle = \langle \rho, f^*(\phi) \rangle$ where $f^* : C^\infty(M) \longrightarrow C^\infty(M)$ denotes the mapping dual to f.

Remark 4.1.4. It is clear that the subspace $Stat(M)$ in the space of all generalized functions is invariant under such mappings, i.e., every positive distribution is mapped to a positive one and the property $\rho(1) = 1$ is preserved.

After this, we extend the problems of state-controllability and reachability to the space $Stat(M)$. First of all, let us notice that the pure-state-controllabilty does not imply the state-controllability in the extended state space. The simplest example is the following: let M be a simplectic manifold and ω be the symplectic form on it, let $\alpha = a \cdot \omega^n$, $2n = \dim(M)$ be the volume form normalized so that $\int_M \alpha = 1$; take the state $\mathcal{E}$ which is a generalized function defined as $\langle \mathcal{E}, \phi \rangle = \int_M \phi \cdot \alpha$, $\phi \in C^\infty(M)$. As already mentioned earlier, for the dynamical polysystem generated by the family of all Hamiltonian vector fields – $Ham(M)$, the space of pure states is controllable but it is not true for the space $Stat(M)$ because the state $\mathcal{E}$ is invariant under the action of Hamiltonian flows. The reasonable classification of the entire space $Stat(M)$ from the point of view of reachability under the dynamical polysystem defined by $Ham(M)$ seems very difficult. As an example, for the application of Orbit Theorem we shall consider a simplified version of this problem. Particularly, we consider the reachability problem for the space of *discrete* mixed states.

The space $Stat(M)$ is a convex space:

$$\rho_1, \rho_2 \in Stat(M) \Longrightarrow c\rho_1 + (1-c)\rho_2 \in Stat(M),\ \forall\, c \in [0,1].$$

Therefore, for any finite set of points $P \subset M$ and a function $s : P \longrightarrow \mathbb{R}$ such that $\varphi > 0$ and $\sum_{x \in P} s(x) = 1$, we can construct the distribution $\rho(P,s) = \sum_{x \in P} s(x)\delta_x \in Stat(M)$. Let us denote the subspace of all such states in $Stat(M)$ by $Stat_0(M)$. From the property $f(\delta_x) = \delta_{f(x)}$, it follows that $Stat_0(M)$ is an invariant subset of $Stat(M)$. Now, we consider the reachability problem for the dynamical polysystem defined by $Ham(M)$ just for the space $Stat_0(M)$. Particularly, we describe the quotient space $Stat_0(M)/\sim$, where two states ρ_1 and ρ_2 from $Stat_0(M)$ are equivalent iff there exists a finite set of Hamiltonians $\{H_1, \ldots, H_k\} \subset C^\infty(M)$ and a set of time intervals $\{t_1, \ldots, t_k\} \subset \mathbb{R}$ such that $\exp(t_1 V_{H_1}) \cdots \exp(t_k V_{H_k})\rho_1 = \rho_2$, where V_{H_i} denotes the Hamiltonian vector field corresponding to the function H_i.

For any $\rho = \rho(P,s) \in Stat(M)$, we call the set of points $P \subset M$ the *support* of the distribution ρ and denote it by $Supp(\rho)$. For any $c \in \mathbb{R}$, we call the number of elements in the set $s^{-1}(c)$ the *multiplicity* of c and denote this number by $m(c)$. For a given $\rho(P,s) \in Stat(M)$, the set of all pairs $\{(c, m(c)) \mid c \in \text{Image}(s)\}$ we call the *spectrum* of the distribution ρ and denote this set by $Spec(\rho)$. It is easy to verify that for any diffeomorphism (or even bijection) $f : M \longrightarrow M$ we have that $f(\rho(P,s)) = \rho(f(P), s \circ f)$. The latter implies that if two states $\rho_1 = \rho(P_1, s_1)$ and $\rho_2 = \rho(P_2, s_2)$ are equivalent then $Spec(\rho_1) = Spec(\rho_2)$. It turns out that the equality of spectra is also a sufficient condition for the equivalency of two states in $Stat_0(M)$. Before we start to prove the sufficiency let us proof the following auxiliary

Lemma 4.1.2. *Let $X_1,\dots,X_k$ be tangent vectors at k different points $\{m_1,\dots,m_k\}\subset M$. There exists such Hamiltonian $H\in C^\infty(M)$ that for the corresponding Hamiltonian vector field V_H we have: $V_H(m_i)=X_i,\ i=1,\dots,k$.*

Proof. For each point m_i take an open neighborhood U_i with canonical coordinate system $\left\{p^i_j,q^i_j \mid j=1,\dots,k\right\}$, so that $U_i\cap U_l=\emptyset$ when $i\neq l$. Since the system of all Hamiltonian vector fields is completely nonholonomic, for each X_i there exists a function $f_i\in C^\infty(M)$ such that $V_{f_i}(m_i)=X_i$ and $f_i(M\setminus U_i)=0$. It is clear that the function f_i does not "affect" the function f_l inside the neighborhood U_l when $i\neq l$. Therefore, the function $H=\sum_{i=1}^{k} f_i$ is the desired Hamiltonian. □

Theorem 4.1.4. *Two states ρ_1 and ρ_2 from $Stat_0(M)$ are equivalent, if and only if $Spec(\rho_1)=Spec(\rho_2)$.*

Proof. Assume that the spectra of $\rho_1=\rho(P_1,s_1)$ and $\rho_2=\rho(P_2,s_2)$ are equal. For ρ_1 consider the ordered set $(x_1,\dots,x_n)$ where n is the number of elements in P_1, $x_i\in P_1$, $i=1,\dots,n$, and $s_1(x_i)\le s_1(x_{i+1})$. In the same way, we construct the ordered set $(y_1,\dots,y_n)$ for ρ_2. From $Spec(\rho_1)=Spec(\rho_2)$ follows that $s_1(x_i)=s_2(y_i)$, $i=1,\dots,n$. Thus, the problem is reduced to the existence of such element g of a Hamiltonian flow that $g(x_i)=g(y_i)$, $i=1,\dots,n$.

For $n\in\mathbb{N}$ let $\widetilde{M}^n$ be the subset of $M^n=\underbrace{M\times\dots\times M}_{n-times}$ defined as

$$\widetilde{M}^n=\left\{(p_1,\dots,p_n)\mid p_i\neq p_j \text{ when } i\neq j,\ i,j=1,\dots,n\right\}$$

$\widetilde{M}^n$ is an open subset of M^n and therefore it is a submanifold. The set $M^n\setminus\widetilde{M}^n$ is a union of submanifolds $M^n_{i,j}=\left\{(p_1,\dots,p_n)\mid p_i=p_j\right\}$, $i\neq j$. It is clear that $\dim(M^n_{i,j})=(n-1)\cdot\dim(M)$ and hence if $\dim(M)\ge 2$ and M is connected then $\widetilde{M}^n$ is also connected. But, recall that M is assumed to be a symplectic manifold; so $\dim(M)\ge 2$. Thus, we have a connected manifold $\widetilde{M}^n$ and the control group generated by the Hamiltonian vector fields on M acts on $\widetilde{M}^n$ as

$$g^m:\widetilde{M}^n\longrightarrow\widetilde{M}^n,\ g^m(x_1,\dots,x_n)=(g(x_1),\dots,g(x_n)),$$

where $g:M\longrightarrow M$ is an element of some Hamiltonian flow. The question is: is this action transitive or not? For a given Hamiltonian $H\in C^\infty(M)$ the action of the corresponding flow on $\widetilde{M}^n$ is $(x_1,\dots,x_n)\mapsto(\exp(tH)x_1\dots,\exp(tH)x_n)$, $t\in\mathbb{R}$; therefore, the corresponding vector field on $\widetilde{M}^n$ is $V^n_H=(V_H,\dots,V_H)$. That is, for any point $(x_1,\dots,x_n)\in\widetilde{M}^n$ the value of this vector field at this point is $(V_H(x_1),\dots,V_H(x_n))$. According to the Lemma 4.1.2 for any tangent vector $(X_1,\dots,X_n)$ at the point $(x_1,\dots,x_n)$ there is a Hamiltonian H such that $V_H(x_1),\dots,V_H(x_n)=(X_1,\dots,X_n)$. This implies that the dynamical polysystem on $\widetilde{M}^n$ generated by the diagonal actions of the Hamiltonian flows on M is completely nonholonomic. Hence, any two states from $Stat_0(M)$ with equal spectra are equivalent. □

Extending this to the entire set of states $Stat(M)$ just by direct analogy, we can see that if two distributions ω_1 and ω_2 are equivalent then there exists such diffeomorphism $g:M\longrightarrow M$ that $g(Supp(\omega_1))=g(Supp(\omega_2))$. The problem in the existence of such diffeomorphism that maps one given closed set to another given closed set, is in itself quite complicated.

4.2. Control Systems on Lie Groups

4.2.1. Homogeneous and Affine Cases

In the case of control systems with manifolds, the main difficulty for controllability and accessibility problems comes from the fact that even for a small number of control parameters the control group is a subgroup of an infinite-dimensional Lie group. For control systems on finite-dimensional Lie groups when we restrict ourselves with right-invariant vector fields, the problem becomes easier sometimes and can be solved completely. In this section, we review the cases of *homogeneous* and *affine* controls which are also discussed in the classical paper of Jurdjevich and Sussmann (see [70]). We start with definitions that address basic notions.

Let G be a finite-dimensional Lie group and $\mathfrak{g}$ be its Lie algebra. According to the general definition (see Section 4.1.3), we need a set of *control parameters* U which is assumed to be $\mathbb{R}^m$ for some positive integer m; a set of *controls* $\mathcal{C}(U)$ which is a subset for the set of mappings $u : \mathbb{R}^+ \longrightarrow \mathbb{R}^m$ and usually is the set of continuous, piecewise-continuous, piecewise-constant or smooth maps; and a continuous map $H : \mathbb{R}^m \longrightarrow \mathcal{F}(G)$. In this case, we consider that such H takes values in the subspace of right-invariant vector fields on G. Thus, we can define H as a continuous map $H : \mathbb{R}^m \longrightarrow \mathfrak{g}$ and call it the *Hamiltonian* (with hope that it will not cause any confusion). These data defines a family of control-dependent differential equations on G

$$\dot{g}(t) = H(u(t))g(t), \quad g(t) \in G,\ g(0) = g_0,\ t \geq 0 \tag{4.2.1}$$

where $u : \mathbb{R}^+ \longrightarrow \mathbb{R}^m$ runs through the space for controls $\mathcal{C}(\mathbb{R}^m)$. The right-hand side of the equation is a time-dependent right-invariant vector field on G defined by the Lie algebra elements $H(u(t))$, $t \in \mathbb{R}^+$. As in the general case, the solution of the equation 4.2.1 is absolutely the continuous map $g : \mathbb{R}^+ \longrightarrow G$ that satisfies this equation for almost every $t \geq 0$. For any given control u and initial state g_0, we denote by $\gamma(g_0,u,t)$, $t \in \mathbb{R}^+$ the solution of the equation 4.2.1.

If for some $\gamma(g_0,u,t) = g$ and with some $u \in \mathcal{C}(\mathbb{R}^m)$ and $t \geq 0$, then we say that the control u *steers* the state g_0 into the state g in time t and the state g is reachable from g_0 in time t. As in this general case, denoted by:

- $\mathcal{R}(g_0,t)$ the set of all reachable in time t elements $g \in G$;
- $\hat{\mathcal{R}}(g_0,T) = \bigcup\limits_{0\leq t\leq T} \mathcal{R}(g_0,t)$ – the set reachable in time $\leq T$;
- $\mathcal{R}(g_0) = \bigcup\limits_{0\leq t\leq \infty} \mathcal{R}(g,t)$ – the set reachable from g_0.

From the fact that this system is a right-invariant easily follows the following properties of the reachable sets

$$\mathcal{R}(g,t) = \mathcal{R}(1,t)\cdot g, \quad \hat{\mathcal{R}}(g,T) = \hat{\mathcal{R}}(1,T)\cdot g, \quad \mathcal{R}(g) = \mathcal{R}(1)\cdot g$$

Therefore, for the right-invariant control systems on Lie groups it is sufficient to study the set reachable from 1.

As already mentioned in the work by Jurdjevich and Sussmann ([70]), this case is considered when the Hamiltonian $H : \mathbb{R}^m \longrightarrow \mathfrak{g}$ has an affine function

$$H(u_1,\ldots,u_m) = X_0 + \sum_{i=1}^{m} u_i \cdot X_i, \quad \text{for fixed } \{X_0,\ldots,X_m\} \subset L \tag{4.2.2}$$

We make a brief review of the results and methods for such control systems, because it turns out that they are useful not only for this specific class of systems but for other classes of control systems.

Let us introduce the following three types of control parameters

- C_u – *unrestricted controls*: locally bounded and measurable mappings from $\mathbb{R}^+ \longrightarrow \mathbb{R}^m$;
- C_r – *restricted controls*: a subset of C_u consisting of the mappings with values in the cube $\{(x_1,\ldots,x_m) \in \mathbb{R}^m \mid |x_i| \leq 1,\ i = 1,\ldots,m\}$;
- C_b – *"bang-bang" controls*: the subset of all piecewise constant mappings (in fact it suffices if $|x_i| = 1,\ i = 1,\ldots,m$);

For these three classes of controls, we use the notation C without subscript when some statement is formulated for each of them.

The following theorem from the classical theory of Lie groups is one of the important tools for studying control systems on Lie groups.

Theorem 4.2.1 (Yamabe's Theorem, see [155]). *Let G be a Lie group and let H be a path-connected subgroup of G. Then H is a Lie subgroup of G.*

The following theorem states that for affine Hamiltonian (see 4.2.2) the solution of the controls system 4.2.1 exists and moreover is complete.

Theorem 4.2.2 (see [70]). *For every control $u \in C$ and $g_0 \in G$, there exists a unique solution $g(t)$, $t \in \mathbb{R}^+$ of 4.2.1 defined for every $t \in [0, \infty]$ such that $g(0) = g_0$.*

Let $\hat{\mathfrak{g}}$ be the Lie subalgebra of L generated by $\{X_0,\ldots,X_m\}$ and $\hat{G}$ be the corresponding connected Lie subgroup of G.

Theorem 4.2.3 (see [70]). *The set $\mathcal{R}(1)$ is a semi-group in G and if it is a group then it coincides with $\hat{G}$.*

Proof. Let $g_1 = \gamma(1,u_1,t_1)$ and $g_2 = \gamma(1,u_2,t_2)$ be two elements of G reachable from 1 applying the controls u_1 and u_2. Define a new control v as

$$v(\tau) = \begin{cases} u_1(\tau) & \text{when } \tau \in [0,\ t_1], \\ u_2(\tau - t_1) & \text{when } \tau > t_1. \end{cases}$$

Since this system is a right-invariant, we then have that $\gamma(1,v,t_1+t_2) = g_2 g_1$.

Now, assume that $\mathcal{R}(1)$ is a group. As the space of controls is path-connected, the subset $\mathcal{R}(1)$ is also path-connected and therefore, according to the Yamabe's theorem (see 4.2.1) it is a Lie subgroup of G. Since $\mathcal{R}(1) \subset \hat{G}$, we have that its Lie algebra V is a subalgebra of $\hat{\mathfrak{g}}$. Conversely, consider any element $v(a_1,\dots,a_m) = X_0 + \sum_{i=1}^{m} a_i X_i \in \hat{\mathfrak{g}}$, $a_i = \pm 1$. By definition of $\mathcal{R}(1)$, we have that $\exp(tv) \in \mathcal{R}(1)$ for $t \geq 0$; but since $\mathcal{R}(1)$ is a group we have that $\exp(tv) \in \mathcal{R}(1)$ for $t \leq 0$, too. The elements $v(a_1,\dots,a_m)$, where $(a_1,\dots,a_m) \in \mathbb{R}^m$, are generators of $\hat{\mathfrak{g}}$ which implies that $\hat{\mathfrak{g}} \subset V$. Since the groups $\mathcal{R}(1)$ and $\hat{G}$ are connected, we have $\mathcal{R}(1) = \hat{G}$. □

An affine control system $X = (X_0, X_1, \dots, X_m)$ is said to be *homogeneous* if $X_0 = 0$. In other words, the Hamiltonian depends on control parameters linearly: $H(u_1,\dots,u_m) = \sum_{i=1}^{m} u_i X_i$.

Theorem 4.2.4 (see [70]). *For homogeneous right-invariant control system on a Lie group G the reachable set $\mathcal{R}(1)$, for any of the three classes of controls, is the connected Lie subgroup of G corresponding to the Lie algebra generated by X. If we use the class of unrestricted controls C_u, then for each $T > 0$ we have that $\mathcal{R}(1,T) = \mathcal{R}(1)$.*

Proof. Since the set $\mathcal{R}(1)$ is a semigroup, according to the previous theorem it is sufficient to show that if $g \in \mathcal{R}(1)$ then therefore $g^{-1} \in \mathcal{R}(1)$. Assume that $g = \gamma(1,u,t)$. Consider a control v defined as

$$v(\tau) = \begin{cases} -u(t-\tau) & \text{when } \tau \in [0,t], \\ u(\tau) & \text{when } \tau > t, \end{cases}$$

it is easy to verify that $\gamma(1,v,t) = g^{-1}$.

The second part of the theorem states that any reachable state can be reached in arbitrary short time interval if we are able to use the class of unrestricted controls. Assume that $g = \gamma(1,u,t)$. For any given time $t_1 > 0$ define a control v as

$$v(\tau) = \frac{t}{t_1} \cdot u\left(\frac{\tau t}{t_1}\right), \quad \tau \in \mathbb{R}^+$$

It is easy to show that $\gamma(1,v,t_1) = g$ which implies that $\mathcal{R}(1,t) \subset \mathcal{R}(1,t_1)$ for arbitrary $t_1 > 0$. □

Summarizing the essential properties of the homogeneous control systems, we have the following:

- The set reachable from 1 is a subgroup of G;
- The set reachable from 1 is one and the same for three classes of controls: every $g \in G$ that can be reached from 1 by means of unrestricted control can also be reached by means of only "bang-bang" control (possibly later);
- If we use the unrestricted controls, then every reachable state $g \in G$ can be reached in arbitrary short time interval.

For the affine case the most important facts are that if the set reachable from 1 is dense in the Lie group $\hat{G}$ corresponding to the Lie algebra generated by the set $\{X_0,\dots,X_m\}$ then $\mathcal{R}(1)=\hat{G}$ and if the subgroup $\hat{G}$ is compact then always $\mathcal{R}(1)=\hat{G}$ and $\exists\, t>0:\ \hat{\mathcal{R}}(1,t)=\mathcal{R}(1)$ (see [70]).

A control system of the type 4.2.1 is said to be *controllable from* $g\in G$ if $\mathcal{R}(g)=G$. It is said to be *controllable* if it is controllable from every $g\in G$.

For the right-invariant control system on Lie group the controllability from 1 is equivalent to the controllability because we know $\mathcal{R}(g)=\mathcal{R}(1)\cdot g$. As it follows from the reachability criteria a necessary condition for the controllability is that the group G is connected and $\hat{\mathfrak{g}}=\mathfrak{g}$. If G is compact or the system is homogeneous then this condition sufficient, too.

Though the class of affine (or homogeneous) Hamiltonians is very restricted, actually the same methods can be used for the reachability problem when the Hamiltonian has a continuous mapping $H:\mathbb{R}^m\longrightarrow\mathfrak{g}$. What will become evident from further discussions; is that, for a compact Lie group this will be more important and not the Hamiltonian itself but the class of controls and the image set of the Hamiltonian.

4.2.2. Control system on Compact Lie Group with Continuous Hamiltonian

This subsection is mainly based on the results in the work by [29], the main idea is that; if we weaken the condition for the Hamiltonian to be affine, or homogeneous, and consider any continuous Hamiltonian, for compact Lie groups many essential results are still true. Moreover, to formally consider a right-invariant control system

$$\dot{g}=H(u)g,\quad g\in G,\ u\in\mathcal{C},\ g(0)=1$$

where G is a compact Lie group, $\mathfrak{g}$ is its Lie algebra, $H:\mathbb{R}^m\longrightarrow\mathfrak{g}$ is a *continuous* map, $\mathcal{C}$ is the set of controls consisting of all piecewise-continuous maps from $\mathbb{R}^+$ to $\mathbb{R}^m$, and $H(u)g$ denotes the tangent vector at $g\in G$ obtained from $H(u)\in\mathfrak{g}$ by the right action of g. Let $\hat{\mathfrak{g}}$ be the Lie subalgebra of $\mathfrak{g}$ generated by $\mathrm{Image}(H)\subset\mathfrak{g}$ and $\hat{G}$ be the corresponding connected Lie subgroup in G. The latter is the maximal integral submanifold for the right-invariant differential system defined by $\hat{\mathfrak{g}}$. Since the control system is right-invariant and each trajectory of this system is tangent to $\hat{\mathfrak{g}}$ at the point $1\in G$, we have that they are contained in $\hat{G}$. Therefore, the set reachable from 1 is a subset of $\hat{G}$. After this, without restriction of generality it can be assumed that $\hat{\mathfrak{g}}=\mathfrak{g}$ and $\hat{G}=G$.

Using this same method as for the case of affine Hamiltonian (see Theorem 4.2.3), it is easy to show that the reachable set $\mathcal{R}(1)$ is a semigroup.

To proceed further, we then need the following theorem.

Theorem 4.2.5 (see [70]). *Let G be a connected Lie group and $X=\{X_1,\dots,X_m\}$ be a set of generators in its Lie algebra $\mathfrak{g}$. Then every $g\in G$ is a finite product of elements in the form* $\exp(tX_i)$, $t\in\mathbb{R}$, $X_i\in X$.

Proof. The set of all finite products for the form $\exp(tX_i)$ is a path-connected subgroup G' of G. According to the Yamabe's theorem (see 4.2.1) G' is a Lie subgroup of G. Since G' contains the elements of the type $\exp(tX_i)$ its Lie algebra $\mathfrak{g}'$ contains the set X, therefore it contains $\mathfrak{g}$. Consequently, we then have $\mathfrak{g}'=\mathfrak{g}$. As the subgroups G and G' are connected with one and the same Lie algebra, they coincide. □

Since $\mathfrak{g}$ is the Lie algebra generated by Image(H), we can take the set of generators X from Image(H). For the reachable set $\mathcal{R}(1)$ the elements $\exp(tX_i)$ are constructed by using only positive time t, while for the elements of the group G can be used any $t \in \mathbb{R}$.

Lemma 4.2.1 (see [29]). *The set reachable from* 1 *is dense subset of* G.

Proof. Consider an element $g \in G$ such that $g = \prod_i \exp(t_i X_{p_i})$ where all t_i-s are negative. We have that $h = g^{-1}$ is an element of $\mathcal{R}(1)$ (because all times become positive). Consider the sequence h^n, $n \in \mathbb{N}$. Since the group G is compact this sequence contains a convergent subsequence $\{h^{n_k} \mid k \in \mathbb{N}\}$. We can assume that the sequence of subscripts $\{n_k\}$ is increasing. Take the sequence $\{h'_k = h^{n_{k+1}-n_k-1}\}$. Since n_k is assumed to be increasing, we have that $n_{k+1} - n_k - 1 \geq 0$ and therefore $\{h'_k\} \subset \mathcal{R}(1)$. But obviously $\lim_{k\to\infty} h'_k = h^{-1} = g$. We obtain that the elements $g = \prod_i \exp(t_i X_{p_i})$ with only negative t_i-s can be approximated by the elements of $\mathcal{R}(1)$ which implies that $\mathcal{R}(1)$ is dense in G. □

The following lemmas help us to go further and show that the set reachable from 1 coincides with G. For $g \in G$ and $V \in \mathfrak{g}$, let $Ad(g)V$ be the element of $\mathfrak{g}$ obtained by the adjoint action of g.

Lemma 4.2.2 (see [30]). *Let* $X = \{X_1, \ldots, X_m\}$ *be a set of generators for the Lie algebra* $\mathfrak{g}$. *If* $m = \dim(\mathfrak{g})$, *then for any* $\varepsilon > 0$ *there exists two elements* X_p *and* X_q *in* X *and a time* τ, $|\tau| < \varepsilon$ *such that the element* $X_{m+1} = Ad(\exp(\tau X_p))X_q$ *is linearly independent from* $\{X_1, \ldots, X_m\}$.

Proof. From $m < \dim(\mathfrak{g})$, it follows that at least for one pair of elements X_p and X_q, their commutator $[X_p, X_q]$ is linearly independent from the set X. If we assume that $Ad(\exp(tX_p))X_q = \sum_{i=1}^{m} a_i(t) X_i$, then the derivation of this equality by t gives $[X_p, X_q] = \sum_{i=1}^{m} \dot{a}_i(0) X_i$ which is a contradiction. □

Applying the same procedure to the extended set of generators $\{X_1, \ldots, X_m, X_{m+1}\}$ and etc., we can formulate the following

Lemma 4.2.3. *The set of generators* $\{X_1, \ldots, X_m\}$ *can be extended to a basis of* $\mathfrak{g}$ – $\{X_1, \ldots, X_m, X_{m+1}, \ldots, X_{m+p}\}$ *so that each element* X_{m+i}, $i = 1, \ldots, p$ *can be written as*

$$X_{m+i} = Ad(\exp(t_q X_{l_q}) \cdots \exp(t_1 X_{l_1}))X_{l_0}$$

where $\{X_{l_0}, \ldots, X_{l_q}\}$ *is a subset of* $\{X_1, \ldots, X_m\}$.

Moreover, in the work by D. D'Alessandro (see [30]) calculated in the upper bound for a number of factors in $\exp(t_q X_{l_q}) \cdots \exp(t_1 X_{l_1})$.

Lemma 4.2.4. *The set reachable from* 1 *contains an open subset of* G.

Proof. Let $\{X_1,\ldots,X_n\}$ be a basis of $\mathfrak{g}$ as a vector space. Consider the mapping $F:\mathbb{R}^n \longrightarrow G$

$$F(t_1,\ldots,t_n)=\prod_{i=1}^{n}\exp(t_iX_i), \quad (t_1,\ldots,t_n)\in\mathbb{R}^n$$

It follows from the inverse function theorem that F is a local diffeomorphism on a neighborhood of $0\in\mathbb{R}^n$. This implies that for sufficiently small $\delta>0$ and some $\varepsilon\in(0,\delta)$, the open neighborhood of $(\delta,\ldots,\delta)\in\mathbb{R}^m$ –

$$\{(t_1,\ldots,t_m)\mid \delta-\varepsilon<t_i<\delta+\varepsilon,\ i=1,\ldots,m\}\subset\mathbb{R}^m$$

is diffeomorphic to some open neighborhood $V\subset G$ of $g=F(\delta,\ldots,\delta)\in G$. According to Lemma 4.2.3 there is a basis of $\mathfrak{g}$ — $\{X_1,\ldots,X_n\}$, where the first m elements are generators from Image(H) and after the subscript m the elements are of the form

$$X_{m+k}=g_kX_{p_k}g_k^{-1}, \quad p_k\in\{1,\ldots,m\},\ g_k\in G,\ k=1,\ldots,n-m$$

Regarding this, we obtain that the mapping F is of the form

$$F(t_1,\ldots,t_n)=e^{t_1X_1}\cdots e^{t_mX_m}\underbrace{g_1}\,e^{t_{m+1}X_{p_1}}\underbrace{g_1^{-1}g_2}e^{t_{m+2}X_{p_2}}g_2^{-1}\cdots$$
$$\cdots g_ke^{t_{m+k}X_{p_k}}\underbrace{g_k^{-1}g_{k+1}}\cdots g_{n-m}e^{t_nX_{p_{n-m}}}\underbrace{g_{n-m}^{-1}}$$

Consider the following terms of this product:

$$g_1'=g_1,\ g_2'=g_1^{-1}g_2,\ldots,g_k'=g_{k-1}^{-1}g_k,\ldots,g_{n-m}'=g_{n-m}^{-1}$$

They are not necessarily in $\mathcal{R}(1)$, but since $\mathcal{R}(1)$ is dense in G we can select elements of $\mathcal{R}(1)$ – $h_1,\ldots,h_{n-m}$ arbitrarily close to $g_1',\ldots,g_{n-m}'$

$$h_1\approx g_1',\ldots,h_{n-m}\approx g_{n-m}'$$

and consider the mapping $\tilde{F}:\mathbb{R}^n\longrightarrow G$

$$\tilde{F}(t_1,\ldots,t_n)=e^{t_1X_1}\cdots e^{t_mX_m}h_1e^{t_{m+1}X_{p_1}}h_2\cdots e^{t_nX_{p_{n-m}}}h_{n-m}$$

The latter is close to F and therefore maps some open neighborhood of $(\delta,\ldots,\delta)$ onto an open neighborhood of $g'=\tilde{F}(\delta,\ldots,\delta)$. But for positive t_i-s we have that $\tilde{F}(t_1,\ldots,t_n)$ is an element of $\mathcal{R}(1)$. Hence, the set $\mathcal{R}(1)$ contains an open subset of G. □

Lemma 4.2.5. *If the set reachable from* 1 *contains some open subset of* G *then it contains an open neighborhood of* 1.

Proof. Let V be an open subset of G contained in $\mathcal{R}(1)$. Consider the set $V^{-1}=\{g^{-1}\mid g\in V\}$. It is clear that V^{-1} is also an open subset in G, though it is not necessarily contained in $\mathcal{R}(1)$. Since $\mathcal{R}(1)$ is dense in G (see 4.2.1), the set V^{-1} contains at least one element $h\in\mathcal{R}(1)$. Consider the set $h\cdot V$. As in $h\in V^{-1}$, clearly the set $h\cdot V$ contains 1 and is open in G. Because $h\in\mathcal{R}(1)$ and $V\subset\mathcal{R}(1)$ and $\mathcal{R}(1)$ is a semigroup, we have $h\cdot V\in\mathcal{R}(1)$. Therefore, $\mathcal{R}(1)$ contains an open neighborhood of 1. □

Hence, we have that $\mathcal{R}(1)$ is a semigroup in G and contains an open neighborhood of $1 \in G$. These implies that $\mathcal{R}(1) = G$.

To summarize, we can conclude that *in the case of continuous Hamiltonian and unrestricted controls on a compact Lie group, the set reachable from* 1 *is the connected Lie subgroup corresponding to the Lie algebra generated by Image*(H) *(regardless of the structure of the Hamiltonian* $H : \mathbb{R}^m \longrightarrow \mathfrak{g}$*)*.

Remark 4.2.1. Actually, during the proofs of the lemmas we use only "bang-bang" controls which is in accordance with the fact that the reachable sets for the unrestricted and "bang-bang" controls are one and the same.

4.2.3. Minimal Set of Generators for $\mathfrak{su}(n)$

As discussed in the previous section, the controllability for right-invariant "*bang-bang*" (or piecewise continuous) systems on compact Lie groups is completely described by the differential systems defined by the set of generators contained in the image of the corresponding Hamiltonian. Hence, the question about a minimal set of generators for a given Lie algebra (and the corresponding Lie group, too) is natural. For this question, the following classical results of Kuranishi (see [91]) and more recent results from [69], [6], [23] and [131] are quite useful. In this section, we give a review for some of them. First, let us recall the general structure of a semi-simple Lie group and the corresponding Lie algebra.

Let G be a compact Lie group and $\mathfrak{g}$ be its Lie algebra. The *rank* of G is defined as the dimension of the maximal abelian subalgebra $\mathfrak{h}$ of $\mathfrak{g}$. This number coincides with the dimension of the maximal torus (i.e., the maximal connected abelian subgroup) H in G.

All irreducible representations of H are 1-dimensional and are described by the elements of the group of *characters* $\hat{H} = Hom(H, \mathbb{C}^*)$, where $\mathbb{C}^*$ denotes the group $\mathbb{C} \setminus \{0\}$. Let $\mathfrak{g}^{\mathbb{C}}$ denote the complexification of $\mathfrak{g}$: $\mathfrak{g}^{\mathbb{C}} = \mathfrak{g} \otimes \mathbb{C}$. Any representation R of G in a complex vector space V defines a representation $\dot{R}$ of $\mathfrak{g}^{\mathbb{C}}$ in V. The restriction of R on H gives the decomposition $V = \bigoplus_{\lambda} V_\lambda$, where λ-s are some elements of $(\mathfrak{h}^{\mathbb{C}})^*$ and $V_\lambda = \left\{ v \in V \mid R(\exp(h))v = \exp(\lambda(h)) \cdot v,\ h \in \mathfrak{h}^{\mathbb{C}} \right\}$.

A vector $\lambda \in (\mathfrak{h}^{\mathbb{C}})^*$ is called a *weight* of the representation R if $\dim(V_\lambda) > 0$. The dimension of V_λ is called the *multiplicity* of the weight λ.

The adjoint action of G on $\mathfrak{g}^{\mathbb{C}}$ is a representation of G on $\mathfrak{g}^{\mathbb{C}}$ the restriction for which on H gives the decomposition

$$\mathfrak{g}^{\mathbb{C}} = \mathfrak{h}^{\mathbb{C}} + \bigoplus_{\alpha \neq 0} \mathfrak{g}_\alpha$$

where $\alpha \in (\mathfrak{h}^{\mathbb{C}})^*$ and

$$\mathfrak{g}_\alpha = \left\{ X \in \mathfrak{g}^{\mathbb{C}} \mid [Y, X] = \alpha(Y) \cdot X,\ \forall\, Y \in \mathfrak{h}^{\mathbb{C}} \right\}$$

The nonzero weights of the adjoint representation are called the *roots* of the group G with respect to the maximal torus H. Let Δ denotes the set of roots.

Let $\langle\ ,\ \rangle$ be a G-invariant scalar product on $\mathfrak{g}$. For instance, it can be the Killing form $\langle X, Y \rangle = tr(ad_X \circ ad_Y)$. This scalar product can be linearly extended to $\mathfrak{g}^{\mathbb{C}}$. The algebra $\mathfrak{g}$ is said to be *semi-simple* if the scalar product $\langle\ ,\ \rangle$ on $\mathfrak{g}$ is non-degenerated. In this case, the group G is said to be *semi-simple*, too. Further, we assume that the group G (and hence, the Lie algebra $\mathfrak{g}$) is semi-simple.

Notice that all roots are pure imaginary: $\Delta \subset i\mathfrak{h}^*$ and for any $\alpha \in \Delta$, we have that $\bar{\mathfrak{g}}_\alpha = \mathfrak{g}_{-\alpha}$. The latter implies that $\Delta = -\Delta$. Besides that $\dim(\mathfrak{g}_\alpha) = 1$. The following theorem summarizes the structure of a semi-simple Lie algebra.

Theorem 4.2.6. *Let G be a connected compact Lie group with dimension n and rank l. Then G has $2m$ roots $\{\pm\alpha_k \mid k = 1, \ldots, m\}$, where $n = l + 2m$. The group G is semi-simple if and only if it has l linearly independent roots. If G is semi-simple, we can choose vectors $e_\alpha \in l_\alpha$, $\alpha \in \Delta$ so that*

$$[e_\alpha, e_{-\alpha}] = h_\alpha \in \mathfrak{h}$$

and

$$[e_\alpha, e_\beta] = \begin{cases} 0 & \text{if } \alpha+\beta \text{ is not a root} \\ c_{\alpha\beta} e_{\alpha+\beta} & \text{if } \alpha+\beta \text{ is a root} \end{cases}$$

The set $\{h_\alpha \mid \alpha \in \Delta\}$ contains a basis for $\mathfrak{h}$ and the vectors e_α are called the root vectors.

An element $h \in \mathfrak{h}$ is said to be *regular* if $\{X \in \mathfrak{g} \mid [h, X] = 0\} = \mathfrak{h}$. Otherwise the element h is said to be *singular*. It is clear that $h \in \mathfrak{h}$ is regular if and only if $\alpha(h) \neq 0$ for each $\alpha \in \Delta$.

An element $h \in \mathfrak{h}$ is said to be *strongly regular* if $\alpha_1(h) \neq \alpha_2(h)$ for any two distinct roots α_1 and α_2. From the property $\Delta = -\Delta$ follows that if h is strongly regular then it is regular, too.

Based on the data about the structure of a semi-simple Lie algebra, we present the proof of the following classical theorem (see [91])

Theorem 4.2.7. *Let L be a semi-simple Lie algebra. There exist two elements in $L^{\mathbb{C}}$ which generate the entire Lie algebra $L^{\mathbb{C}}$.*

Proof. Let A be a maximal abelian subalgebra of L and $L^{\mathbb{C}} = A^{\mathbb{C}} + \bigoplus_{\alpha\in\Delta} l_\alpha$ be the corresponding root space decomposition of L. Let one element that we are looking for be any strongly regular element $h \in A$ and the other one be the vector $e = \sum_{\alpha\in\Delta} e_\alpha$. Consider the vectors

$$\begin{aligned} e_0 &= e \\ e_1 &= [h, e_0] = \sum_{\alpha\in\Delta} \alpha(h) e_\alpha \\ &\vdots \\ e_{k+1} &= [h, e_k] = \sum_{\alpha\in\Delta} \alpha(h)^k e_\alpha \\ &\vdots \\ e_{2m-1} &= [h, e_{2m-2}] = \sum_{\alpha\in\Delta} \alpha(h)^{2m-1} e_\alpha \end{aligned}$$

where $2m$ is the number of elements in Δ. Since the numbers $\alpha(h)$, $\alpha \in \Delta$ are different, the matrix

$$\begin{pmatrix} 1 & \cdots & 1 \\ \alpha_1(h) & \cdots & \alpha_{2m}(h) \\ \vdots & \cdots & \vdots \\ \alpha_1(h)^{2m-1} & \cdots & \alpha_{2m}(h)^{2m-1} \end{pmatrix}$$

where $\{\alpha_1,\ldots,\alpha_{2m}\}=\Delta$, is non-degenerate and therefore, the set of vectors $\{e_0,\ldots,e_{2m-1}\}$ is a basis for $\bigoplus_{\alpha\in\Delta} l_\alpha$. Thus, from the vectors $e_0,\ldots,e_{2m-1}$ we can construct the root vectors e_α, $\alpha\in\Delta$ and then use the vectors $[e_\alpha,e_{-\alpha}]$, $\alpha\in\Delta$ to construct the basis for $A^{\mathbb{C}}$. □

Notice that this theorem concerns not the Lie algebra L itself, but its complexification. Now, we consider the ways for constructing the pair of generators for a Lie algebra itself. We concentrate on the Lie algebra $\mathfrak{su}(n)$. First, we review the structure of its complexification in the framework of *root space decomposition*.

Let A be any element of $\mathfrak{su}(n)$. We can take A as a diagonal matrix, otherwise we can diagonalize it. So, we have $A = diag(\lambda_1,\ldots,\lambda_n)$, $\sum_{i=1}^{n}\lambda_j = 0$, $\lambda_j\in i\cdot\mathbb{R}$. Let H be the maximal abelian subalgebra of $\mathfrak{su}(n)^{\mathbb{C}}$ containing A, and let H' be the zero eigenspace of ad_A:

$$H' = \left\{X\in\mathfrak{su}(n)^{\mathbb{C}} \mid [A,X]=0\right\}$$

It is clear that

$$H = H' \Longleftrightarrow \lambda_p\neq\lambda_q,\ \text{when } p\neq q$$

Therefore, the element A is regular if and only if its eigenvalues λ_j, $j=1,\ldots,n$ are distinct. We assume that the latter is true for A, and thus H is the subgroup of all diagonal (with respect to the basis in which A is diagonal) traceless matrices. We have n linear functions $\lambda_j : H\longrightarrow\mathbb{C}$, $j=1,\ldots,n$, where for any $X\in H$ we have $X = diag(\lambda_1(X),\ldots,\lambda_n(X))$. Let $\mathcal{E}_{pq}$ be the matrix with (p,q)-entry equal to 1 and other entries equal to 0. It is clear that for any $X\in H$ we have $[X,\mathcal{E}_{pq}] = \alpha_{pq}(X)\cdot\mathcal{E}_{pq}$, where $\alpha_{pq}(X) = \lambda_p(X)-\lambda_q(X)$. Therefore, the matrices $\mathcal{E}_{pq}\in\mathfrak{su}(n)^{\mathbb{C}}$ can be considered as the root vectors and the functionals $\alpha_{pq} = \lambda_p-\lambda_q$ as the roots. Each root vector has the multiplicity 1 if and only if $\alpha_{pq}\neq\alpha_{ls}$ for $(p,q)\neq(l,s)$. This is so when H contains such matrix B that $\lambda_p(B)-\lambda_q(B)\neq\lambda_l(B)-\lambda_s(B)$ for $(p,q)\neq(l,s)$. Recall that such element B is said to be strongly regular. We assume that the matrix A itself is strongly regular.

The roots $\alpha_{12},\alpha_{23},\ldots,\alpha_{n-1n}$ are called the *fundamental roots* since the others are obtained as sums of pairs of them. The basis for $\mathfrak{su}(n)^{\mathbb{C}}$ corresponding to the roots is

$$\{E_k = \mathcal{E}_{kk}-\mathcal{E}_{k+1k+1} \mid k=1,,\ldots,n-1\}\cup\{\mathcal{E}_{pq}\mid p,q=1,\ldots,n,\ p\neq q\}$$

This basis is known as the Wayl basis for $\mathfrak{su}(n)^{\mathbb{C}}$. It is clear that the elements of this basis are not in $\mathfrak{su}(n)$ but such basis can be constructed from the Wayl basis as

$$\{h_k = i\cdot E_k\}\cup\{U_{pq}=\mathcal{E}_{pq}-\mathcal{E}_{qp}\}\cup\{V_{pq}=i\cdot(\mathcal{E}_{pq}+\mathcal{E}_{qp})\}$$

It is worth noticing that for the following commutating relations: for any diagonal $A = i\cdot h\in\mathfrak{su}(n)$ we have

$$\begin{aligned}&[A,U_{pq}] = \alpha_{pq}(A)V_{pq},\ [A,V_{pq}] = -\alpha_{pq}(A)U_{pq}\\ &[U_{pp+1},V_{pp+1}] = 2h_p\end{aligned}\tag{4.2.3}$$

Our further discussion in this section is in accordance with [6]. Hence, we have a matrix $A\in\mathfrak{su}(n)$ which is a regular element (i.e., its eigenvalues have multiplicity 1) and therefore, the associated Cartian subalgebra in $\mathfrak{su}(n)$ is the set of all traceless diagonal matrices. Let B be another element of $\mathfrak{su}(n)$. Consider the graph $\Gamma_{A,B}$ defined as follows: $\Gamma_{A,B}$ has n nodes $\{|1\rangle,\ldots,|n\rangle\}$, and the nodes $|p\rangle$ and $|q\rangle$ are joined if and only if the (p,q)-entry of the matrix B is nonzero. The graph is said to be *connected* if for all pairs of nodes there exists an oriented path connecting them. The following two theorems are very important for the controllability of affine systems on $\mathfrak{su}(n)$ (see [6])

Theorem 4.2.8. *For $A,B\in\mathfrak{su}(n)$ if A is diagonal, a necessary condition that the pair (A,B) generates the entire $\mathfrak{su}(n)$ is that the graph $\Gamma_{A,B}$ is connected.*

Theorem 4.2.9. *Given a pair of matrices $A,B\in\mathfrak{su}(n)$, assume that the graph $\Gamma_{A,B}$ is connected. If A is strongly regular then the pair (A,B) generates the entire Lie algebra $\mathfrak{su}(n)$.*

The first step of the proof (as in the case of the Theorem 4.2.7) is to consider the set of matrices $A,\ [A,B],\ldots,\underbrace{[A,\ldots,[A,B]]}_{n^2-n-1}$ and then use the commutating relations 4.2.3 to construct the remaining $n-1$ elements of the basis by the commutators between them.

4.3. Recursive Construction of Quantum Gates

Regardless, in quantum computation it is important to construct the set of generators (gates) recursively: assuming that we have the complete controllability of some subsystem for a large system and the task is to construct an optimal extension of the controllability to the entire system. Our further discussions can exact more in regard to this question.

Let G be a compact Lie group with the Lie algebra $\mathfrak{g}$; $\mathfrak{g}_0$ shall be a Lie subalgebra of $\mathfrak{g}$ and G_0 be the corresponding connected Lie subgroup in G. Assume that we have a set of generators for $\mathfrak{g}_0$, thus the set reachable from $1\in G$ when using these generators is G_0. The problem is to find a minimal extension in the given set of generators for $\mathfrak{g}_0$ to a set of generators for $\mathfrak{g}$. This problem is actually for control systems that are used in quantum computations, because usually systems of universal quantum gates are constructed hierarchically; by adding new gates to existing lower-dimensional subsystems.

Consider the adjoint action of the subgroup G_0 on the Lie algebra $\mathfrak{g}$. Since $\mathfrak{g}_0$ is the Lie algebra of G_0, it is invariant under this action and thus we have the following decomposition $\mathfrak{g}=\mathfrak{g}_0\oplus L_1\oplus\cdots\oplus L_p$ where each L_i is invariant and irreducible for the adjoint action of G_0.

Remark 4.3.1. For instance, we can take the orthogonal complement of $\mathfrak{g}_0$ under a scalar product invariant for the adjoint action. This space itself is invariant for the adjoint action of G_0, and then take its decomposition on irreducible components.

Lemma 4.3.1. *Consider a set $A=\mathfrak{g}_0\cup\{X_1,\ldots,X_p\}$ where $X_i\in L_i\setminus\{0\}$, $i=1,\ldots,p$. The set A generates the entire Lie algebra $\mathfrak{g}$.*

Proof. Let $\mathfrak{g}'$ be the Lie algebra generated by the set A. Since A is a subset of $\mathfrak{g}'$, the space $\mathfrak{g}'$ is invariant under the adjoint action of the subgroup G_0. For each $i=1,\ldots,p$ the intersection $\mathfrak{g}'\cap L_i$ is nontrivial, because it contains at least X_i. Since the subspaces $\mathfrak{g}'$ and L_i are invariant for the adjoint action of G_0, their intersection is also invariant. But by assumption, the adjoint action of G_0 on each L_i is irreducible, which implies that L_i is a subspace of $\mathfrak{g}'$ for each $i = 1, \ldots, p$ and consequently $\mathfrak{g}' = \mathfrak{g}$. □

After this proof, it is still not clear if the set $\{X_1, \ldots, X_p\}$ is a minimal complement for $\mathfrak{g}_0$ to a set of generators for $\mathfrak{g}$ or not. Although this is the case when $p = 1$, but in general the answer is negative. To illustrate this, consider the case when $\dim(L_i) = 1,\ i = 1, \ldots, p$. Let us select nonzero vectors $e_i \in L_i,\ i = 1, \ldots, p$. The set $\{e_1, \ldots, e_p\}$ is a basis for $L = \bigoplus_{i=1}^{p} L_i$, and we have that $[x, e_i] = \lambda_i(x) \cdot e_i,\ x \in \mathfrak{g}_0$, where λ_i is the character of the representation of G_0 on L_i via the adjoint action. Consider the vector $e = \sum_{i=1}^{p} e_i$. If we are able to find such elements $x_1, \ldots, x_p$ in $\mathfrak{g}_0$ that the matrix $\left(a_i^j = \lambda_i(x_j)\right)_{i,j=1}^{p}$ is non-degenerate then the vectors $[x_1, e], \ldots, [x_p, e]$ form a basis for L; and therefore, we need just one element e which together with $\mathfrak{g}_0$ gives a set of generators for $\mathfrak{g}$. Of course, this is a minimal complement for any set of generators for $\mathfrak{g}_0$ that has a set of generators for $\mathfrak{g}$.

The method described in this Lemma 4.3.1 is in agreement with the standard method of constructing the universal set of gates for quantum computation. Now, we will describe a method of recursive construction for generators which follow the ideas discussed in [24]. Let X and Y be finite-dimensional Hilbert spaces. Assuming that we have universal sets of generators for the unitary groups $U(X)$ and $U(Y)$, the problem is to extend them to the universal set of generators for the unitary group $U(X \otimes Y)$. First, let us recall some useful facts from the representation theory.

Let V be a finite-dimensional Hilbert space and $\pi : G \longrightarrow U(V)$ be a unitary representation of a Lie group G. For $g \in G$ and $v \in V$, we shall use the notation gv for the action $\pi(g)(v)$. Let us denote by $C(\pi)$ the set of all such linear operators $A : V \longrightarrow V$ that $A \circ \pi(g) = \pi(g) \circ A,\ \forall\ g \in G$. It is clear that $C(\pi)$ is a subalgebra of the algebra of all endomorphisms of V.

Lemma 4.3.2. *The representation π is irreducible if and only if the algebra $C(\pi)$ does not contain a projection operator different from 0 and 1.*

Proof. If the representation π is reducible then $V = V_1 \oplus V_2$ where V_1 and V_2 are proper non-trivial invariant subspaces. It is clear that the projector on the subspace V_1 (or V_2) commutes with every $\pi(g),\ g \in G$. Conversely, if $P : V \longrightarrow V$ is a non-trivial projector on a proper subspace which commutes with all $\pi(g)$, then we have

$$\forall\ x \in \mathrm{Im}(P),\ \forall\ g \in G:\ P(gx) = gP(x) = gx \Longrightarrow gx \in \mathrm{Im}(P)$$

Hence, the subspace $\mathrm{Im}(P)$ is invariant for every $\pi(g)$ and therefore the representation π is not irreducible. □

Lemma 4.3.3. *The representation π is irreducible if the algebra $C(\pi)$ does not contain a self-adjoint operator different from 0 and 1.*

Proof. If π is not irreducible then according to the previous lemma the algebra $C(\pi)$ contains a projection different from 0 and 1, which is a self-adjoint operator. Conversely, if

$C(\pi)$ contains a self-adjoint operator A different from 0 and 1, consider the spectral decomposition of A: $A = \sum \lambda_i P_i$, where P_i-s are orthogonal projectors. From $Ag = gA$ follows that if $x \in \mathrm{Im}(P_i)$, then we have

$$A(gx) = gA(x) = \lambda_i \cdot gx \Longrightarrow gx \in \mathrm{Im}(P_i)$$

This implies that each subspace $\mathrm{Im}(P_i)$ is invariant for the representation π. □

The last lemma implies the following

Theorem 4.3.1. *The representation π is irreducible if and only if the algebra $C(\pi)$ consists of only scalar operators (i.e., operators for the form $\lambda \cdot \mathbf{1}$, $\lambda \in \mathbb{C}$).*

Proof. Let $A \in C(\pi)$ be a non-scalar operator. Then at least one of the following two self-adjoint operators $B_1 = A + A^*$ and $B_2 = i(A - A^*)$ is also non-scalar. According to the previous lemma this implies that the representation π is not irreducible which contradicts to the assumption of the theorem. □

For two representations $\pi_1 : G_1 \longrightarrow U(V_1)$ and $\pi_2 : G_2 \longrightarrow U(V_2)$, where G_1 and G_2 are Lie groups and V_1 and V_2 are Hilbert spaces, let us denote by $\pi_1 \otimes \pi_2$ the representation of $G_1 \times G_2$ in $V_1 \otimes V_2$ defined as

$$(\pi_1 \otimes \pi_2)(g_1, g_2)(x, y) = g_1 x \otimes g_2 y,\ g_1 \in G_1,\ g_2 \in G_2,\ x \in V_1,\ y \in V_2$$

Theorem 4.3.2. *If the spaces V_1 and V_2 are finite-dimensional, then $C(\pi_1 \otimes \pi_2) = C(\pi_1) \otimes C(\pi_2)$.*

Proof. Any operator $T : V_1 \otimes V_2 \longrightarrow V_1 \otimes V_2$ can be represented as $T = \sum^{p} A_p \otimes B_p$ where A_p is an endomorphism of V_1 and B_p is an endomorphism of V_2. The right-hand side of this equality can be reduced to the form when the operators B_p are linearly independent. If $T \in C(\pi_1 \otimes \pi_2)$, then we have the following

$$[T, g_1 \otimes g_2] = 0, \quad \forall\, g_1 \in G_1 \text{ and } \forall\, g_2 \in G_2$$

This implies that

$$[T, g \otimes 1] = 0, \quad \forall\, g \in G_1$$

Expanding T, we obtain

$$\sum^{p} [A_p, g] \otimes B_p = 0, \quad \forall\, g \in G_1$$

Since the operators B_p are linearly independent, we have that each $[A_p, g]$ is 0. Therefore, each $A_p \in C(\pi_1)$, which implies that $T \in End(V_1) \otimes C(\pi_2)$. In the same manner, we obtain that $T \in C(\pi_1) \otimes End(V_2)$. Thus, we have that $T \in C(\pi_1) \otimes End(V_2) \bigcap End(V_1) \otimes C(\pi_2)$ and therefore $T \in C(\pi_1) \bigcap C(\pi_2)$. □

Corollary 4.3.1. *if the representations π_1 and π_2 are irreducible then the representation $\pi_1 \otimes \pi_2$ is also irreducible.*

Now consider the unitary group $U(V_1 \otimes V_2)$ where V_1 and V_2 are finite-dimensional Hilbert spaces and $\dim(V_1) = m$, $\dim(V_2) = n$. First of all, notice that its Lie algebra is

$$\mathfrak{u}(V_1 \otimes V_2) = \mathfrak{u}(V_1) \otimes i\mathfrak{u}(V_2)$$

because (see [24]), obviously the Lie algebra $\mathfrak{u}(V_1) \otimes i\mathfrak{u}(V_2)$ with the commutator defined as

$$[a_1 \otimes ib_1, a_2 \otimes ib_2] = a_2a_1 \otimes b_2b_1 - a_1a_2 \otimes b_1b_2$$

is a subalgebra of $\mathfrak{u}(V_1 \otimes V_2)$. Also, we have that

$$\dim(\mathfrak{u}(V_1 \otimes V_2)) = (mn)^2 = \dim(V_1)^2 \dim(V_2)^2 = \dim(\mathfrak{u}(V_1) \otimes i\mathfrak{u}(V_2))$$

Consider the subgroup of $U(V_1 \otimes V_2)$ generated by the elements for the form $g_1 \otimes g_2$, $g_1 \in U(V_1)$, $g_2 \in U(V_2)$. Let us denote this subgroup by $U(V_1) \otimes U(V_2)$. The Lie algebra of this subgroup is the subalgebra of elements for the form

$$u_1 \otimes 1 + 1 \otimes u_2, \quad u_1 \in \mathfrak{u}(V_1)\ u_2 \in \mathfrak{u}(V_2)$$

Such element is known as the Kronecker sum of u_1 and u_2.

Remark 4.3.2. The Kronecker sum of u_1 and u_2 can be written as $u_1 \otimes i(-i1) - i1 \otimes iu_2$ in accordance with the equality $\mathfrak{u}(V_1 \otimes V_2) = \mathfrak{u}(V_1) \otimes i\mathfrak{u}(V_2)$.

Consider the decomposition

$$\mathfrak{u}(V_1 \otimes V_2) = (i\mathbb{R} \cdot 1 \oplus \mathfrak{su}(V_1)) \otimes (\mathbb{R} \cdot 1 \oplus i \cdot \mathfrak{su}(V_2)) =$$

$$= (i\mathbb{R} \otimes 1) \oplus (\mathfrak{su}(V_1) \otimes 1) \oplus (1 \otimes \mathfrak{su}(V_2)) \oplus (\mathfrak{su}(V_1) \otimes i \cdot \mathfrak{su}(V_2)) =$$

$$= (\mathfrak{u}(V_1) \otimes 1 + 1 \otimes \mathfrak{u}(V_2)) \oplus (\mathfrak{su}(V_1) \otimes i \cdot \mathfrak{su}(V_2))$$

The first summand is obviously the Lie algebra of $U(V_1) \otimes U(V_2) \equiv G$. Let us denote this Lie algebra by L. Consider the adjoint action of the group G on the Lie algebra $\mathfrak{u}(V_1 \otimes V_2)$. It is clear that L is invariant under this action. Since the adjoint action of $U(V_1)$ on $\mathfrak{su}(V_1)$ is irreducible and the same is true for $U(V_2)$ and $i \cdot \mathfrak{su}(V_2)$, we have according to the corollary 4.3.1 that the action of G on $\mathfrak{su}(V_1) \otimes i \cdot \mathfrak{su}(V_2)$ is irreducible, too. This together with Lemma 4.3.1 implies that if we have a set of generators for L, we need only one new element from $\mathfrak{su}(V_1) \otimes i \cdot \mathfrak{su}V_2$ to generate the entire Lie algebra $\mathfrak{u}(V_1 \otimes V_2)$ (actually, any one element *outside* for L is enough). Hence, an optimal (in the sense of quantity) extension for the set of generators with the Lie algebra for $U(V_1) \otimes U(V_2)$ is obtained by adding one and only one element outside of the Lie algebra of $U(V_1) \otimes U(V_2)$.

The set of generators for the Lie algebra is used to construct a control system in this corresponding set reachable from 1 which is the entire Lie group. But the dynamics of a control system is a continuous process while the conventional computational process (quantum too) is discrete. In other words, the latter uses not 1-parameter flows of the form $\exp(tX_i)$, $t > 0$ as gates to generate group elements, but discrete "flows" of the form g_i^n, $n = 1, 2, \ldots$, where g_i-s are so-called *gates*. The generators of Lie algebra can be used to construct such gates.

Let $\mathcal{X} = \{X_1, \ldots, X_m\}$ be a set of generators for the Lie algebra L of a compact connected Lie group G. For any $X_i \in \mathcal{X}$ consider the corresponding 1-parameter flow $\exp(tX_i)$, $t \in \mathbb{R}$. If $\exp(tX_i)$ is periodic and therefore $\exp(TX_i) = 1$ for some $T \in \mathbb{R}^+$ then take the element $g_i = \exp(\alpha_i TX_i)$, where $\alpha_i \in \mathbb{R}^+$ is an irrational number. The set $S(g_i) = \{g_i^n \mid n \in \mathbb{N}^+\}$ is dense in the orbit $\exp(tX_i)$, $t \in \mathbb{R}$. If for some generator $X_j \in \mathcal{X}$ the corresponding orbit $\exp(tX_j)$, $t \in \mathbb{R}^+$ is not periodic then consider the maximal torus $\mathbb{T}_k \subset G$ where it is contained. Replace the element X_j in the set of generators $\mathcal{X}$ by periodic generators of the Lie algebra of the torus $\mathbb{T}^k$. The obtained set is again a set of generators for L because X_j is a linear combination for the generators of the torus. In this way, we can achieve that all the generators for L are periodic, and then construct elements of the group G: $g_1, \ldots, g_m$, such that for each group $S(g_k) = \{g_k^n \mid n \in \mathbb{N}^+\}$, $k = 1, \ldots, m$ is dense in the 1-parameter subgroup $\exp(tX_k)$, $t \in \mathbb{R}$. Since the subgroups $\exp(tX_k)$, $k = 1, \ldots, m$ generate the group G, the subgroups $S(g_k)$, $k = 1, \ldots, m$ generate a dense subgroup in G.

Lemma 4.3.4. *If G is a Lie group with representation in some Hilbert space $\mathcal{H}$ and $\mathcal{V} \subset \mathcal{H}$ is a subspace invariant for some dense subset $S \subset G$ then $\mathcal{V}$ is invariant for the entire group G.*

Proof. Consider the continuous map $F : G \longrightarrow Hom(\mathcal{V}, \mathcal{V}^{\perp})$ defined as $F(g) = P \circ g|_{\mathcal{V}}$ where $P : \mathcal{H} \longrightarrow \mathcal{V}^{\perp}$ is the operator of orthogonal projection on $\mathcal{V}^{\perp}$. Since $\mathcal{V}$ is invariant for S, we have that $F(S) = 0$. Since S is dense in G and F is continuous, we have that $F(G) = 0$ which means that $\mathcal{V}$ is invariant for the entire group G. □

Let L_1 be a Lie subalgebra of L and $G_1 \subset G$ be the corresponding connected Lie group. Let $L = L_1 \oplus A$ be such decomposition of L that the subspace A is invariant under the adjoint action of G_1 and the action of G_1 on A is irreducible. As it follows from Lemma 4.3.1, if we take only one element $X \in A$, then the set $L_1 \cap \{X\}$ generates the entire Lie algebra L. If X is such element that the 1-parameter subgroup $\{\exp(tX) \mid t \in \mathbb{R}\}$ is periodic, then for some $\alpha \in \mathbb{R}$, the element $g = \exp(\alpha X)$ generates a dense subgroup in $\{\exp(tX) \mid t \in \mathbb{R}\}$. In this situation the set $G_1 \cap \{g\}$ generates a *dense subgroup* in G. Moreover, we then follow this theorem:

Theorem 4.3.3. *Under these conditions the set $G_1 \cap \{g\}$ generates the entire group G.*

Proof. First, let us show that if $g_0 = g^m$ is a normalizer of the subgroup G_1, then $m = 0$. Next, we assume that $m > 0$ and $g_0 G_1 g_0^{-1} = G_1$. It is clear that in this case, the subgroup $\{g_0^n \mid n \in \mathbb{Z}\}$ is also a normalizer of G_1. But this subgroup is dense too in $\{\exp(tX) \mid t \in \mathbb{R}\} \equiv S$. Consider the adjoint action of S on the Lie algebra L. Since $\{g_0^n \mid n \in \mathbb{Z}\}$ is a normalizer of G_1, we have that group S contains a dense subgroup for which L_1 is invariant. This implies according to the previous lemma that L_1 is invariant for the adjoint action of the group S. Therefore, we have that $[X, L_1] \subset L_1$. Since A is invariant under the adjoint action of G_1 and $X \in A$, we have that $[X, L_1] \subset A$. The result is that $[X, L_1] = 0$, which contradicts to the assumption that the action of G_1 on A is irreducible. Hence, we obtain that there is no nontrivial normalizer for G_1 in $\{g^n \mid n \in \mathbb{Z}\}$.

Now, take any element $g_1 = g^m$, $m \neq 0$ and consider the group $G_2 = g_1 G_1 g_1^{-1}$. Since g_1 cannot be a normalizer of G_1, we have then $G_2 \neq G_1$. Therefore, the group $\hat{G}$ generated by $G_1 \cap G_2$ contains G_1 as a proper subgroup. The group $\hat{G}$ is connected because G_1 is such.

Therefore, its Lie algebra $\hat{L}$ contains L_1 as a proper subalgebra. From this, it then follows that $\hat{L} \cap A \neq 0$. But as the adjoint action of G_1 on A is irreducible and $\hat{L}$ is invariant under this action, we have that $\hat{L} \cap A = A$. The latter implies $\hat{L} = L$. Since G and its subgroup $\hat{G}$ are connected and they have a common Lie algebra, they must be equal: $\hat{G} = G$. □

Regarding this theorem in the context of inductive construction for generators in this unitary group of tensor product, we have the following situation: consider the unitary group $U(V_1 \otimes \cdots \otimes V_m)$, where $V_1, \ldots, V_m$ are some finite-dimensional Hilbert spaces; if we are able to generate the subgroups $U(V_1), \ldots, U(V_m)$, then we can construct a set of generators for the entire group step-by-step by adding only one element at each step. For instance, consider the group $U(V_1 \otimes V_2 \otimes V_3 \otimes V_4 \otimes V_5$. If we have the groups $U(V_1)$, $U(V_2)$, $U(V_3)$, $U(V_4)$ and $U(V_5)$, then to construct $U(V_1 \otimes V_2)$ and $U(V_3 \otimes V_4)$; we need only 2 more elements. Then with $U(V_1 \otimes V_2)$ and $U(V_3 \otimes V_4)$, we can construct $U(V_1 \otimes V_2 \otimes V_3 \otimes V_4)$ by adding 1 more element. And finally, with $U(V_1 \otimes V_2 \otimes V_3 \otimes V_4)$ and $U(V_5)$, we can construct the entire group $U(V_1 \otimes V_2 \otimes V_3 \otimes V_4 \otimes V_5)$ by adding again 1 more element. Therefore, besides the subgroups $U(V_1)$, $U(V_2)$, $U(V_3)$, $U(V_4)$ and $U(V_5)$, we need 4 more elements to construct the group $U(V_1 \otimes V_2 \otimes V_3 \otimes V_4 \otimes V_5)$.

This construction is, of course, in accordance with [24], and the key point here is that an element outside of the subgroup is not a normalizer of this subgroup.

4.4. Vector Bundle with Connection on Grassmann Manifold

4.4.1. The Differential Structure on Complex Grassmann Manifold

Let $\mathcal{H}$ be a complex Hilbert space, infinite or finite-dimensional, as required. For any positive integer number m, let us denote by $Gr_m(\mathcal{H})$ the set of all m-dimensional complex subspaces of the space $\mathcal{H}$. It is clear that any m-dimensional subspace X of the Hilbert space $\mathcal{H}$ can be uniquely defined by the corresponding operator of orthogonal projection $\mathcal{P}(X) : \mathcal{H} \longrightarrow X \subset \mathcal{H}$. The operator $\mathcal{P}(X)$ is characterized by the properties: $\mathcal{P}(X)^* = \mathcal{P}(X)$ and $\mathrm{trace}(\mathcal{P}(X)) = \dim(X)$.

Let us denote by $\mathcal{P}_m(\mathcal{H})$ the set

$$\mathcal{P}_m(\mathcal{H}) = \left\{P : \mathcal{H} \longrightarrow \mathcal{H} \,|\, P^2 = P,\, P^* = P,\, \mathrm{trace}(P) = m\right\} \tag{4.4.1}$$

We have then the mapping $\mathcal{P} : Gr_m(\mathcal{H}) \longrightarrow \mathcal{P}_m(\mathcal{H})$ is a bijection (see [44], [45], [46]) and defines an injection of the set $Gr_m(\mathcal{H})$ in the vector space of Hermitian operators on the Hilbert space $\mathcal{H}$. This injection induces a topology and a differential structure on the set $Gr_m(\mathcal{H})$. The set $Gr_m(\mathcal{H})$, together with this differential structure is known as the Grassmann manifold for the m-dimensional complex subspaces of the Hilbert space $\mathcal{H}$. As it follows from our earlier discussion, we can identify the following two objects: $\mathcal{P}_m(\mathcal{H})$ and $Gr_m(\mathcal{H})$.

Any fixed element $X \in Gr_m(\mathcal{H})$ defines a mapping

$$\Gamma_X : Hom(X, X^{\perp}) \longrightarrow Gr_m(\mathcal{H})$$

where, for $\varphi \in Hom(X, X^{\perp})$: $\Gamma_X(\varphi) = \{x + \varphi(x) \,|\, x \in X\}$

In other words, the subspace corresponding to the linear mapping $X \xrightarrow{\varphi} X^{\perp}$ is the graph of the mapping φ. In some cases, we shall ommit the subscript $_X$ in the expression Γ_X and write just Γ. This mapping is an injection of the vector space $Hom(X,X^{\perp})$ into the set $Gr_m(\mathcal{H})$ and its image is the subset

$$\left\{Y \in Gr_m(\mathcal{H})\,|\,Y \cap X^{\perp} = \{0\}\right\} \equiv \mathcal{W}_X$$

It follows that from varying the element $X \in Gr_m(\mathcal{H})$, we can cover the Grassmann manifold by the open subsets $\mathcal{W}_X$. Actually, it is sufficient to take only a finite number of the elements $X \in Gr_m(\mathcal{H})$ to cover the total manifold by the sets of the type $\mathcal{W}_X$.

Hence, we can state that a pair $(Hom(X,X^{\perp}),\Gamma_X)$, for $X \in Gr_m(\mathcal{H})$, defines a local coordinate system in the neighborhood of the point X (see [104]).

Now, let us find the expression for the diffeomorphism

$$\mathcal{P} : Gr_m(\mathcal{H}) \longrightarrow \mathcal{P}_m(\mathcal{H})$$

in this local coordinate system. In other words, the problem is to find the coordinate expression for the the following composition map

$$\Phi_X : Hom(X,X^{\perp}) \xrightarrow{\Gamma_X} Gr_m(\mathcal{H}) \xrightarrow{\mathcal{P}} \mathcal{P}_m(\mathcal{H})$$

For any $f \in Hom(X,X^{\perp})$, the graph of the operator $-f^* \in Hom(X^*,X)$ is the orthogonal subspace of the space $\Gamma_X(f)$. Indeed, for $x \in X$ and $u \in X^{\perp}$ we have the following

$$\langle x+f(x), u-f^*(u)\rangle = \underbrace{\langle x,u\rangle}_{0} - \langle x, f^*(u)\rangle + \langle f(x),u\rangle - \underbrace{\langle f(x), f^*(u)\rangle}_{0} =$$
$$= \langle f(x),u\rangle - \langle x, f^*(u)\rangle = 0 \text{ (by the definition of dual operator).}$$

For $f \in Hom(X,X^{\perp})$, consider the following operator on the Hilbert space $\mathcal{H}$:

$$\widetilde{f} = \begin{bmatrix} \mathbf{1} & -f^* \\ f & \mathbf{1} \end{bmatrix} : X \oplus X^{\perp} \longrightarrow X \oplus X^{\perp}$$

This operator is automorphism and it is easy to verify that $\widetilde{f}(X) = \Gamma(f)$ and $\widetilde{f}(X^{\perp}) = \Gamma(-f^*) = \Gamma(f)^{\perp}$. It is clear that if P_U is the projection operator on a subspace $U \subset W$ corresponding to a decomposition of a vector space $W = U \oplus V$ and $A : W \longrightarrow W$ as an automorphism, then the projection operator on the subspace $A(U)$; corresponding to the decomposition $W = A(U) \oplus A(V)$ is $P_{A(U)} = AP_UA^{-1}$. Therefore, the projection operator, corresponding to the decomposition $\mathcal{H} = \Gamma(f) \oplus \Gamma(f)^{\perp} = \widetilde{f}(X) \oplus \widetilde{f}(X^{\perp})$ is $\mathcal{P}(\Gamma(f)) = \widetilde{f}\mathcal{P}(X)\widetilde{f}^{-1}$. The explicit expression for the operator $\widetilde{f}^{-1}$, for the decomposition $\mathcal{H} = X \oplus X^{\perp}$ is

$$\begin{bmatrix} \mathbf{1} & -f^* \\ f & \mathbf{1} \end{bmatrix}^{-1} = \begin{bmatrix} (1+f^*f)^{-1} & f^*(1+ff^*)^{-1} \\ -f(1+f^*f)^{-1} & (1+ff^*)^{-1} \end{bmatrix}$$

Remark 4.4.1. The operator $1+f^*f : X \longrightarrow X$ is invertible, because

$$(1+f^*f)(x)=0 \implies (f^*f)(x)=-x \implies \langle (f^*f)(x),x\rangle = -\|x\|^2$$
$$\implies \|f(x)\|^2 = -\|x\|^2 \implies x=0$$

The same is true for the operator $1+ff^* : X^{\perp} \longrightarrow X^{\perp}$.

Therefore, the explicit expression for the projection operator $\mathcal{P}(\Gamma(f))$, corresponding to the decomposition $\mathcal{H} = X \oplus X^{\perp}$ is

$$\mathcal{P}(\Gamma(f)) = \begin{bmatrix} \mathbf{1} & -f^* \\ f & \mathbf{1} \end{bmatrix} \begin{bmatrix} 1 & 0 \\ 0 & 0 \end{bmatrix} \begin{bmatrix} (1+f^*f)^{-1} & f^*(1+ff^*)^{-1} \\ -f(1+f^*f)^{-1} & (1+ff^*)^{-1} \end{bmatrix} =$$
$$= \begin{bmatrix} (1+f^*f)^{-1} & f^*(1+ff^*)^{-1} \\ f(1+f^*f)^{-1} & ff^*(1+ff^*)^{-1} \end{bmatrix} \tag{4.4.2}$$

The action in the group $U(\mathcal{H})$ (the group with the unitary transformations for the Hilbert space $\mathcal{H}$, on $\mathcal{H}$) induces the action of this group on the Grassmann manifold. This action is transitive and for any point $X \in Gr_m(\mathcal{H})$, the corresponding stabilizer subgroup is $U(X) \times U(X^{\perp})$. Therefore, the manifold $Gr_m(\mathcal{H})$ can be considered as the homogeneous space (see, for example, [44], [45], [46])

$$Gr_m(\mathcal{H}) \cong \frac{U(\mathcal{H})}{U(X)\times U(X^{\perp})}$$

For any unitary transformation $u : \mathcal{H} \longrightarrow \mathcal{H}$, the corresponding diffeomorphism of the Grassmann manifold $gr(u) : Gr_m(\mathcal{H}) \longrightarrow Gr_m(\mathcal{H})$ induces a transformation of the local coordinate systems

$$\widetilde{u} : Hom(X,X^{\perp}) \longrightarrow Hom(Y,Y^{\perp})$$

where $Y = u(X)$ (and therefore $Y^{\perp} = u(X^{\perp})$). This transformation is defined by the condition: $\Gamma(\widetilde{u}(f)) = u(\Gamma(f))$. We have that

$$u(\Gamma(f)) = \{ux + uf(x) \,|\, x \in X\} = \left\{y + uf(u^{-1}y) \,|\, y \in u(X) \equiv Y\right\}$$

which implies that $u(\Gamma(f)) = \Gamma(ufu^{-1})$. Hence, the transformation of the local coordinate system is of the form

$$\widetilde{u} : Hom(X,X^{\perp}) \longrightarrow Hom(u(X),u(X^{\perp}))$$
$$\widetilde{u}(f) = ufu^{-1} \tag{4.4.3}$$

The tangent space of the manifold $Gr_m(\mathcal{H})$ at a point $X \in Gr_m(\mathcal{H})$ can be identified with the vector space $Hom(X,X^{\perp})$ (see [104]). The identification of the manifold $Gr_m(\mathcal{H})$

with the manifold of projections $\mathcal{P}_m(\mathcal{H})$, gives another representation of the tangent space $T_X(Gr_m(\mathcal{H}))$, which follows from 4.4.1:

$$T_X(Gr_m(\mathcal{H})) \cong \{\Phi : \mathcal{H} \longrightarrow \mathcal{H} \,|\, \Phi \text{ is linear and } \mathcal{P}(X)\circ\Phi + \Phi\circ\mathcal{P}(X) = \Phi,\ \Phi^* = \Phi,\ \text{trace}(\Phi) = 0\} \tag{4.4.4}$$

This is exactly the set of operators $\Phi : \mathcal{H} \longrightarrow \mathcal{H}$, that the decomposition of Φ, corresponding to the decomposition of the Hilbert space $\mathcal{H} = X \oplus X^{\perp}$ is in the form

$$\Phi = \begin{bmatrix} 0 & \varphi^* \\ \varphi & 0 \end{bmatrix}$$

The differential of the mapping $f \mapsto \mathcal{P}(\Gamma(f))$ (see 4.4.2), at the point $0 \in Hom(X, X^{\perp})$, gives the isomorphism between the two representations of the tangent space $T_X(Gr_m(\mathcal{H}))$:

$$Hom(X, X^{\perp}) \ni \varphi \mapsto \mathcal{P}'_X(\varphi) = \begin{bmatrix} 0 & \varphi^* \\ \varphi & 0 \end{bmatrix} \tag{4.4.5}$$

Recall that an action of a Lie group G on a smooth manifold M, gives rise to the homomorphism from the Lie algebra of G to the Lie algebra of vector fields on the manifold M: for $v \in \mathfrak{g}$, where $\mathfrak{g}$ denotes the Lie algebra of the Lie group G, the corresponding vector field $\widetilde{v}$, induced by the action of the group on M, is defined as (see [110])

$$\widetilde{v}(x) = \frac{\partial F}{\partial G}(1, x)(v)$$

where $F : G \times M \longrightarrow M$ is the mapping that corresponds to the action of the group G on the manifold M.

So, the action of the group $U(\mathcal{H})$ on the manifold $Gr_m(\mathcal{H})$, gives rise of the homomorphism from the Lie algebra of $U(\mathcal{H})$, to the Lie algebra of vector fields on the manifold $Gr_m(\mathcal{H})$. To describe this homomorphism, recall that if $\mathcal{P}(X)$ is the orthogonal projection corresponding to the element $X \in Gr_m(\mathcal{H})$, then for any orthogonal transformation $g \in U(\mathcal{H})$, the orthogonal projection corresponding to $g(X)$ is $\mathcal{P}(g(X)) = g\mathcal{P}(X)g^{-1}$. Therefore, for any fixed point $X \in Gr_m(\mathcal{H})$, we have a mapping

$$U(\mathcal{H}) \ni g \mapsto g\mathcal{P}(X)g^{-1} \in \mathcal{P}_m(\mathcal{H}) \cong Gr_m(\mathcal{H})$$

This implies that the tangent vector corresponding to an element $u \in \mathfrak{u}(\mathcal{H})$ in the Lie algebra of the Lie group $U(\mathcal{H})$, at a point $X \in Gr_m(\mathcal{H})$ is $\widetilde{u}_X = [u, \mathcal{P}(X)]$, where $[\cdot,\cdot]$ denotes the commutator of two operators. If the decomposition of the linear operator $u : \mathcal{H} \longrightarrow \mathcal{H}$, corresponding to the decomposition $\mathcal{H} = X \oplus X^{\perp}$ is

$$u = \begin{bmatrix} U_{XX} & -U^*_{XX^{\perp}} \\ U_{XX^{\perp}} & U_{X^{\perp}X^{\perp}} \end{bmatrix}$$

then it is clear that

$$\widetilde{u}_X = [u, \mathcal{P}(X)] = \begin{bmatrix} 0 & U^*_{XX^{\perp}} \\ U_{XX^{\perp}} & 0 \end{bmatrix}$$

This, together with the formula 4.4.5, implies that the vector field corresponding to the Lie algebra element $u \in \mathfrak{u}(\mathcal{H})$ via the action of the group $U(\mathcal{H})$ on $Gr_m(\mathcal{H})$ is

$$\widetilde{u}_X = U_{XX^{\perp}} = (1 - \mathcal{P}(X)) \circ u \in Hom(X, X^{\perp}) \cong T_X(Gr_m(\mathcal{H})) \quad (4.4.6)$$

4.4.2. Some Operations over Connections on Vector Bundles

In this section, we shall review some facts from the theory of vector bundles and the connections found within the context of this book.

For any vector bundle $\pi : E \longrightarrow M$, we denote by $S(E)$ the space of smooth sections in this vector bundle.

Let $V \xrightarrow{p} M$ and $W \xrightarrow{q} M$ be two vector bundles over a smooth manifold M. Let $S(V) \xrightarrow{\nabla^V} S(T^*(M) \otimes V)$ and $S(W) \xrightarrow{\nabla^W} S(T^*(M) \otimes W)$ be connections (covariant derivations) on them. We define the following operations over these connections:

The direct (or Whitney) sum of two connections. Define a connection $\nabla = \nabla^V \oplus \nabla^W$ on the Whitney sum of the vector bundles (V, M, p) and (W, M, q) is as follows: for a vector field ξ on the manifold M and a section $s = s_1 \oplus s_2$ of the vector bundle $V \oplus W$ let

$$\nabla_\xi(s) = \nabla^V_\xi(s_1) + \nabla^W_\xi(s_2) \quad (4.4.7)$$

The tensor product of two connections. Define a connection $\nabla = \nabla^V \otimes \nabla^W$ on the tensor product of the vector bundles (V, M, p) and (W, M, q) as

$$\nabla_\xi(s_1 \otimes s_2) = \nabla^V_\xi(s_1) \otimes s_2 + s_1 \otimes \nabla^W_\xi(s_2) \quad (4.4.8)$$

The connection on the dual vector bundle. Define a connection ∇^{V^*} on the vector bundle $p_* : V^* \longrightarrow M$ a fiber of which at a point $x \in M$ is the dual vector space of the vector space $p^{-1}(x)$

$$\xi\langle s, \tau\rangle = \langle \nabla^V_\xi(s), \tau\rangle + \langle s, \nabla^{V^*}_\xi(\tau)\rangle \quad (4.4.9)$$

where s is a section in the vector bundle V and τ is a section in the dual vector bundle V^*, and $\langle s, \tau\rangle$ is the function on the manifold M obtained by the pairing of the sections s and τ.

The connection on the bundle of homomorphisms. Let

$$Hom(V, W) \xrightarrow{h} M$$

be the fiber bundle, the fiber of which at a point $x \in M$ is the space of all homomorphisms from the vector space $p^{-1}(x)$ to the vector space $q^{-1}(x)$. In the case of finite dimensional fibers, the fiber bundle is canonically isomorphic to the fiber bundle $h : V^* \otimes W \longrightarrow M$. We have defined the connections on the dual spaces bundle and the tensor products bundle. Therefore, the covariant derivation $\nabla = Hom(\nabla^V, \nabla^W) = \nabla^{V^*} \otimes \nabla^W$ on the space of sections $S(Hom(V, V))$ can be defined as $\nabla_\xi(\tau \otimes s) = \nabla^{V^*}_\xi(\tau) \otimes s + \tau \otimes \nabla^W_\xi(s)$ for $\tau \in S(V^*)$ and

$s \in S(W)$. This implies that for $t \in S(V)$, we have the following

$$\nabla_\xi(\tau \otimes s)(t) = \langle \nabla_\xi^{V^*}(\tau), t\rangle \cdot s + \tau(t) \cdot \nabla_\xi^{W}(s) =$$

$$= \xi(\tau(t)) \cdot s + \tau(t) \cdot \nabla_\xi^{W}(s) - \langle \tau, \nabla_\xi^{V}(t)\rangle \cdot s =$$

$$= \nabla_\xi^{W}(\tau(t) \cdot s) - \langle \tau, \nabla_\xi^{V}(t)\rangle \cdot s$$

which suggests a more elegant expression for the covariant derivation $\nabla = Hom(\nabla^{V}, \nabla^{W})$ on the space of sections $S(Hom(V,W))$:

$$\text{for } f \in S(Hom(V,W)) \text{ and } t \in S(V) \text{ let}$$
$$\nabla_\xi(f)(t) = \nabla_\xi^{W}(f(t)) - f(\nabla_\xi^{V}(t)) \tag{4.4.10}$$

It is easy to verify that the results of all the previously defined operations satisfy the conditions required for covariant derivations on vector bundles.

Let $p : V \longrightarrow M$ and $q : W \longrightarrow M$ be such vector bundles over a smooth manifold M that their Whitney sum is a trivial vector bundle: $V \oplus W \cong M \times H$. This situation gives rise to a connection on the vector bundles (V,M,p) and (W,M,q) as follows. Any section of the vector bundle V can be regarded as a function on M with values in the vector space H. Let us denote this function under $\widetilde{s} : M \longrightarrow H$. For a vector field ξ on the manifold M, define the covariant derivation of the section s as $\nabla_\xi(s)(x) = P_{V_x}(\widetilde{s}'_x(\xi_x))$, $\forall x \in M$. Here, P_{V_x} is the projection operator $P_{V_x} : H \longrightarrow V_x$ corresponding to the decomposition $H = V_x \oplus W_x$. The covariant derivation on the sections of the vector bundle W can be defined analogically. It can be verified directly that this defined operation satisfies the conditions required for covariant derivations.

Remark 4.4.2. If V_1, V_2, W_1 and W_2 are vector bundles over the manifold M such that $V_1 \oplus W_1 \cong M \times H_1$ and $V_2 \oplus W_2 \cong M \times H_2$; ∇^1 and ∇^2 are the connections on V_1 and V_2, accordingly, as defined earlier, then the covariant derivation on the vector bundle $V_1 \otimes V_2$ induced by the decomposition

$$\left(V_1 \otimes V_2\right) \oplus \left((V_1 \otimes W_2) \oplus (W_1 \otimes V_2) \oplus (W_1 \otimes W_2)\right) \cong M \times \left(H_1 \otimes H_2\right)$$

is the same as $\nabla = \nabla^1 \otimes \nabla^2$ (see the formula 4.4.8).

Definition 4.4.1. *Let $\pi : T(M) \longrightarrow M$ be the tangent vector bundle for a smooth manifold M and $p : V \longrightarrow M$ be such vector bundle that the Whitney sum $T(M) \oplus V$ is a trivial bundle $M \times H$. We call the decomposition of a trivial fiber bundle, $M \times H = T(M) \oplus V$, integrable if the induced covariant derivation ∇ on the sections of the tangent bundle (i.e., vector fields on M) satisfies the condition*

$$\nabla_X(Y) - \nabla_Y(X) = [X,Y] \tag{4.4.11}$$

for any two vector fields X and Y on the manifold M.

If the vector space H, which is the fiber of the Whitney sum, is a Hilbert space; and, for each point $x \in M$, the vector space V_x (the fiber of the vector bundle V at the point x) is the orthogonal complement of the subspace $T_x(M) \subset H$, then the covariant derivation ∇ satisfies the condition

$$X\langle Y,Z\rangle = \langle \nabla_X(Y),Z\rangle + \langle Y,\nabla_X(Z)\rangle$$

Hence, we can state that the connection ∇, induced by the decomposition $M \times H = T(M) \oplus V$, is the covariant derivation corresponding to the Levi-Civita connection on the manifold M, for the metric induced by the embedding $T(M) \subset M \times H$.

Remark 4.4.3. If $\Phi : M \longrightarrow H$ is a smooth mapping, such that for each point $x \in M$ the linear mapping $\Phi'_x : T_x(M) \longrightarrow H$ is a monomorphism, then the vector bundle $p : \Phi^{\perp}(H) \longrightarrow M$, where $p^{-1}(x) = \mathrm{Im}(\Phi'_x)^{\perp}$, is such that the Whitney sum $T(M) \oplus \Phi^{\perp}(H)$ is trivial: $M \times H$; and the connection (covariant derivation) induced by the decomposition $M \times H = T(M) \oplus \Phi^{\perp}(H)$ is the Levi-Civita connection on $T(M)$.

Let $\pi : P \longrightarrow M$ be a principal bundle with a structure group G, acting from right on the total space P. Let F be a vector space on which the Lie group G acts from left. The associated vector bundle over the manifold M is defined as the vector bundle with the total space $(P \times V)/G \equiv P_V$, where the quotient is taken under the right action of the group G on the manifold $P \times V$: $(p,v) \mapsto (pg, g^{-1}v)$, for each $(p,v) \in P \times G$ and $g \in G$ (see [110]). The projection mapping $\pi^V : P_V \longrightarrow M$ for this vector bundle is defined as

$$\pi^V([p,v]) = \pi(p), \quad \forall\, [p,v] \in (P \times V)/G$$

It follows from the definition that any section of the associated vector bundle can be regarded as a fuction $\varphi : P \longrightarrow V$, such that $\varphi(pg) = g^{-1}\varphi(p)$, $\forall p \in P$, $\forall g \in G$. Let us denote the space of such functions on the total space of the principal bundle P, by $C^\infty(P,V)_G$.

A connection form A on the principle bundle (P,M,π) is a differential 1-form, with values in $\mathfrak{g}$ – the Lie algebra of the Lie group G, such that for any $g \in G$: $R_g^*(A) = Ad(g^{-1})(A)$; and for any $u \in \mathfrak{g}$: $A(\widetilde{u}) = u$; where $\widetilde{u}$ denotes the tangent vector

$$\widetilde{u} \in T_p(P), \quad \widetilde{u} = \frac{d}{dt}(p \cdot \exp(t \cdot u))$$

The connection form A defines a distribution of horizontal subspaces on the total space P:

$$H_p = \ker(A_p) \subset T_p(P), \quad \forall\, p \in P$$

This distribution is carried to the total space of the associated vector bundle, by the quotient mapping: $P \times V \longrightarrow P_V = (P \times V)/G$.

A vector field X on the manifold M can be lifted to a "horizontal" vector field $\widetilde{X}$ on the total space P, which is uniquely defined by the following conditions

$$\widetilde{X}_p \in \ker(A_p),\ \forall\, p \in P \text{ and } \pi'(\widetilde{X}) = X$$

As mentioned, a section of the associated vector bundle $s \in \Gamma(P_V)$ can be identified with an element $\varphi_s \in C^\infty(P,V)_G$. The covariant derivation for the section s is the section of the vector bundle (P_V, M, π_V), corresponding to the function $\varphi'_s(\widetilde{X}) \in C^\infty(P,V)_G$.

4.4.3. The Canonical Vector Bundle on Grassmann Manifold

Consider the set $\Pi_m(\mathcal{H}) = \{(X,x)\,|\,X \in Gr_m(\mathcal{H}),\, x \in X\}$, which is the subset of $Gr_m(\mathcal{H}) \times \mathcal{H}$. The set $\Pi_m(\mathcal{H})$ together with the differential structure induced from $Gr_m(\mathcal{H}) \times \mathcal{H}$ is a differential manifold. The mapping

$$p : \Pi_m(\mathcal{H}) \longrightarrow Gr_m(\mathcal{H}), \quad p(X,x) = X$$

is a vector bundle over the Grassmann manifold $Gr_m(\mathcal{H})$. The fiber at a point $X \in Gr_m(\mathcal{H})$ is the vector space $X \subset \mathcal{H}$, itself. This vector bundle is known as the *canonical vector bundle* over the Grassmann manifold $Gr_m(\mathcal{H})$. For any point $X \in Gr_m(\mathcal{H})$, the local trivialization of this vector bundle, in the neighborhood of the point X, naturally arises from the coordinate system:

$$\Phi_X : Hom(X,X^{\perp}) \times X \longrightarrow \Pi_m(\mathcal{H}), \quad \Phi_X(\varphi,x) = (\Gamma(\varphi), x + \varphi(x))$$

If we regard the manifold $Gr_m(\mathcal{H})$ as the manifold of orthonormal projectors, $\mathcal{P}_m(\mathcal{H})$ (see 4.4.1), we can describe the total space $\Pi_m(\mathcal{H})$, as a submanifold of $\mathcal{P}_m(\mathcal{H}) \times \mathcal{H}$ defined by the equation $P(x) = x$, $(P,x) \in \mathcal{P}_m(\mathcal{H}) \times \mathcal{H}$. The latter implies that the tangent space of the manifold $\Pi_m(\mathcal{H})$ is defined by the equation $dP(x) + P(dx) = dx$. More precisely: for any point $Q = (P,x) \in \Pi_m(\mathcal{H})$, where $P \in \mathcal{P}_m(\mathcal{H}) \cong Gr_m(\mathcal{H})$ is a projector and $x \in \mathrm{Im}(P) = X$, the tangent space $T_Q(\Pi_m(\mathcal{H}))$ is the subspace of the space $Hom(X,X^{\perp}) \times \mathcal{H}$ defined by the linear equation: $\varphi(x) = (1-P)u$, $\varphi \in Hom(X,X^{\perp})$, $u \in \mathcal{H}$. That is

$$T_{(X,x)}(\Pi_m(\mathcal{H})) = \left\{(\varphi,u) \in Hom(X,X^{\perp}) \times \mathcal{H} \mid \varphi(x) = \mathcal{P}(X^{\perp})(u)\right\} \tag{4.4.12}$$

From this description of the tangent space in the total space of the canonical fiber bundle, this easily follows that the vertical tangent subspace at a point $(X,x) \in \Pi_m(\mathcal{H})$ is

$$Vert(X,x) = \left\{(0,u) \in Hom(X,X^{\perp}) \times \mathcal{H}\,|\,u \in X\right\}$$

One natural choice of a horizontal complement of the vertical subspace could be

$$Hor(X,x) = \left\{(\varphi,\varphi(x))\,|\,\varphi \in Hom(X,X^{\perp})\right\}$$

The projection operator on the vertical tangent subspace, along the horizontal tangent subspace is

$$P_{Vert} : T_{(X,x)}(\Pi_m(\mathcal{H})) \longrightarrow Vert(X,x), \quad P_{Vert}(\varphi,w) = (0, w - \varphi(x)) \tag{4.4.13}$$

Proposition 4.4.1. *The distribution of horizontal subspaces*

$$Hor(X,x) \subset T_{(X,x)}(\Pi_m(\mathcal{H})),\ (X,x) \in \Pi_m(\mathcal{H})$$

is a connection on the canonical vector bundle $\Pi_m(\mathcal{H})$. The corresponding covariant derivation acts on the space of sections for $\Pi_m(\mathcal{H})$ as follows: any section $s \in S(\Pi_m(\mathcal{H}))$ can be regarded as a function $s : Gr_m(\mathcal{H}) \longrightarrow \mathcal{H}$; for any tangent vector $\varphi \in T_X(Gr_m(\mathcal{H})) \cong Hom(X,X^{\perp})$, the covariant derivative of s by φ is $\nabla_{\varphi}(s) = \mathcal{P}(X)(s'_X(\varphi)) = s'_X(\varphi) - \varphi(s(X))$.

Proof. As it follows from the formula for the projection on the vertical tangent subspace (see 4.4.13), the action of the differential for the function s on the tangent vector φ and then its projection on the vertical tangent space is $s'_X(\varphi) - \varphi(s(X))$. As it follows from the definition for the tangent space of $\Pi_m(\mathcal{H})$ (see 4.4.12), it is the same as the projection of $s'_X(\varphi)$ on the subspace X along its orthogonal complement $X^\perp$.
It can be verified by direct calculations that the operation $\nabla_\varphi(s) = S'(\varphi) - \varphi \circ s$ has the properties of covariant derivation. □

Now, consider the vector bundle $q : \Pi_m^\perp(\mathcal{H}) \longrightarrow M$ the fiber of which at a point $X \in Gr_m(\mathcal{H})$ is $X^\perp$ – the orthogonal complement of X. It is clear that the Whitney sum of two vector bundles $\Pi_m(\mathcal{H}) \oplus \Pi_m^\perp(\mathcal{H})$ is the trivial vector bundle: $Gr_m(\mathcal{H}) \times \mathcal{H}$. As in the case of $\Pi_m(\mathcal{H})$, we can describe the total space $\Pi_m^\perp(\mathcal{H})$ as a submanifold in $\mathcal{P}_m(\mathcal{H}) \cong Gr_m(\mathcal{H}) \times \mathcal{H}$:

$$\Pi_m^\perp(\mathcal{H}) = \{(P,y) \in \mathcal{P}_m(\mathcal{H}) \times \mathcal{H} \,|\, P(y) = 0\}$$

Differentiating the equation $P(y) = 0$, we obtain the equation for the tangent space of $\Pi_m^\perp(\mathcal{H})$: $dP(y) + P(dy) = 0$. This implies that the tangent space of $\Pi_m^\perp(\mathcal{H})$ at a point (X,y), $X \in Gr_m(\mathcal{H})$, $y \in X^\perp$ is the subspace of $Hom(X,X^\perp) \times \mathcal{H}$:

$$T_{(X,y)}(\Pi_m^\perp(\mathcal{H})) = \left\{(\varphi,u) \in Hom(X,X^\perp) \times \mathcal{H} \,|\, \varphi^*(y) + \mathcal{P}(X)(u) = 0\right\} \tag{4.4.14}$$

From this description of the tangent space for $\Pi_m^\perp(\mathcal{H})$, it follows that the vertical tangent subspace for the fiber bundle $q : \Pi_m^\perp(\mathcal{H}) \longrightarrow M$ at a point (X,y) is

$$Vert(X,y) = \left\{(0,u) \in Hom(X,X^\perp) \times \mathcal{H} \,|\, u \in X^\perp\right\} \tag{4.4.15}$$

One choice for the complementar ("horizontal") subspace is

$$Hor(X,y) = \left\{(\varphi, -\varphi^*(y)) \in Hom(X,X^\perp) \times \mathcal{H}\right\}$$

The discussion analogical to that is in the proof of the proposition 4.4.1, which shows that the covariant derivation corresponding to this distribution of horizontal subspaces is

$$\nabla_\varphi^\perp(s) = s'_X(\varphi) + \varphi^*(s(X)) = \mathcal{P}(X^\perp)(s'_X(\varphi)))$$

for $\varphi \in Hom(X,X^\perp) \cong T_X(Gr_m(\mathcal{H}))$ and s is a section of the vector bundle $\Pi_m^\perp(\mathcal{H})$ (in this case regarded as a function $s : Gr_m(\mathcal{H}) \longrightarrow \mathcal{H})$.

It is clear that $\Pi_m(\mathcal{H}) \oplus \Pi_m^\perp(\mathcal{H}) = M \times \mathcal{H}$. Using the definition of the covariant derivation on the sum of two vector bundles (see the formula 4.4.7), we obtain a covariant derivation on the trivial vector bundle $M \times \mathcal{H}$: $\nabla^{\mathcal{H}} = \nabla \oplus \nabla^\perp$

$$\nabla_\varphi^{\mathcal{H}}(s) = s'_X(\varphi) - \varphi(s_1(X)) + \varphi^*(s_2(X)) \tag{4.4.16}$$

where $X \in Gr_m(\mathcal{H})$, $\varphi \in Hom(X,X^\perp) \cong T_X Gr_m(\mathcal{H})$, s_1 is a section of $\Pi_m(\mathcal{H})$, s_2 is a section of $\Pi_m^\perp(\mathcal{H})$ and $s = s_1 + s_2$ is a section of $Gr_m(\mathcal{H}) \times \mathcal{H}$ (that is, a function $Gr_m(\mathcal{H}) \longrightarrow \mathcal{H})$.

The formula 4.4.16 for the covariant derivation on $Gr_m(\mathcal{H}) \times \mathcal{H}$ implies that the corresponding connection form $\mathcal{F}$ is a 1-form on the manifold $Gr_m(\mathcal{H})$ with values in the Lie algebra for the group of unitary transformations on the Hilbert space $\mathcal{H}$ (which is the Lie algebra antisymmetric operators on $\mathcal{H}$):

$$\mathcal{F}_X(\varphi) = \begin{bmatrix} 0 & \varphi \\ -\varphi^* & 0 \end{bmatrix} \in \mathfrak{u}(\mathcal{H})$$

for $X \in Gr_m(\mathcal{H})$ and $\varphi \in T_X(Gr_m(\mathcal{H})) \cong Hom(X, X^{\perp})$

As mentioned (see the formula 4.4.5), the projective representation of the Grassmann manifold gives rise to the 1-form with values in the space of symmetric operators

$$d\mathcal{P}(\varphi) = \begin{bmatrix} 0 & \varphi^* \\ \varphi & 0 \end{bmatrix}, \ \varphi \in T_X(Gr_m(\mathcal{H})) \cong Hom(X, X^{\perp})$$

Using this form, the connection form $\mathcal{F}$, can be written as a 1-form on $\mathcal{P}_m(\mathcal{H}) \cong Gr_m(\mathcal{H})$, as

$$\mathcal{F} = \mathcal{P}d\mathcal{P} + (\mathcal{P} - 1)d\mathcal{P} = 2\mathcal{P}d\mathcal{P} - d\mathcal{P} \tag{4.4.17}$$

Its differential (the curvature form) is $d\mathcal{F} = 2d\mathcal{P}d\mathcal{P}$, which is a 2-form with values in the Lie algebra of antisymmetric operators on the Hilbert space $\mathcal{H}$

$$(d\mathcal{F})(\varphi, \psi) = 2 \begin{bmatrix} \varphi^*\psi - \psi^*\varphi & 0 \\ 0 & \varphi\psi^* - \psi\varphi^* \end{bmatrix}, \quad \varphi, \psi \in Hom(X, X^{\perp}).$$

Let us denote by $\mathfrak{S}(\mathcal{H})$ the space of symmetric operators on the Hilbert space $\mathcal{H}$. The empedding $\mathcal{P} : Gr_m(\mathcal{H}) \longrightarrow \mathfrak{S}(\mathcal{H})$ of the Grassmann manifold as the set of projectors, induces a mapping $\mathcal{P}' : T(Gr_m(\mathcal{H})) \longrightarrow \mathfrak{S}(\mathcal{H}) \times \mathfrak{S}(\mathcal{H})$

$$\mathcal{P}'_X(\varphi) = (\mathcal{P}(X), \widetilde{\varphi}),$$

where

$$\widetilde{\varphi} = \begin{bmatrix} 0 & \varphi^* \\ \varphi & 0 \end{bmatrix}.$$

This mapping, according to the remark 4.4.3 induces a connection on the tangent bundle $T(Gr_m(\mathcal{H}))$, which itself, coincides with the connection $\nabla^* \otimes \nabla^{\perp}$, where ∇ is the connection on the canonical vector bundle $\Pi_m(\mathcal{H})$ and $\nabla^{\perp}$ is the connection on $\Pi_m^{\perp}(\mathcal{H})$.

4.4.4. Symplectic Structure

Let L be a finite-dimensional Lie algebra over the field of complex or real numbers and L^* be its dual vector space. The linear mapping

$$B : L \wedge L \longrightarrow L, \quad B(x \wedge y) = [x, y]$$

induces the dual mapping between the dual spaces

$$B^*: L^* \longrightarrow L^* \wedge L^*, \quad B^*(\varphi)(x \wedge y) = \varphi([x,y]).$$

The latter can be regarded as an antisymmetric covariant tensor field on the linear space L^*. Such a field defines a bracket on the algebra of smooth functions $C^\infty(L^*)$:

$$\{f,g\}(\varphi) = (df|_\varphi \wedge dg|_\varphi)(B^*(\varphi)) = \varphi([df|_\varphi, dg|_\varphi]),$$

where the elements of the space L^{**}: $df|_\varphi$ and $dg|_\varphi$ are considered as elements of the vector space L which is canonically isomorphic to L^{**}. The algebra $C^\infty(L^*)$ together with this defined bracket is a Poisson algebra. Particularly, if we take any two elements $x, y \in L$, they can be regarded as linear functions on the vector space L^*: $x(\varphi) = \varphi(x)$, $\forall\, \varphi \in L^*$; and the Poisson bracket of these two linear functions is $\{x,y\} = [x,y]$.

If the vector space L is equipped with a scalar product $\langle \cdot\,, \cdot \rangle$, we can carry the Poisson structure from the space L^* to L by the linear mapping $u \mapsto \langle u, \cdot \rangle : L \longrightarrow L^*$. Under these conditions, an element $x \in L$ can be considered as a linear function on the vector space L: $\widetilde{x}(u) = \langle x, u \rangle$, $\forall u \in L$. The Poisson bracket for these two functions is

$$\{\widetilde{x}, \widetilde{y}\}(u) = \langle [x,y], u \rangle = \widetilde{[x,y]}(u), \ \forall u \in L.$$

Now, assume that L is a Lie algebra of a Lie group G and the scalar product on the vector space L is invariant under the adjoint action of the group G:

$$\langle gug^{-1}, gvg^{-1} \rangle = \langle u, v \rangle, \quad \forall u, v \in L, \ \forall g \in G \tag{4.4.18}$$

Suppose that $g = \exp(tw)$, $w \in L, t \in \mathbb{R}$ is a one-parameter subgroup generated by an element w. Differentiating the equality 4.4.18 by the parameter t at the point $t = 0$, we obtain the following infinitezimal version of the equality 4.4.18

$$\langle [w,u], v \rangle + \langle u, [w,v] \rangle = 0 \tag{4.4.19}$$

For any function $f \in C^\infty(L)$, we denote by $\mathrm{ham}(f)$ the Hamiltonian vector field corresponding to the function f for the Poisson structure defined by $\langle\ ,\ \rangle$. For any element $x \in L$ and the corresponding linear function $\widetilde{x} = \langle x, \cdot \rangle$ on L, we have the following

$$d\widetilde{y}(\mathrm{ham}(\widetilde{x})) = \{\widetilde{x}, \widetilde{y}\} = \widetilde{[x,y]} \Longrightarrow$$
$$\Longrightarrow \forall u \in L: \widetilde{y}(\mathrm{ham}(\widetilde{x}_u)) = \widetilde{[x,y]}(u) = \langle [x,y], u \rangle.$$

Keeping in mind the equality 4.4.19, we obtain the following

$$\mathrm{ham}(\widetilde{x})_u = [x,u], \ \forall u \in L$$

That is: for a Poisson structure defined by an adjoint-invariant metric on a Lie algebra L, the Hamiltonian vector field corresponding to a linear function $\widetilde{x} = \langle x, \cdot \rangle$, $x \in L$ is $\mathrm{ham}(\widetilde{x})_u = [x,u]$, $\forall u \in L$. Hence, we obtain that for any point $u \in L$, the subspace of the tangent space $T(L) \cong L$, generated by the Hamiltonian vector fields at the point u is

$$Ham_u = \{[x,u] \,|\, x \in L\}$$

The distribution of the subspaces for the tangent spaces: $u \mapsto Ham_u$, $u \in L$, is integrable and the integral submanifolds of this distribution are the orbits of the adjoint action of the corresponding Lie group G. These orbits that are exactly symplectic leaves of the Poisson structure and as it should be, the restriction of the Poisson structure on each of them is nondegenerated. The corresponding symplectic form on any orbit of the adjoint action is

$$\omega(x;u,v) = \langle x, [u,v] \rangle,\ x,u,v \in L \tag{4.4.20}$$

Any curve $u : \mathbb{R} \longrightarrow L$ can be regarded as a time-dependent Hamiltonian: $\widetilde{u}_t(x) = \langle u(t), x \rangle$, $t \in \mathbb{R}, x \in L$; and the corresponding Hamiltonian equation is $\dot{x}(t) = [u(t), x(t)]$. As the symplectic leaves of the Poisson structure are the orbits of the adjoint action, we have that any solution of the Hamiltonian equation lies in some orbit of the adjoint action.

Now, consider the situation when the Lie algebra L is the Lie algebra of antisymmetric operators on the Hilbert space $\mathcal{H}$. This Lie algebra is the same as the Lie algebra of the Lie group $U(\mathcal{H})$ – the group of unitary transformations of the Hilbert space $\mathcal{H}$. We denote this Lie algebra by $\mathfrak{u}(\mathcal{H})$. The mapping $\mathcal{P}_m(\mathcal{H}) \longrightarrow \mathfrak{u}(\mathcal{H}),\quad P \mapsto -iP$ is an injection of the Grassmann manifold in the Lie algebra $\mathfrak{u}(\mathcal{H})$. The equality $\mathcal{P}(g(X)) = g\mathcal{P}(X)g^{-1}$, $\forall X \in Gr_m(\mathcal{H})$, $\forall g \in U(\mathcal{H})$, implies that the image of this mapping – $i\mathcal{P}_m(\mathcal{H})$, is an orbit for the adjoint action of the Lie group $U(\mathcal{H})$ on its Lie algebra – $\mathfrak{u}(\mathcal{H})$. Therefore, the Grassmann manifold can be regarded as an orbit for an adjoint action of a Lie group on its Lie algebra.

Consider the following scalar product on the vector space $\mathfrak{u}(\mathcal{H})$:

$$\langle u,v \rangle = \mathrm{trace}(u^* \circ v), \quad u,v \in \mathfrak{u}(\mathcal{H})$$

For any unitary transformation $g \in U(\mathcal{H})$, we have

$$\langle gug^{-1}, gvg^{-1} \rangle = \mathrm{trace}(gu^*vg^{-1}) = \mathrm{trace}(u^*v) = \langle u,v \rangle,\ \forall u,v \in \mathfrak{u}(\mathcal{H})$$

which implies that the scalar product on the space $\mathfrak{u}(\mathcal{H})$ is invariant under the adjoint action of the group $U(\mathcal{H})$. Hence, we have the symplectic form on any orbit of the adjoint action, defined by the formula 4.4.20. Particularly, for the embedded Grassmann manifold in $\mathfrak{u}(\mathcal{H})$, we have

$$\omega(iP;\Phi,\Psi) = \mathrm{trace}(iP \circ [\Phi,\Psi]) \tag{4.4.21}$$

To obtain more explicit expression for the differential 2-form ω, at a point $P \in \mathcal{P}_m(\mathcal{H}) \cong Gr_m(\mathcal{H})$, consider the decomposition of the operators Φ and Ψ corresponding to the decomposition $\mathcal{H} = \mathrm{Im}(P) \oplus \mathrm{Im}(P)^{\perp}$:

$$\Phi = \begin{bmatrix} 0 & i\varphi^* \\ i\varphi & 0 \end{bmatrix}, \qquad \Psi = \begin{bmatrix} 0 & i\psi^* \\ i\psi & 0 \end{bmatrix}$$

where $\varphi, \psi : \mathrm{Im}(P) \longrightarrow \mathrm{Im}(P)^{\perp}$. From the formula 4.4.21, this easily follows the expression for the symplectic form:

$$\begin{gathered} \omega(X;\varphi,\psi) = i \cdot \mathrm{trace}(\varphi^*\psi + \psi^*\varphi)) \\ \text{for } X \in Gr_m(\mathcal{H}) \text{ and } \varphi,\psi \in T_X(Gr_m(\mathcal{H})) \cong Hom(X, X^{\perp}) \end{gathered} \tag{4.4.22}$$

which can be written as $w = i\text{trace}(P \cdot dP \wedge dP)$.

As mentioned for the case of a general Lie algebra, any element $u \in \mathfrak{u}(\mathcal{H})$ can be considered as a function $\widetilde{u} : Gr_m(\mathcal{H}) \longrightarrow \mathbb{R}$

$$\widetilde{u}(P) = i \cdot \text{trace}(u^* \circ P) = -i \cdot \text{trace}(u \circ P), \quad P \in \mathcal{P}_m(\mathcal{H}) \cong Gr_m(\mathcal{H})$$

and the corresponding Hamiltonian vector field on the manifold $Gr_m(\mathcal{H})$ is

$$\text{ham}(\widetilde{u})_P = [u, P],\ P \in \mathcal{P}_m(\mathcal{H}) \cong Gr_m(\mathcal{H}) \tag{4.4.23}$$

This vector field is the same as that one generated by the action for the group $U(\mathcal{H})$.

4.4.5. The Canonical Principal Bundle and Canonical Connection

Consider the set of all orthonormal m-frames in the Hilbert space $\mathcal{H}$. This set is known as the Stiefel manifold (see [104]) for the complex Hilbert space $\mathcal{H}$. It is clear that the Stiefel manifold of orthonormal m-frames can be identified with the set of all isometric mappings from the complex vector space $\mathbb{C}$ to the Hilbert space $\mathcal{H}$. We denote this set of all isometric mappings from $\mathbb{C}$ to $\mathcal{H}$ by $Isom(\mathbb{C}^m, \mathcal{H})$. This set equipped with the topology and the differential structure for the subset of the vector space $Hom(\mathbb{C}^m, \mathcal{H})$, is a manifold, and can be described as follows:

$$Isom(\mathbb{C}^m, \mathcal{H}) = \left\{ \varphi \in Hom(\mathbb{C}^m, \mathcal{H}) \,|\, \varphi^* \varphi = 1 \right\}$$

There is a natural mapping from the manifold $Isom(\mathbb{C}^m, \mathcal{H})$ to the manifold $Gr_m(\mathcal{H})$ defined as

$$\pi : Isom(\mathbb{C}^m, \mathcal{H}) \longrightarrow Gr_m(\mathcal{H}), \qquad \pi(\varphi) = \text{Im}(\varphi)$$

Recall that for any $X \in Gr_m(\mathcal{H})$, the symbol $\mathcal{P}(X)$ denotes the operator on the Hilbert space $\mathcal{H}$ that is the orthogonal projection on the subspace X.

Lemma 4.4.1. *For any $\varphi \in Isom(\mathbb{C}^m, \mathcal{H})$, we have that $\mathcal{P}(Im(\varphi)) = \varphi \circ \varphi^*$ (see [44], [45], [46])*

Proof. The proof consists of the following two steps: first, we check that for any $x \in \text{Im}(\varphi)$: $(\varphi\varphi^*)(x) = x$, and then we check that for any $y \in \text{Im}(\varphi)^\perp$: $(\varphi\varphi^*)(y) = 0$.

For $x \in \text{Im}(\varphi)$, we have the following:

$$x = \varphi(u) \implies (\varphi\varphi^*)(x) = (\varphi \underbrace{\varphi^* \varphi}_{1})(u) = \varphi(u) = x$$

For $y \in \text{Im}(\varphi)^\perp$, we have:

$$\forall u \in \mathcal{H} :\ \langle (\varphi\varphi^*)(y), u \rangle = \langle \varphi^*(y), \varphi^*(u) \rangle = \langle y, \underbrace{(\varphi\varphi^*)(y)}_{\in \text{Im}(\varphi)} \rangle = 0 \implies$$

$$\implies (\varphi\varphi^*)(y) = 0$$

□

This implies that as the manifold $Gr_m(\mathcal{H})$ is identified with the space of projectors $\mathcal{P}_m(\mathcal{H})$, in some cases, we can substitute the mapping

$$\pi : Isom(\mathbb{C}^m, \mathcal{H}) \longrightarrow Gr_m(\mathcal{H})$$

with the mapping

$$\pi : Isom(\mathbb{C}^m, \mathcal{H}) \longrightarrow \mathcal{P}_m(\mathcal{H}), \qquad \varphi \mapsto \varphi\varphi^*$$

The latter, we also denote by π.

For any $X \in Gr_m(\mathcal{H})$, consider any isometric map $\varphi : \mathbb{C}^m \longrightarrow X$. It is clear that $\pi(\varphi) = X$. Hence, the mapping π is subjective and for each $X \in Gr_m(\mathcal{H})$, we then have

$$\pi^{-1}(X) = \{\varphi \in Isom(\mathbb{C}^m, \mathcal{H}) \,|\, \mathrm{Im}(\varphi) = X\}$$

Lemma 4.4.2. *Let $\{u_1, \cdots, u_m\}$ be any basis of a Hermitian vector space X. There exists one and only one orthogonal basis $\{e_1, \cdots, e_m\}$ in the space X, such that for any $i = 1, \cdots, m$: $Span < e_1, \cdots, e_i >= Span < u_1, \cdots, u_i >$ and $e_1 \wedge \cdots \wedge e_i = k \cdot u_1 \wedge \cdots \wedge u_i$, where $k > 0$ (i.e., the orientations of the frames $\{e_1, \cdots, e_i\}$ and $\{u_1, \cdots, u_i\}$ coincide).*

Proof. We prove this lemma by induction. For $m = 1$, it is clear that the basis e_1 should be $u_1/\|u_1\|$. If the statement is true for $m-1$, then consider a vector of the type $e' = x_1e_1 + \cdots + x_{m-1}e_{m-1} + u_m$. The system of equations $\langle e', e_i\rangle = 0$, $i = 1, \ldots, m-1$ gives the values of the numbers $x_1, \ldots, x_{m-1}$. After this, we can obtain the vector e_m by normalizing the vector e': $e_m = c \cdot e'$; so that $\|e_m\| = 1$ and the orientation of the frame $\{e_1, \ldots, e_m\}$ be the same as the orientation of $\{u_1, \ldots, u_m\}$ and it is clear that this could be done in exactly one way. □

The statement of this lemma implies that for any isomorphism $f : \mathbb{C}^m \longrightarrow X$, there is exactly one linear transformation $I(f) : X \longrightarrow X$, such that $\Phi(f) = I(f) \circ f : \mathbb{C}^m \longrightarrow X$ is an isometric mapping and $Span < \Phi(f)(e_1), \ldots, \Phi(f)(e_i) >= Span < f(e_1), \ldots, f(e_i) >$ and the orientations of the frames $\{\Phi(f)(e_1), \ldots, \Phi(f)(e_i)\}$ and $\{f(e_1), \ldots, f(e_i)\}$ coincide, for $i = 1, \ldots, m$, where $e_1, \ldots, e_m$ is the natural basis in the complex vector space $\mathbb{C}^m$. We call the mapping $\Phi(f)$ the *isometrization* of the isomorphism f.

By using of this construction, we can define a local trivialization of $\pi : Isom(\mathbb{C}^m, \mathcal{H}) \longrightarrow Gr_m(\mathcal{H})$ as follows: for any point $X \in Gr_m(\mathcal{H})$, consider the mapping

$$\Phi_X : Isom(\mathbb{C}^m, X) \times Hom(X, X^{\perp}) \longrightarrow Isom(\mathbb{C}^m, \mathcal{H})$$

$$\Phi_X(\varphi, f) = \Phi((1+f)\varphi)$$

It is clear that the mapping of $(1+f)\varphi$ is a monomorphism, the image of which is the subspace $\Gamma(f) \in Gr_m(\mathcal{H})$, and $\Phi((1+f)\varphi)$ is the isometrization in the mapping of $(1+f)\varphi$.

Hence, we obtain that $\pi : Isom(\mathbb{C}^m, \mathcal{H}) \longrightarrow Gr_m(\mathcal{H})$ is a locally trivial fiber bundle.

Consider the following right action of $U(m)$ – the group of unitary transformations of $\mathbb{C}^m$, on the space $Isom(\mathbb{C}^m, \mathcal{H})$:

$$\forall g \in U(m) \text{ and } \forall \varphi \in Isom(\mathbb{C}^m, \mathcal{H}) \text{ let } \varphi \mapsto \varphi \circ g$$

It is clear that $\pi(\varphi g) = \pi(\varphi)$.

If φ and ψ are in one and the same fiber $\pi^{-1}(X)$ for $X \in Gr_m(\mathcal{H})$, then $\varphi\varphi^* = \psi\psi^*$, which implies that $\varphi = \psi \underbrace{\psi^* \varphi}_{g} = \psi g$. For $g = \psi^* \varphi$, we have $g^* g = \varphi^* \underbrace{\psi\psi^*}_{1} \varphi = \varphi^* \varphi = 1$, which implies that g is the element of the unitary group $U(m)$. We obtain that the right action for the group $U(m)$ on the fiber of the bundle $\pi : Isom(\mathbb{C}^m, \mathcal{H}) \longrightarrow Gr_m(\mathcal{H})$ is transitive and effective. In other words, this fiber bundle is a principal bundle with the structure group $U(m)$. The associated vector bundle $(Isom(\mathbb{C}^m, \mathcal{H}) \times \mathbb{C}^m)/U(m)$ where

$$(\varphi, x) \sim (\varphi \circ g, g^{-1}(x)),\ \forall (\varphi, x) \in Isom(\mathbb{C}^m, \mathcal{H}) \times \mathbb{C}^m \text{ and } \forall g \in U(m)$$

is isomorphic to the canonical vector bundle $\Pi_m(\mathcal{H})$ via the mapping

$$[\varphi, x] \mapsto \varphi(x)$$

The subset $Isom(\mathbb{C}^m, \mathcal{H})$ in the space $Hom(\mathbb{C}^m, \mathcal{H})$ is defined by the equation $\varphi^* \varphi = 1$, which implies that the tangent space of $Isom(\mathbb{C}^m, \mathcal{H})$ can be described by the equation

$$d\varphi^* \varphi + \varphi^* d\varphi = 0 \tag{4.4.24}$$

Otherwise

$$T_\varphi(Isom(\mathbb{C}^m, \mathcal{H})) = \{u \in Hom(\mathbb{C}^m, \mathcal{H}) \,|\, u^* \varphi + \varphi^* u = 0\} \tag{4.4.25}$$

Differentiating the mapping $\pi(\varphi) = \varphi\varphi^*$ from $Isom(\mathbb{C}^m, \mathcal{H})$ to $Gr_m(\mathcal{H})$, we obtain

$$d\pi = d\varphi\varphi^* + \varphi d\varphi^* \tag{4.4.26}$$

Proposition 4.4.2. *The vertical tangent subspace of the fiber bundle*

$$\pi : Isom(\mathbb{C}^m, \mathcal{H}) \longrightarrow Gr_m(\mathcal{H})$$

at a point $\varphi \in Isom(\mathbb{C}^m, \mathcal{H})$ is the subspace of the tangent space T_φ consisting of such $v : \mathbb{C}^m \longrightarrow \mathcal{H}$ that $Im(v) \subset Im(\varphi)$.

Proof. By definition, the vertical tangent space at a point $\varphi \in Isom(\mathbb{C}^m, \mathcal{H})$ is the kernel for the mapping $\pi'(\varphi)$. Hence, it is described by the following system of linear equations

$$\begin{cases} v^* \varphi + \varphi^* v = 0 & \text{– Tangent} \\ v\varphi^* + \varphi v^* = 0 & \text{– Vertical} \end{cases}$$

The multiplication of the second one by φ from the right, gives $v + \varphi v^* \varphi = 0$. Substitute the term $v^* \varphi$ by $-\varphi^* v$, from the first equation, gives the following equation $v - \varphi\varphi^* v = 0$, or equivalently $(1 - \mathcal{P}(\text{Im}(\varphi)))v = 0$. But it is clear that the operator $(1 - \mathcal{P}(\text{Im}(\varphi)))$ is the orthogonal projection operator on the subspace $\text{Im}(\varphi)^\perp$. This implies that $\text{Im}(v) \subset \text{Im}(\varphi)$. □

For a given subspace $X \subset \mathcal{H}$, any linear mapping $w : \mathbb{C}^m \longrightarrow \mathcal{H}$ is uniquely decomposed as a direct sum: $w = v + u$, with $\mathbb{C}^m \xrightarrow{v} X$ and $\mathbb{C}^m \xrightarrow{u} X^\perp$. This description of the vertical tangent space for the fiber at a point $X \in Gr_m(\mathcal{H})$ suggests the following natural choice of its complementar (horizontal) subspace

$$Hor_X = \left\{ u \in Hom(\mathbb{C}^m, \mathcal{H}) \,|\, \mathrm{Im}(u) \subset X^\perp \right\}$$

For any $\varphi \in Isom(\mathbb{C}^m, \mathcal{H})$ with $\mathrm{Im}(\varphi) = X$, we have that $\ker(\varphi) = X^\perp$, therefore: $\varphi^* u = 0$, which implies that $(\varphi^* u)^* = 0$ and finally: $u^* \varphi + \varphi^* u = 0$. Therefore, the space Hor_X automatically is subspace of the tangent space $T_\varphi(Isom(\mathbb{C}^m, \mathcal{H}))$.

The distribution of the horizontal subspaces $\varphi \mapsto Hor_\varphi$ can be described by the equation $\varphi^* d\varphi = 0$. Consider the following differential form $\mathcal{A} = \varphi^* d\varphi$ on the manifold $Isom(\mathbb{C}^m, \mathcal{H})$.

Proposition 4.4.3. *The differential form $\mathcal{A} = \varphi^* d\varphi$ takes its values in the Lie algebra of the Lie group $U(m)$, and is a connection form on the total space of the $U(m)$-principal bundle $\pi : Isom(\mathbb{C}^m, \mathcal{H}) \longrightarrow Gr_m(\mathcal{H})$.*

Proof. From the equation for the tangent space of $Isom(\mathbb{C}^m, \mathcal{H})$: $\varphi^* d\varphi + d\varphi^* \varphi = 0$, it then follows that the differential form $\mathcal{A}$ which takes its values in the Lie algebra of antisymmetric matrices: $\mathcal{A}^* = d\varphi^* \varphi = -\varphi^* d\varphi = -\mathcal{A}$.

The next step is to verify the transformation rule for the differential form $\mathcal{A}$ under the right action of the unitary group $U(m)$. For any $g \in U(m)$, we have the following:

$$R_g(\varphi) = \varphi g \implies R_g^*(\mathcal{A}) = g^* \varphi^* d(\varphi g) = g^{-1} \varphi^* d\varphi g = g^{-1} \mathcal{A} g = Ad(g^{-1})(\mathcal{A})$$

which shows that the transformation rule for the differential form $\mathcal{A}$ is in accordance with the requirement for the connection forms.

For any fixed point $\varphi \in Isom(\mathbb{C}^m, \mathcal{H})$, consider the mapping

$$m_\varphi : U(m) \longrightarrow Isom(\mathbb{C}^m, \mathcal{H}), \quad m_\varphi(g) = \varphi g, \; \forall g \in U(m)$$

Its differential at the point $g = 1$ is the linear mapping from the Lie algebra $\mathfrak{u}(m)$ to the tangent space $T_\varphi(Isom(\mathbb{C}^m, \mathcal{H}))$: $m'_\varphi(1)(u) = \varphi u$, $\forall u \in \mathfrak{u}(m)$. Consider the value of the differential form $\mathcal{A}$ on the vertical tangent vector $m'_\varphi(1)(u) = \varphi u$. We obtain $\mathcal{A}(\varphi u) = \underbrace{\varphi^* \varphi}_{1} u = u$. The latter equality is also in accordance with the requirement for the connection forms. □

Let $\xi \in T_\varphi(Isom(\mathbb{C}^m, \mathcal{H}))$ be a horizontal tangent vector. That is: $\xi \in Hom(\mathbb{C}^m, \mathcal{H})$ and $\mathrm{Im}(\xi) \subset \mathrm{Im}(\varphi)^\perp$. Its image in $T_{\pi(\varphi)} Gr_m(\mathcal{H})$ by the differential of the projection map π is

$$\pi'(\varphi)(\xi) = \xi \varphi^* + \varphi \xi^* = \begin{bmatrix} 0 & \varphi \xi^* \\ \xi \varphi^* & 0 \end{bmatrix} : \mathrm{Im}(\varphi) \oplus \mathrm{Im}(\varphi)^\perp \longrightarrow \mathrm{Im}(\varphi) \oplus \mathrm{Im}(\varphi)^\perp$$

Otherwise, we can consider ξ as a linear mapping from $\mathbb{C}^m$ to $\mathrm{Im}(\varphi)^\perp$, φ as a linear mapping from $\mathbb{C}^m$ to $\mathrm{Im}(\varphi)$, and $\pi'(\varphi)(\xi)$, as the composition

$$\xi \circ \varphi^* : \mathrm{Im}(\varphi) \xrightarrow{\varphi^*} \mathbb{C}^m \xrightarrow{\xi} \mathrm{Im}(\varphi)^\perp$$

which is the element of $Hom(\mathrm{Im}(\varphi), \mathrm{Im}(\varphi)^{\perp}) \cong T_{\mathrm{Im}(\varphi)}(Gr_m(\mathcal{H}))$. Vice versa, any tangent vector $\mu \in Hom(X, X^{\perp})$ at a point $X \in Gr_m(\mathcal{H})$, could be lifted to the corresponding horizontal space at a point $\varphi \in Isom(\mathbb{C}^m, \mathcal{H})$ as $\widetilde{mu} = \mu\varphi$, which is a mapping from $\mathbb{C}^m$ to $X^{\perp}$ (i.e., the element of the horizontal subspace in $T_{\varphi}(Isom(\mathbb{C}^m, \mathcal{H}))$). It is clear that $\widetilde{\xi\varphi^*} = \xi \underbrace{\varphi^*\varphi}_{1} = \xi$.

4.4.6. Holonomy Algebra of the Connection form $\varphi^* d\varphi$

Let us recall the following well known result from the theory on connections and their holonomies on principal bundles.

Theorem 4.4.1. ***(Ambrous, Singer).*** *Let A be a connection form on a principle bundle $\pi : P \longrightarrow M$. For any point $p \in P$, the holonomy algebra Φ_p at the point p is the algebra spanned on the set of elements for the type $\Omega(X,Y)$, where Ω is the curvature form for the connection form A and X and Y are horizontal tangent vectors of P, at such points q that can be connected with the point p by a horizontal curve.*

We use the statement of this theorem to investigate the holonomy algebra for the connection form $\mathcal{A} = \varphi^* d\varphi$ on the canonical principal bundle over the Grassmann manifold.

Proposition 4.4.4. *For any point $\varphi \in Isom(\mathbb{C}^m, \mathcal{H})$, the holonomy algebra of the connection $\mathcal{A} = \varphi^* d\varphi$ at this point coincides with the Lie algebra for the Lie group $U(\mathcal{H})$.*

Proof. As shown, the horizontal subspace corresponding to the connection form $\mathcal{A}$, at a point $\varphi \in Isom(\mathbb{C}^m, \mathcal{H})$ is

$$Hor(\varphi) = \left\{ u \in Hom(\mathbb{C}^m, \mathcal{H}) \,|\, \mathrm{Im}(u) \subset \mathrm{Im}(\varphi)^{\perp} \right\}$$

and the value of the curvature form on a pair of horizontal vectors (u, v) at the point φ is

$$\Omega(u, v) = (d\mathcal{A})(u, v) = \frac{1}{2}(u^* v - v^* u)$$

We show that for any antisymmetric operator $w : \mathbb{C}^m \longrightarrow \mathbb{C}^m$, there can be found such a pair of linear maps $u, v : \mathbb{C}^m \longrightarrow \mathrm{Im}(\varphi)^{\perp}$ that $w = \frac{1}{2}(u^* v - v^* u)$. In this case, we assume that the Hilbert space is infinite-dimensional (or, at least has a dimension as big as we need). So, we can choose an m-dimensional complex subspace in $\mathrm{Im}(\varphi)^{\perp}$. Fix any basis in X, after which it is identified with $\mathbb{C}^m$. If we take $u = 1 : \mathbb{C}^m \longrightarrow X \cong \mathbb{C}^m$, then from the equation $w = \frac{1}{2}(v - v^*)$, it easily follows the solution $v = w$. □

Hence, for the Grassmann manifold of an infinite-dimensional Hilbert space, it is not necessary to use the full strength of the theorem 4.4.6, because the $\mathrm{Im}(\Omega)$ is only in one point of the total space $Isom(\mathbb{C}^m, \mathcal{H})$, and covers all the Lie algebra $\mathfrak{u}(m)$, even without spanning. The same is true for a finite-dimensional case, but when $\dim(\mathrm{Im}(\varphi)^{\perp}) \geq m$. In the case when the Hilbert space is finite-dimensional and $\dim(\mathrm{Im}(\varphi)^{\perp}) < m$, things are not so simple.

Proposition 4.4.5. *For any integer $n > m$ and any $\psi \in Isom(\mathbb{C}^m, \mathbb{C}^n)$, the holonomy algebra of the connection form $\mathcal{A} = \varphi^* d\varphi$ at the point ψ, coincides with the entire Lie algebra $\mathfrak{u}(m)$.*

Proof. Let us assume, that the antisymetric operator $w : \mathbb{C}^m \longrightarrow \mathbb{C}^m$ is diagonal

$$w = \begin{pmatrix} a_1 & 0 & \cdots & 0 \\ \cdot & \cdot & \cdots & \cdot \\ \cdot & \cdot & \cdots & \cdot \\ 0 & 0 & \cdots & a_m \end{pmatrix}, \qquad \text{where } a_i \in \mathbf{i} \cdot \mathbb{R}, i = 1, \ldots, m$$

It is sufficient to consider the case when $n = m + 1$. We have that $w = \sum_{i=1}^{m} w_i$, where

$$w_i = \begin{pmatrix} 0 & \cdot & \cdot & \cdot & \cdot & 0 \\ 0 & \cdot & \cdot & \cdot & \cdot & 0 \\ 0 & \cdot & a_i & \cdot & \cdot & 0 \\ 0 & \cdot & \cdot & \cdot & \cdot & 0 \\ 0 & \cdot & \cdot & \cdot & \cdot & 0 \\ 0 & \cdot & \cdot & \cdot & \cdot & 0 \end{pmatrix}, \qquad i = 1, \cdots, m$$

For each w_i consider the pair of linear mappings $u_i, v_i : \mathbb{C}^m \longrightarrow \mathbb{C}$, where $u_i = (\underbrace{0, \cdots, 0}_{i-1}, 1, 0, \cdots, 0)$ and $v_i = (\underbrace{0, \cdots, 0}_{i-1}, a_i, 0, \cdots, 0)$. For these mappings, we have then $w_i = \frac{1}{2}(u_i^* v_i - v_i^* u_i)$. Hence, we obtain that any diagonal matrix w can be represented as $w = \sum_i \Omega(u_i, v_i)$, where $u_i, v_i : \mathbb{C}^m \longrightarrow \mathbb{C}$. Now, recall that for any antisymmetric operator $w : \mathbb{C}^m \longrightarrow \mathbb{C}^m$ there exists a unitary transformation $\tau : \mathbb{C}^m \longrightarrow \mathbb{C}^m$, such that $\tau w \tau^{-1}$ is diagonal. If $\tau w \tau^{-1} = \sum_i \Omega(u_i, v_i)$, then consider the operators $\widetilde{u_i} = u_i \tau$ and $\widetilde{v_i} = v_i \tau$. We have $w = \sum_i \Omega(\widetilde{u_i}, \widetilde{v_i})$. □

Let V be a complex subspace of the Hilbert space $\mathcal{H}$. As before, $U(\mathcal{H})$ is the group of unitary transformations for the Hilbert space $\mathcal{H}$. And, $\mathfrak{u}(V)$ is the Lie algebra for the Lie group of unitary transformations for the subspace V. Consider the following $\mathfrak{u}(V)$-valued 1-form on the Lie group $U(\mathcal{H})$:

$$\mathcal{B} = \iota^*(g^* dg)\iota$$

where ι denotes the natural immersion mapping $\iota : V \hookrightarrow \mathcal{H}$ and ι^* is its dual $\iota^* : \mathcal{H} \longrightarrow V$. The value of the form $\mathcal{B}$ can be interpreted as the $(g^{-1}dg)_{VV}$ component of the operator $g^{-1}dg \in \mathfrak{u}(\mathcal{H})$ in the decomposition

$$g^{-1}dg = \begin{bmatrix} (g^{-1}dg)_{VV} & (g^{-1}dg)_{V^\perp V} \\ (g^{-1}dg)_{VV^\perp} & (g^{-1}dg)_{V^\perp V^\perp} \end{bmatrix}$$

corresponding to the decomposition of the Hilbert space $\mathcal{H} = V \oplus V^\perp$. The differential form $\mathcal{B}$ is left-invariant: for any $a \in U(\mathcal{H})$, we have the following:

$$L_a^*(\mathcal{B}) = \iota^*(g^* a^* a dg)\iota = \iota^*(g^* dg)\iota = \mathcal{B}$$

Consider the distribution of tangent subspaces on the manifold $U(\mathcal{H})$ defined by the equation $\mathcal{B} = 0$, i.e., the subspace of the tangent space at a point $u \in U(\mathcal{H})$ is the kernel of the 1-form $\mathcal{B}$ at point u. As the differential form $\mathcal{B}$ is left-invariant, the distribution X is left-invariant, too: $X_{gu} = L'_g(X_u)$. Therefore, the distribution X is of a constant rank. But X is not an involutive distribution. It follows from the definition of the form $\mathcal{B}$: the subset of the tangent space $X_1 \in T_1(U(\mathcal{H}))$, consists of the operators with the form

$$f = \begin{bmatrix} 0 & -\varphi^* \\ \varphi & 0 \end{bmatrix} : \mathcal{H} = V \oplus V^\perp \longrightarrow \mathcal{H} = V \oplus V^\perp$$

If we have two such operators

$$f_1 = \begin{bmatrix} 0 & -\varphi_1^* \\ \varphi_1 & 0 \end{bmatrix} \text{ and } f_2 = \begin{bmatrix} 0 & -\varphi_2^* \\ \varphi_2 & 0 \end{bmatrix}$$

and $u_1, u_2 \in X$ are the left-invariant vector fields on $U(\mathcal{H})$, generated by f_1 and f_2, respectively, we have that

$$[u_1, u_2]_1 = [f_1, f_2] = \begin{bmatrix} \varphi_2^*\varphi_1 - \varphi_1^*\varphi_2 & 0 \\ 0 & \varphi_2\varphi_1^* - \varphi_1\varphi_2^* \end{bmatrix}$$

Let us denote the family of vector fields defined by the distribution X, also by X and suppose that the subspace V in the Hilbert space $\mathcal{H}$ is finite-dimensional.

Proposition 4.4.6. *The family of vector fields X on the manifold $U(\mathcal{H})$ is completely nonholonomic.*

Proof. Let u_0 and u_1 be any two elements of $U(\mathcal{H})$. We have to show that there exists such control $(t_1, X_1)\cdots(t_n, X_n)$, $t_i \in \mathbb{R}$, $X_i \in X$, that $\exp(t_1X_1)\cdots\exp(t_nX_n)u_0 = u_1$.

As the distribution X is left-invariant, we can assume that $u_0 = 1$. Consider the fiber bundle $\pi : Isom(V, \mathcal{H}) \longrightarrow Gr_m(\mathcal{H})$, where $m = \dim(V)$, and the connection form $\mathcal{A} = \varphi^* d\varphi$ on it. This follows from the theorem about the holonomy algebra of the connection $\mathcal{A}$, any two points $\varphi_0, \varphi_1 \in Isom(V, \mathcal{H})$, can be connected by a piece-wise smooth curve

$$\Phi : [0,1] \longrightarrow Isom(V, \mathcal{H}), \qquad \Phi(0) = \varphi_0, \Phi(1) = \varphi_1$$

Assume that φ_0 and φ_1 are such that $\varphi_0(v) = v$ and $\varphi_1(v) = u_1(v), \forall v \in V$. Consider the fiber bundle $\pi^\perp : Isom(V^\perp, \mathcal{H}) \longrightarrow Gr_m(\mathcal{H})$, the fiber of which at a point $W \in Gr_m(\mathcal{H})$ is $Isom(V^\perp, W^\perp)$. Consider two points $\varphi_0^\perp, \varphi_1^\perp \in Isom(V^\perp, \mathcal{H})$, such that $\varphi_0^\perp(x) = x$ and $\varphi_1^\perp(x) = u_1(x), \forall x \in V^\perp$. It is clear that $\varphi_0 \oplus \varphi_0^\perp = \mathbf{1}$ and $\varphi_1 \oplus \varphi_1^\perp = u_1$. On the manifold $Gr_m(\mathcal{H})$, we have a curve $\alpha : [0,1] \longrightarrow Gr_m(\mathcal{H})$, $\alpha = \pi\Phi$ for $\alpha(0) = V$ and $\alpha(1) = u_1(V)$. Let $\Phi^\perp : [0,1] \longrightarrow Isom(V^\perp, \mathcal{H})$ be the lifting of curve α that $\Phi^\perp(0) = \varphi_0^\perp$ and $\Phi^\perp(1) = \varphi_1^\perp$. It is clear that $\{u(t) = (\Phi + \Phi^\perp)(t) \,|\, t \in [0,1]\}$ is a piece-wise smooth family for unitary transformations of the Hilbert space $\mathcal{H}$, such that $u(t)\varphi_0 = \Phi(t), u(0) = 1, u(1) = u_1$, $t \in [0,1]$. As the curve Φ is horizontal, we obtain that $\varphi_0^* u^* \dot{u} \varphi_0 = 0$. Hence, we obtain that the piece-wise smooth curve $u(t), t \in [0,1]$, connects the points $u_0 = 1$ and u_1 in $U(\mathcal{H})$, and at the same time is an integral curve of the dynamical polysystem X on the Lie group $U(\mathcal{H})$. □

4.5. The Geometric Phase

4.5.1. Schrödinger Equation and Berry Phase

Any element g of the unitary group $U(\mathcal{H})$ acts on the total space $Isom(\mathbb{C}^m,\mathcal{H})$ as $\varphi \mapsto g\varphi$, and on the base $Gr_m(\mathcal{H})$ as $X \mapsto g(X)$ (or, by the language of projectors: $P \mapsto gPg^{-1}$). These two actions are correlated:

$$\pi(g\varphi) = g\pi(\varphi),\, \forall\, \varphi \in Isom(\mathbb{C}^m,\mathcal{H}) \text{ and } \forall\, g \in U(\mathcal{H})$$

It other words, the unitary transformation of the Hilbert space $\mathcal{H}$ defines an automorphism of the principal bundle $\pi : Isom(\mathbb{C}^m,\mathcal{H}) \longrightarrow Gr_m(\mathcal{H})$. The connection form $\mathcal{A} = \varphi^* d\varphi$ is invariant under the action of such automorpism

$$(g\varphi)^* d(g\varphi) = \varphi^* g^{-1} g d\varphi = \varphi^* d\varphi$$

For any element u of the Lie algebra $\mathrm{u}(\mathcal{H})$, we have a one-parameter subgroup of the group $U(\mathcal{H})$ generated by u — $\{\exp(tu),\, t \in \mathbb{R}\}$. The action of this group on the total space $Isom(\mathbb{C}^m,\mathcal{H})$ defines a vector field $\widehat{u}$, which can be described as $\widehat{u}_\varphi = u \circ \varphi$, $\forall\, \varphi \in Isom(\mathbb{C}^m,\mathcal{H})$. The action of this group on the base $Gr_m(\mathcal{H})$, defines a vector field $\check{u}_P = [u,P]$, $\forall\, P \in \mathcal{P}_m(\mathcal{H}) \cong Gr_m(\mathcal{H})$ which is the Hamiltonian vector field corresponding to the function $\widetilde{u} : Gr_m(\mathcal{H}) \longrightarrow \mathbb{R}$ defined as $\widetilde{u}(P) = \mathbf{i} \cdot \mathrm{trace}(u^* \circ P)$ (see 4.4.23). The tangent vector $\check{u}$ at a point $X \in Gr_m(\mathcal{H})$, can be described as $\check{u}_X = u_{XX^\perp}$, where $u_{XX^\perp} \in Hom(X,X^\perp)$ is a component in the decomposition

$$u = \begin{bmatrix} u_{XX} & -u^*_{XX^\perp} \\ & \\ u_{XX^\perp} & u_{X^\perp X^\perp} \end{bmatrix} : \mathcal{H} = X \oplus X^\perp \longrightarrow \mathcal{H} = X \oplus X^\perp$$

As the actions of the group $U(\mathcal{H})$ on the total space and the base are compatible, we have that $\pi'(\hat{u}) = \check{u}$.

Remark 4.5.1. This can be checked directly:

$$\pi(\varphi) = \varphi\varphi^* \implies d\pi = d\varphi\varphi^* + \varphi d\varphi^* \implies$$

$$\implies d\pi(\hat{u}) = u\varphi\varphi^* + \varphi(u\varphi)^* = u\underbrace{\varphi\varphi^*}_{\pi(\varphi)} - \underbrace{\varphi\varphi^*}_{\pi(\varphi)} u = [u,\pi(\varphi)]$$

As the action of the group $U(\mathcal{H})$ preserves the connection form $\mathcal{A}$, we have that $L_{\hat{u}}(\mathcal{A}) = 0$, where $L_{\hat{u}}$ denotes the Lie derivation by the vector field $\hat{u}$.

Remark 4.5.2. It can be checked directly:

$$L_{\hat{u}}(\mathcal{A}) = d(\mathcal{A}(\hat{u})) + (d\mathcal{A})(\hat{u},\cdot) =$$

$$= d(\varphi^* u\varphi) + (u\varphi)^* d\varphi - d\varphi^*(u\varphi) = d\varphi^* u\varphi + \varphi^* u d\varphi - \varphi^* u d\varphi - d\varphi^* u\varphi = 0$$

The vertical component of the vector field $\hat{u}$ is measured by the connection form $\mathcal{A}$ and is

$$\varphi\mathcal{A}(\hat{u}) = \varphi\varphi^* u\varphi = \pi(\varphi)u\varphi \tag{4.5.1}$$

Therefore, the horizontal component of the vector field $\hat{u}$ at the point $\varphi \in Isom(\mathbb{C}^m, \mathcal{H})$ is

$$u\varphi - \pi(\varphi)u\varphi = (1 - \pi(\varphi))u\varphi \tag{4.5.2}$$

It is clear that in the expression 4.5.1, the term $\pi(\varphi)$ is the projection operator on the subspace $\mathrm{Im}(\varphi) \subset \mathcal{H}$ and the term $1 - \pi(\varphi)$, in the expression 4.5.2, is the projection operator on the subspace $\mathrm{Im}(\varphi)^{\perp} \subset \mathcal{H}$. Let us describe the vertical and horizontal components of the tangent vector $u\varphi$ more explicitly. For any point $\varphi \in Gr_m(\mathcal{H})$, let

$$u = \begin{bmatrix} u_{11} & -u_{21}^* \\ u_{21} & u_{22} \end{bmatrix}$$

be the decomposition of the operator $u \in U(\mathcal{H})$, corresponding to the decomposition of the Hilbert space $\mathcal{H} = \mathrm{Im}(\varphi) \oplus \mathrm{Im}(\varphi)^{\perp}$. From the formulas 4.5.1 and 4.5.2, it follows that the vertical component of the vector $u \circ \varphi$ is

$$u_{11} \circ \varphi : \mathbb{C}^m \longrightarrow \mathrm{Im}(\varphi)$$

and the horizontal component is

$$u_{21} \circ \varphi : \mathbb{C}^m \longrightarrow \mathrm{Im}(\varphi)^{\perp}$$

Let $H(t), t \in [0,T], T > 0$ be a time-dependent Hamiltonian, i.e., for any $t \in [0,T]$, $H(t)$ is an antisymmetric operator on the Hilbert space $\mathcal{H}$. This can be regarded as a time-dependent C^{∞}-class function (Hamiltonian)

$$\tilde{H(t)} : Gr_m(\mathcal{H}) \cong \mathcal{P}_m(\mathcal{H}) \longrightarrow \mathbb{R}, \qquad \tilde{H(t)}(P) = \mathbf{i} \cdot \mathrm{trace}(H(t)^* \circ P)$$

which, together with the symplectic form (see the formula 4.4.22) $\omega(X;\varphi,\psi) = i \cdot \mathrm{trace}(\varphi^*\psi + \psi^*\varphi))$, defines a time-dependent Hamiltonian vector field on the Grassmann manifold $Gr_m(\mathcal{H})$: $ham(\tilde{H(t)})_P = [H(t), P], \forall P \in \mathcal{P}_m(\mathcal{H})$ (or equivalently: $ham(\tilde{H(t)})_X = H(t)_{XX^{\perp}} \in Hom(X, X^{\perp}), \forall X \in Gr_m(\mathcal{H})$), and the corresponding time-dependent vector field $\check{H(t)}$ on the total space $Isom(\mathbb{C}^m, \mathcal{H})$: $\check{H(t)}_{\varphi} = H(t) \circ \varphi, \forall \varphi \in Isom(\mathbb{C}^m, \mathcal{H})$. The corresponding differential equation on the space $Isom(\mathbb{C}^m, \mathcal{H})$ is

$$\dot{\varphi}(t) = H(t) \circ \varphi(t), t \in [0,T] \tag{4.5.3}$$

and the differential equation (Hamiltonian (or Schrödinger) equation) on the Grassmann manifold is

$$\dot{P}(t) = [H(t), P(t)], t \in [0,T] \tag{4.5.4}$$

Let $P(t), t \in [0,T], P(0) \equiv P_0$ be a solution of the equation 4.5.4 and $\varphi(t), t \in [0,T], \varphi(0) \cong \varphi_0 \in \pi^{-1}(P_0)$ be the corresponding solution for the equation 4.5.3, in the total space $Isom(\mathbb{C}^m, \mathcal{H})$. It is clear that $\pi(\varphi(t)) = P(t), \forall t \in [0,T]$, but the curve $\varphi(t)$, in general, is not the horizontal lifting of the curve $P(t)$, defined by the connection form $\mathcal{A} = \varphi^* d\varphi$, because, the tangent vector $\dot{\varphi}(t)$ has the vertical component — $H(t)_{11} \circ \varphi(t)$, which, in general, is different from zero. This, pure geometric fact has a close relation with the effect, known in Physics as *Berry Phase*. The integral curve $P(t), t \in [0,T]$ on the Grassmann manifold defines two isomorphisms between the fibers $\pi^{-1}(P_0)$ and $\pi^{-1}(P(T) \equiv P_1)$:

1. For a point $\sigma \in \pi^{-1}(P_0)$, let $\psi_\sigma(t), t \in [0,T]$ be the horizontal curve such that $\psi_\sigma(0) = \sigma$ and $\pi(\psi_\sigma(t)) = P(t), t \in [0,T]$. Define an isomorphism $\mathfrak{B}_1 : \pi^{-1}(P_0) \longrightarrow \pi^{-1}(P_1)$ as $\mathfrak{B}_1(\sigma) = \psi_\sigma(T)$

2. For a point $\sigma \in \pi^{-1}(P_0)$, let $\varphi_\sigma(t), t \in [0,T]$ be the solution for the equation 4.5.3, such that $\varphi_\sigma(0) = \sigma$. Define a mapping $\mathfrak{B} : \pi^{-1}(P_0) \longrightarrow \pi^{-1}(P_1)$ as $\mathfrak{B}(\sigma) = \varphi_\sigma(T)$.

When the Hamiltonian curve $P(t)$ on the Grassmann manifold is closed: $P(0) = P(T)$, then the isomorphism $\mathfrak{B} : \pi^{-1}(P_0) \longrightarrow \pi^{-1}(P_0)$ is known as the *Berry phase* and the isomorphism $\mathfrak{B}_1 : \pi^{-1}(P_0) \longrightarrow \pi^{-1}(P_0)$ is known as the *geometric Berry phase.*

We call a Hamiltonian curve $P(t) \in Gr_m(\mathcal{H}), t \in [0,T]$ *geometric* if the corresponding curve $\varphi(t) \in Isom(\mathbb{C}^m, \mathcal{H}), t \in [0,T]$ in the total space, which is the solution for the equation 4.5.3 is horizontal; that is for each $t \in [0,T]$ the tangent vector $H(t) \circ \varphi(t)$ is horizontal. As the vertical component of this tangent vector is $H(t)_{11} \circ \varphi$, the curve $P(t)$ is geometric if and only if $H(t)_{11} = 0$ for each $t \in [0,T]$, where $H(t)_{11} : X(t) \longrightarrow X(t)$ is the upper left component in the decomposition

$$H(t) = \begin{bmatrix} H(t)_{11} & -H(t)_{12}^* \\ H(t)_{12} & H(t)_{22} \end{bmatrix} : \mathcal{H} = X(t) \oplus X(t)^\perp \longrightarrow \mathcal{H} = X(t) \oplus X(t)^\perp$$

Proposition 4.5.1. *For any curve $Q : [0,T] \longrightarrow Gr_m(\mathcal{H})$ there exists such a time-dependent Hamiltonian $H^Q(t) : \mathcal{H} \longrightarrow \mathcal{H}, t \in [0,T]$ that the curve Q is a Hamiltonian curve for $H^Q(t)$ and is geometric.*

Proof. The tangent vector $\dot{Q}(t)$, for any $t \in [0,T]$ is an element in the tangent space of the Grassmannian: $\dot{Q}(t) \in Hom(Q(t), Q(t)^\perp)$. Consider the time-dependent Hamiltonian

$$H^Q(t) = \begin{bmatrix} 0 & -\dot{Q}(t)^* \\ \dot{Q}(t) & 0 \end{bmatrix}, \qquad t \in [0,T]$$

it is clear that the curve $Q : [0,T] \longrightarrow Gr_m(\mathcal{H}) \cong \mathcal{P}_m(\mathcal{H})$ satisfies the Hamiltonian (Schrödinger) equation $\dot{Q}(t) = [H^Q(t), Q(t)]$. This implies that the curve $Q(t)$ is Hamiltonian. At the same time, for any $t \in [0,T]$, the component H_{11}^Q is automatically 0. Therefore, the tangent vectors $H^Q \circ \varphi(t)$ for the points $\varphi(t) \in \pi^{-1}(Q(t)), \quad t \in [0,T]$ are "pure horizontal." □

It is clear that for geometric curves on the base manifold, the two mappings $\mathfrak{B}$ and $\mathfrak{B}_1$ coincide. In other words, *for geometric curves the geometric Berry phase and the Berry phase coincide.*

4.5.2. Cyclic Trajectory in Grassmann Manifold and Berry Phase

Let $\mathcal{H}$ be a finite-dimensional Hilbert; $\mathrm{U}(\mathcal{H})$ be the Lie group of unitary transformations of $\mathcal{H}$ and $\mathfrak{u}(\mathcal{H})$ be the corresponding Lie group. For any positive integer k the Grassmann manifold $Gr_k(\mathcal{H})$ is defined as the set of all k-dimensional subspaces of $\mathcal{H}$. This manifold can also be defined as the set of corresponding orthogonal projectors

$$Gr_k(\mathcal{H}) = \left\{P : \mathcal{H} \longrightarrow \mathcal{H} \mid P \text{ is linear }, P^\dagger = P, \operatorname{tr}(P) = k\right\}.$$

As already established, it is well-known for any given Hamiltonian $H \in \mathfrak{u}(\mathcal{H})$ the corresponding Schrödinger equation is defined as the equation in the form

$$\dot{\psi}(t) = H(\psi(t)), \quad \psi(t) \in \mathcal{H},\ t \in \mathbb{R},\ \psi(0) = \psi_0 \tag{4.5.5}$$

The corresponding one-parameter family of transformations for $\mathcal{H}$ is the on-parameter subgroup of unitary transformations for $\mathcal{H}$

$$\Phi_H = \{\exp(tH) \mid t \in \mathbb{R}\}$$

Obviously, the equation (4.5.5) defines a dynamical system on the Grassmann manifold $Gr_k(\mathcal{H})$:

$$\dot{P}(t) = [H, P(t)], \quad t \in \mathbb{R} \tag{4.5.6}$$

and the corresponding one-parameter group for diffeomorphisms of $Gr_k(\mathcal{H})$ is defined by the action of the group Φ_H on $Gr_k(\mathcal{H})$. For any given element $V \in Gr_k(\mathcal{H})$, which is a k-dimensional subspace of $\mathcal{H}$, the action of $\exp(tH)$ on V is defined as

$$V \mapsto \exp(tH)V,$$

and the corresponding action for the projector representation of $Gr_k(\mathcal{H})$, is

$$P \mapsto \exp(tH)P\exp(-tH).$$

For a given k-dimensional subspace $V_0 \in \mathcal{H}$, we are interested in Hamiltonians $H \in \mathfrak{u}(\mathcal{H})$ such that, after the time period $t = 1$, the one-parameter group Φ_H brings V_0 to itself. In other words, for a given point $P_0 \in Gr_k(\mathcal{H})$ we are looking for Hamiltonians $H \in \mathfrak{u}(\mathcal{H})$ such that the trajectory of the equation (4.5.6) through the point P is closed:

$$\exp(H)P_0\exp(-H) = P_0.$$

Since the transformation $\exp(H)$ brings the subspace V_0 to itself, it defines a unitary transformation

$$g_0 = \exp(H)|_{V_0} : V_0 \longrightarrow V_0.$$

Remark 4.5.3. In fact, the unitary transformation $g_0 : V_0 \longrightarrow V_0$, induced by the one-parameter flow $\{\exp(tH) \mid t \in \mathbb{R}\}$, is the well-known Berry phase and can be decomposed in so-called "dynamical" and "geometrical" factors. Here, we don't concern this decomposition and consider the Berry phase as a single whole.

After this, we can reformulate our problem as

Problem 4.5.1. for a given k-dimensional subspace $V_0 \in \mathcal{H}$ and a unitary transformation $g_0 : V_0 \longrightarrow V_0$, find a skew-hermitian operator $H : \mathcal{H} \longrightarrow \mathcal{H}$ such that $\exp(H)V_0 = V_0$ and $\exp(H)|_{V_0} = g_0$.

Remark 4.5.4. In [144], a similar problem was considered, but for the "geometric" factor of the Berry phase corresponding to the cyclic trajectory on the Grassmannian $Gr_k(\mathcal{H})$ defined by $\exp(tH)$, $t \in [0,1]$.

Further, we will discuss the solution to this Problem 4.5.1.

Let $m = \dim(V_0)$ and $\mathcal{E}_0 = \{e_0, \ldots, e_m\}$ be an orthonormal basis of V_0 consisting of eigenvectors with the operator g_0:

$$g_0(e_k) = a_k \cdot e_k, \quad a_k \in \mathbb{C},\ |a_k| = 1,\ k = 1, \ldots, m.$$

Consider an orthonormal extension of the basis $\mathcal{E}_0$ to the basis of the entire Hilbert space $\mathcal{H}$:

$$\mathcal{E} = \mathcal{E}_0 \bigcap \mathcal{E}_1, \quad \mathcal{E}_1 = \{e_{m+1}, \ldots, e_n\} \subset V_0^{\perp},$$

where $n = \dim(\mathcal{H})$, and define the unitary operator $g : \mathcal{H} \longrightarrow \mathcal{H}$ as

$$g|_{V_0} = g_0, \quad g(e_{m+1}) = a_m \cdot e_{m+1} \quad \text{and} \quad g(e_p) = e_p \quad \text{for} \quad m+2 \le p \le n.$$

In other words, we set that $e_1, \ldots, e_m, e_{m+1}, \ldots, e_n$ are eigenvectors of g, the restriction for the operator g to the subspace V_0 coincides with g_0, the eigenvalues of g on e_m and e_{m+1} are equal and its eigenvalues on the vectors $e_{m+2}, \ldots, e_n$ are equal to 1. Let W be the subspace of $\mathcal{H}$ spanned by the vectors e_m and e_{m+1}. For any u which is a unitary transformation of W define a new orthonormal basis of $\mathcal{H}$ as

$$\mathcal{E}_u = \left\{e_1, \ldots, e_{m-1}, e'_m = u(e_m), e'_{m+1} = u(e_{m+1}), e_{m+2}, \ldots, e_n\right\}.$$

Since the restriction of g to the subspace W, is just a scalar operator: $g(v) = a_m \cdot v$, it is clear that the matrix of the operator g in the basis $\mathcal{E}_u$ is the same as in the basis $\mathcal{E}$ and has the form

$$\begin{bmatrix} a_1 & 0 & & \cdots & & & 0 \\ 0 & \ddots & \ddots & & & & \\ & \ddots & a_m & 0 & & & \vdots \\ \vdots & & 0 & a_m & \ddots & & \\ & & & \ddots & 1 & & \\ & & & & & \ddots & 0 \\ 0 & & \cdots & & & 0 & 1 \end{bmatrix}$$

Assume $a_1 = e^{i\lambda_1}, \ldots, a_m = e^{i\lambda_m}$, $\lambda_k \in \mathbb{R}$, $k = 1, \ldots, m$. Obviously, the number a_m can also be written as $a_m = e^{i(\lambda_m + 2\pi n)}$, $n \in \mathbb{Z}$. Consider the skew-hermitian operator H defined as

$$H(e_1) = i\lambda_1 e_1, \ldots, H(e_{m-1}) = i\lambda_{m-1} e_{m-1},$$

$$H(e'_m) = i\lambda_m e'_m, \quad H(e'_{m+1}) = i(\lambda_m + 2\pi n) e'_{m+1},$$

$$H(e_{m+2}) = 0, \ldots, H(e_n) = 0$$

We have $\exp(H) = g$, and therefore determine $\exp(H)|_{V_0} = g_0$.

Let ω be the matrix of the operator u in the basis $\{e_m, e_{m+1}\}$. Since the matrix of $H|_W$ in the basis $\{e'_m, e'_{m+1}\}$ is

$$\begin{pmatrix} i\lambda_m & 0 \\ 0 & i(\lambda_m + 2\pi n) \end{pmatrix},$$

the matrix of $H|_W$ in the basis $\{e_m, e_{m+1}\}$ is

$$h = \omega \cdot \begin{pmatrix} i\lambda_m & 0 \\ 0 & i(\lambda_m + 2\pi n) \end{pmatrix} \cdot \omega^{-1},$$

which is not necessarily diagonal. Hence, the matrix of H in the basis $\mathcal{E}$ is

$$A = \begin{bmatrix} i\lambda_1 & 0 & & \cdots & & & & & 0 \\ 0 & \ddots & \ddots & & & & & & \\ & \ddots & i\lambda_{m-1} & 0 & & & & & \\ & & 0 & h_1^1 & h_2^1 & & & & \vdots \\ \vdots & & & h_1^2 & h_2^2 & 0 & & & \\ & & & & 0 & 1 & \ddots & & \\ & & & & & \ddots & \ddots & 0 & \\ 0 & & & \cdots & & & 0 & 1 \end{bmatrix},$$

where

$$\begin{pmatrix} h_1^1 & h_2^1 \\ h_1^2 & h_2^2 \end{pmatrix}$$

is the matrix h. If P_0 is the orthogonal projector corresponding to the subspace V_0 and $P(0) = P_0$ is the initial condition for the Schrödinger equation (4.5.6), then we have

$$P(t) = P_0, \quad \forall\, t \in [0,\, 1] \quad \Leftrightarrow \quad [H,\, P_0] = 0,$$

and when the matrix of H in the basis $\mathcal{E}$ is A, then the equality $[H,\, P_0] = 0$ is equivalent to the condition that the matrix h is diagonal.

4.6. Canonical Parametrization Density Matrices

In this section, we review some properties for the set of density matrices of an n-level quantum system and, by using a canonical coset parameterization for $n \times n$ unitary matrices (see [3]), we present a coset parameterization for n-level density matrices. The state space of an $n-$level quantum system is identified by the set of all $n \times n$ Hermitian positive semidefinite complex matrices of trace unity, and comprise an $(n^2 - 1)$-dimensional convex set $\mathcal{M}_n$. The total number of independent variables needed to parameterize a density matrix ρ is equal to $n^2 - 1$, provided no degeneracy occurs.

Let us denote the set of all diagonal density matrices of an n-level system with $\mathcal{D}_n$. An arbitrary element $D \in \mathcal{D}_n$ can be written as

$$D = \operatorname{diag}\{\lambda_1, \lambda_2, \cdots, \lambda_n\}, \qquad 0 \le \lambda_i \le 1, \qquad \sum_{i=1}^{n} \lambda_i = 1. \tag{4.6.1}$$

This means that the set of all diagonal density matrices forms an $(n-1)$-dimensional simplex $\mathcal{S}_{n-1}$. A generic density matrix in an arbitrary basis can be obtained as the orbit of points $D \in \mathcal{D}_n$ under the action of the unitary group $U(n)$ as

$$\rho = UDU^{\dagger}. \tag{4.6.2}$$

Let H be a maximum stability subgroup, i.e., a subgroup of $U(n)$ that consists of all the group elements h that will leave the diagonal state D invariant,

$$hDh^{\dagger} = D, \qquad h \in H, \qquad D \in \mathcal{D}_n, \tag{4.6.3}$$

that is, H contains all elements of $U(n)$ that commute with D. For every element $U \in U(n)$, there is a unique decomposition of U into a product of two group elements, one in H and the other in the quotient G/H [48], i.e.,

$$U = \Omega\, h, \qquad U \in U(n), \qquad h \in H, \qquad \Omega \in U(n)/H. \tag{4.6.4}$$

This decomposition implies that the action of an arbitrary group element $U \in U(n)$ on the point $D \in \mathcal{D}_n$ is given by

$$\rho = UDU^{\dagger} = \Omega hDh^{\dagger}\Omega^{\dagger} = \Omega D\Omega^{\dagger}. \tag{4.6.5}$$

This means that in order to characterize the space $\mathcal{M}_n$, it is sufficient to consider the orbit of points $D \in \mathcal{D}_n$ under the action of the quotient $U(n)/H$. Since $\mathcal{D}_n$ consists of points with a different degree of degeneracy, the maximum stability subgroup will differ for different $D \in \mathcal{D}_n$ [161]. Let m_i denote the degree of degeneracy for eigenvalue λ_i of matrix D. This kind of spectrum follows that D remains invariant under the action of arbitrary unitary transformation performed in each of the m_i-dimensional eigensubspace. Therefore, $H = U(m_1) \otimes U(m_2) \otimes \cdots U(m_k)$ is maximum stability subgroup for D, and the quotient space $U(n)/H$ is a complex flag manifold

$$\mathcal{F} = \frac{U(n)}{U(m_1) \otimes U(m_2) \otimes \cdots U(m_k)}, \qquad m_1 + m_2 + \cdots + m_k = n. \tag{4.6.6}$$

There are two special kinds of degeneracy in the spectrum of D; defined as follows: i) Let D represent the maximally mixed state $\rho_* = \mathrm{diag}\{\frac{1}{n}, \frac{1}{n}, \cdots, \frac{1}{n}\}$. In this case, the stability subgroup H is $U(n)$, and the orbit of point ρ_* is only one point, i.e., $\rho = \rho_*$. ii) On the other hand, if the spectrum of D is non-degenerate, then the stability subgroup is n-dimensional torus $T^n = U(1)^{\otimes n}$, and the orbit of the point D is

$$\rho = \Omega D \Omega^{\dagger}, \qquad \Omega \in U(n)/T^n. \tag{4.6.7}$$

The maximal torus T^n is itself a subgroup of all maximum stability subgroups, therefore the orbit of points $D \in \mathcal{D}_n$ under the action of quotient $U(n)/T^n$ generates all points of the space $\mathcal{M}_n$. The diagonal matrix D is defined up to a permutation of its entries. One can divide the simplex $\mathcal{S}_{n-1}$ into $n!$ identical simplexes and take any of them. Each part can identify points of $\mathcal{S}_{n-1}$ which have the same coordinates, but with different ordering, and can be considered as the homomorphic image of simplex $\mathcal{S}_{n-1}$ relative to the discreet permutation group P_n, i.e., $\mathcal{S}_{n-1}/P_n$. Therefore, the points of $\mathcal{M}_n$ can be characterized as the orbit of diagonal matrices $D \in \mathcal{S}_{n-1}/P_n$ under the action of quotient $U(n)/T^n$.

Further insight into the space of density matrices can be obtained by writing the arbitrary element $U \in U(n)$ as [48]

$$U = \Omega_n^{(n)} \Omega_{n-1}^{(n)} \cdots \Omega_2^{(n)} \Omega_1^{(n)}, \tag{4.6.8}$$

where $\Omega_1^{(n)} \in T^n$, and

$$\Omega_k^{(n)} \in \frac{U(k) \otimes T^{n-k}}{U(k-1) \otimes T^{n-k+1}}, \qquad k = 2, \cdots, n. \tag{4.6.9}$$

Comparing this with the decomposition $U = \Omega h$, where $h \in T^n$, gives the following decomposition for an arbitrary element Ω of quotient $U(n)/T^n$

$$\Omega = \Omega_n^{(n)} \Omega_{n-1}^{(n)} \cdots \Omega_2^{(n)}. \tag{4.6.10}$$

A typical coset representative $\Omega_k^{(n)}$ can be written as [48]

$$\Omega_k^{(n)} = \left(\begin{array}{c|c} SU(k)/U(k-1) & O \\ \hline O^T & I_{n-k} \end{array} \right), \tag{4.6.11}$$

where O, O^T, and I_{n-k} represent, respectively. The $k \times (n-k)$ zero matrix is transposed and the $(n-k) \times (n-k)$ identity matrix. The $2(k-1)$-dimensional coset space $SU(k)/U(k-1)$ have the following $k \times k$ matrix representation [48]

$$SU(k)/U(k-1) = \left(\begin{array}{c|c} \cos\sqrt{B^{(k)}[B^{(k)}]^\dagger} & B^{(k)} \frac{\sin\sqrt{[B^{(k)}]^\dagger B^{(k)}}}{\sqrt{[B^{(k)}]^\dagger B^{(k)}}} \\ \hline -\frac{\sin\sqrt{[B^{(k)}]^\dagger B^{(k)}}}{\sqrt{[B^{(k)}]^\dagger B^{(k)}}} [B^{(k)}]^\dagger & \cos\sqrt{[B^{(k)}]^\dagger B^{(k)}} \end{array} \right), \tag{4.6.12}$$

where $B^{(k)}$ represents $(k-1) \times 1$ complex matrix and $[B^{(k)}]^\dagger$ is its adjoint. In the next section, we consider the $n = 2, 3$ cases, explicitly.

4.6.1. Two-Level System

In this subsection, we give the coset parameterization for a two-level quantum system [3]. Let us consider a diagonal two-level density matrix $D = \text{diag}\{\lambda_1, \lambda_2\}$. Every $h \in H = T^2$ leaves the density matrix D invariant, i.e., $hDh^\dagger = D$ for $h \in H$. Any group element $U \in U(2)$ can be decomposed, uniquely, as $U = \Omega h$ where $\Omega \in U(2)/T^2$ and $h \in T^2$ [48]. The coset space with respect to the stability subgroup $H = T^2$ will provide the unitary transformations to construct a generic density matrix ρ as an element of the orbit of the diagonal matrix D, i.e.,

$$\rho = \Omega D \Omega^\dagger, \qquad \Omega \in U(2)/T^2. \tag{4.6.13}$$

A typical coset representative in the coset space $U(2)/T^2$ is

$$\Omega = \begin{pmatrix} \cos\alpha & e^{i\phi}\sin\alpha \\ -e^{-i\phi}\sin\alpha & \cos\alpha \end{pmatrix}, \tag{4.6.14}$$

where α, ϕ are real. The range of parameters λ_1, λ_2 can be determined as follows. The set of all diagonal 2×2 matrices D forms a 1-dimensional simplex $\mathcal{S}_1$, which can be divided into two identical parts $0 \le \lambda_1 \le \frac{1}{2}$, $\frac{1}{2} \le \lambda_1 \le 1$ (see figure 1a). It can be easily seen that each part can be obtained as the orbit of the other part under the action of the group element $\Omega = \Omega(\alpha = \phi = \frac{\pi}{2})$ which is an element of the coset space $U(2)/T^2$. This means that we can easily consider the diagonal matrix D as

$$D = \begin{pmatrix} \cos^2\theta & 0 \\ 0 & \sin^2\theta \end{pmatrix}, \qquad 0 \le \theta \le \frac{\pi}{4}. \tag{4.6.15}$$

By this parameterization any 2×2 density matrix can be written explicitly as

$$\rho = \begin{pmatrix} \sin^2\alpha\sin^2\theta + \cos^2\alpha\cos^2\theta & -\frac{1}{2}e^{i\phi}\sin 2\alpha\cos 2\theta \\ -\frac{1}{2}e^{-i\phi}\sin 2\alpha\cos 2\theta & \sin^2\alpha\cos^2\theta + \cos^2\alpha\sin^2\theta \end{pmatrix}. \tag{4.6.16}$$

4.6.2. Three-Level System

The density matrices for the three-level system comprise eight-dimensional convex set. Let $D = \mathrm{diag}\{\lambda_1, \lambda_2, \lambda_3\}$ be a 3×3 diagonal density matrix of a three-level system. The density matrix D is invariant under the action of every group element $h \in H = T^3$, i.e., $hDh^\dagger = D$ for $h \in H$. The coset decomposition $U = \Omega h$ where $\Omega \in U(3)/T^3$, $h \in T^3$ provides a parameterization for a generic ρ as

$$\rho = UDU^\dagger = \Omega h D h^\dagger \Omega^\dagger = \Omega D \Omega^\dagger. \tag{4.6.17}$$

On the other hand, the decomposition

$$U = \Omega_3^{(3)}\Omega_2^{(3)}\Omega_1^{(3)}, \tag{4.6.18}$$

with

$$\begin{aligned} &\Omega_3^{(3)} \in U(3)/U(2)\otimes U(1), \\ &\Omega_2^{(3)} \in (U(2)\otimes U(1))/(U(1)\otimes U(1)\otimes U(1)), \\ &\Omega_1^{(3)} \in U(1)\otimes U(1)\otimes U(1), \end{aligned} \tag{4.6.19}$$

follows that

$$\Omega = \Omega_3^{(3)}\Omega_2^{(3)}. \tag{4.6.20}$$

The coset representatives $\Omega_2^{(3)}$ and $\Omega_3^{(3)}$ can be parameterized respectively as

$$\Omega_2^{(3)} = \begin{pmatrix} \cos\alpha & e^{i\phi}\sin\alpha & 0 \\ -e^{-i\phi}\sin\alpha & \cos\alpha & 0 \\ 0 & 0 & 1 \end{pmatrix}, \tag{4.6.21}$$

and

$$\Omega_3^{(3)} = \begin{pmatrix} 1 + \frac{\beta_1^2}{\beta^2}(\cos\beta - 1) & \frac{\beta_1\beta_2}{\beta^2}e^{i(\psi_1-\psi_2)}(\cos\beta - 1) & \frac{\beta_1}{\beta}e^{i\psi_1}\sin\beta \\ \frac{\beta_1\beta_2}{\beta^2}e^{-i(\psi_1-\psi_2)}(\cos\beta - 1) & 1 + \frac{\beta_2^2}{\beta^2}(\cos\beta - 1) & \frac{\beta_2}{\beta}e^{i\psi_2}\sin\beta \\ -\frac{\beta_1}{\beta}e^{-i\psi_1}\sin\beta & -\frac{\beta_2}{\beta}e^{-i\psi_2}\sin\beta & \cos\beta \end{pmatrix}, \tag{4.6.22}$$

where $\beta = \sqrt{\beta_1^2 + \beta_2^2}$.

The two-dimensional simplex $\mathcal{S}_2$ of eigenvalues for ρ is divided into 3! parts. Let us take one of the parts (e.g., the shaded one) from this partition. It is easy to see that all other parts can be obtained from this one by applying the permutation group P_3. On the other hand, the elements of the permutation group P_3 can be obtained from the coset representative (4.6.20) up to a phase:

$$\begin{aligned}
&\Omega_3^{(3)}(\beta_1 = \beta_2 = 0)\Omega_2^{(3)}(\alpha = 0) = (Id),\\
&\Omega_3^{(3)}(\beta_1 = \beta_2 = 0)\Omega_2^{(3)}(\alpha = \phi = \tfrac{\pi}{2}) = i(12),\\
&\Omega_3^{(3)}(\beta_1 = \tfrac{\pi}{2}, \beta_2 = 0, \psi_1 = \tfrac{\pi}{2})\Omega_2^{(3)}(\alpha = 0) = i(13),\\
&\Omega_3^{(3)}(\beta_1 = 0, \beta_2 = \tfrac{\pi}{2}, \psi_2 = \tfrac{\pi}{2})\Omega_2^{(3)}(\alpha = 0) = i(23),\\
&\Omega_3^{(3)}(\beta_1 = \tfrac{\pi}{2}, \beta_2 = 0, \psi_1 = \tfrac{\pi}{2})\Omega_2^{(3)}(\alpha = \phi = \tfrac{\pi}{2}) = i(123),\\
&\Omega_3^{(3)}(\beta_1 = 0, \beta_2 = \tfrac{\pi}{2}, \psi_2 = \tfrac{\pi}{2})\Omega_2^{(3)}(\alpha = \phi = \tfrac{\pi}{2}) = i(321).
\end{aligned}$$

Therefore, the ranges of the eigenvalues of D can be determined as $\frac{1}{3} \leq \lambda_1 \leq 1, 0 \leq \lambda_2 \leq \frac{1}{2}$ and $0 \leq \lambda_3 \leq \frac{1}{3}$, or equivalently, one can parameterize the diagonal matrix D as

$$D = \begin{pmatrix} \cos^2\theta_1 & 0 & 0 \\ 0 & \sin^2\theta_1\cos^2\theta_2 & 0 \\ 0 & 0 & \sin^2\theta_1\sin^2\theta_2 \end{pmatrix}, \tag{4.6.23}$$

where $0 \leq \theta_1 \leq \cos^{-1}\frac{1}{\sqrt{3}},\ \frac{\pi}{6} \leq \theta_2 \leq \frac{\pi}{4}$.

4.7. Quantum Computation with Holonomy

In quantum computation, one constructs a product of various unitary gates to implement a quantum algorithm. As noted previously, a natural idea is to use the holonomy to produce unitary gates necessary for quantum computation. To implement an n-qubit resistor we take a quantum system whose ground state is k-fold degenerate where $k = 2^n$. Then by changing the control parameter adiabatically, we will obtain any unitary gate as a resultant holonomy. Of course, we need to design an appropriate control loop λ to implement a particular unitary gate.

It is easy, in principle, to compute the holonomy for a given loop.

In contrast, to find a loop λ which produces a specified unitary matrix Γ as its holonomy is far from trivial. Moreover, to build a working quantum computer it is strongly desired to reduce the time required to manipulate the computer since a sequence of operations should be carried out before decoherence extinguishes quantum information from the system. At the same time, the control parameter must be changed as slowly as possible to keep adiabaticity intact. Therefore our task is to find a control loop as short as possible to fulfill these seemingly opposed conditions. This kind of problem is known as *isoholonomic* problem, which is first formulated by Montgomery [105]. The isoholonomic problem is one of generalizations for the isoperimetric (also known as Dido's) problem. The isoholonomic problem is formulated here. Assume that we have a principal fiber bundle (P, M, π, G) with a connection. The base space M is assumed to be a Riemannian manifold. The isoholonomic problem asks to find the shortest possible piecewise smooth loop in M with a given base point $x_0 \in M$, that produces a given element g_0 in the structure group G as its associated holonomy.

The isoholonomic problem. Following the work of [144], we formulate the isoholonomic problem. The state vector $\psi(t) \in \mathbb{C}^N$ evolves according to the Schrödinger equation

$$i\hbar \frac{d}{dt}\psi(t) = H(t)\psi(t). \tag{4.7.1}$$

The Hamiltonian admits a spectral decomposition

$$H(t) = \sum_{l=1}^{L} \varepsilon_l(t) P_l(t) \tag{4.7.2}$$

with projection operators $P_l(t)$. Therefore, the set of energy eigenvalues $(\varepsilon_1, \ldots, \varepsilon_L)$ and orthogonal projectors $(P_1, \ldots, P_L)$ constitutes a complete set of control parameters for the system. Now, we concentrate on the eigenspace associated with the lowest energy, which is assumed to be identically zero, $\varepsilon_1 \equiv 0$. Suppose that the degree of degeneracy $k = \mathrm{tr}(P)_1(t)$ is constant. For each t, there exists $V(t) \in S_{N,k}(\mathbb{C})$ such that $P_1(t) = V(t)V^\dagger(t)$, where $S_{N,k}(\mathbb{C})$ is the Stiefel manifold: the set of orthonormal k-frames in $\mathbb{C}^N$,

$$S_{N,k}(\mathbb{C}) = \{V \in M(N,k;\mathbb{C}) \,|\, V^\dagger V = I_k\}. \tag{4.7.3}$$

Here, $M(N,k;\mathbb{C})$ is the set of $N \times k$ complex matrices and I_k is the k-dimensional unit matrix.

By adiabatic approximation with the mean substitution of $\psi(t) \in \mathbb{C}^N$ by a reduced state vector $\phi(t) \in \mathbb{C}^k$ as

$$\psi(t) = V(t)\phi(t). \tag{4.7.4}$$

Since $H(t)\psi(t) = \varepsilon_1\psi(t) = 0$, the Schrödinger equation (4.7.1) becomes

$$\frac{d\phi}{dt} + V^\dagger \frac{dV}{dt}\phi(t) = 0 \tag{4.7.5}$$

and its formal solution is written as

$$\phi(t) = \mathcal{P}\exp\left(-\int V^\dagger dV\right)\phi(0). \tag{4.7.6}$$

Therefore, $\psi(t)$ is written as

$$\psi(t) = V(t)\mathcal{P}\exp\left(-\int V^\dagger dV\right)V^\dagger(0)\psi(0). \tag{4.7.7}$$

In particular, when the control parameter comes back to the initial point as $P_1(T) = P_1(0)$, the holonomy $\Gamma \in U(k)$ is defined via

$$\psi(T) = V(0)\Gamma\phi(0) \tag{4.7.8}$$

and it is given explicitly as

$$\Gamma = V(0)^\dagger V(T)\,\mathcal{P}\exp\left(-\int V^\dagger dV\right). \tag{4.7.9}$$

If the condition

$$V^{\dagger}\frac{dV}{dt}=0 \tag{4.7.10}$$

is satisfied, the curve $V(t)$ in $S_{N,k}(\mathbb{C})$ is called a horizontal lift for the curve $P_1(t)=\pi(V(t))$ in $G_{N,k}(\mathbb{C})$. Then the holonomy (4.7.9) is reduced to

$$\Gamma=V^{\dagger}(0)V(T)\in U(k). \tag{4.7.11}$$

In the present context, isoholonomic problem formulates the question:

Given a specified unitary gate $U_{\text{gate}}\in U(k)$ and a fixed point $P_1(0)\in G_{N,k}(\mathbb{C})$, find the shortest loop $P(t)$ in $G_{N,k}(\mathbb{C})$ with the base points $P_1(0)=P(T)=P_0$ whose horizontal lift $V(t)$ in $S_{N,k}(\mathbb{C})$ produces a holonomy Γ that coincides with U_{gate}.

In this case, [144] reformulates the isoholonomic problem as a *variational problem.* The length of the horizontal curve $V(t)$ is evaluated by the functional

$$S[V,\Omega]=\int_0^T\left\{\operatorname{tr}\left(\frac{dV^{\dagger}}{dt}\frac{dV}{dt}\right)-\operatorname{tr}\left(\Omega V^{\dagger}\frac{dV}{dt}\right)\right\}dt, \tag{4.7.12}$$

where $\Omega(t)\in\mathrm{u}(k)$ is a Lagrange multiplier to impose the horizontal condition (4.7.10) on the curve $V(t)$. The value of the functional S is equal to the length of the projected curve $P(t)=\pi(V(t))$,

$$S=\int_0^T\frac{1}{2}\operatorname{tr}\left(\frac{dP}{dt}\frac{dP}{dt}\right)dt. \tag{4.7.13}$$

Thus, the problem is formulated as:

Find a curve $V(t)$ that attains an extremal value of the functional (4.7.12) *and satisfies the boundary condition* (4.7.11).

Euler-Lagrange equation for the variational problem of the functional (4.7.12) is

$$\frac{d}{dt}(\dot{V}V^{\dagger}-V\dot{V}^{\dagger}+V\Omega V^{\dagger})=0. \tag{4.7.14}$$

Thus, the isoholonomic problem is reduced to the system of equations (4.7.10) and (4.7.14) with respect to $V(t),\Omega(t)$.

The equation (4.7.14) is integrated to yield

$$\dot{V}V^{\dagger}-V\dot{V}^{\dagger}+V\Omega V^{\dagger}=\text{const}=X\in\mathrm{u}(N). \tag{4.7.15}$$

From (4.7.10), and (4.7.15), we obtain

$$\dot{V}+V\Omega=XV. \tag{4.7.16}$$

By multiplying $V^{\dagger}$ on (4.7.16) from the left, we obtain

$$\Omega=V^{\dagger}XV. \tag{4.7.17}$$

The equation (4.7.16) implies $\dot{V}=XV-V\Omega$, and hence the time derivative of $\Omega(t)$ becomes

$$\dot{\Omega}=V^{\dagger}X\dot{V}+\dot{V}^{\dagger}XV=V^{\dagger}X(XV-V\Omega)+(-V^{\dagger}X+\Omega V^{\dagger})XV=[\Omega,\Omega]=0. \tag{4.7.18}$$

Therefore, $\Omega(t)$ is actually a constant. Thus, the solution of (4.7.16) and (4.7.17) is

$$V(t) = e^{tX} V_0 e^{-t\Omega}, \qquad \Omega = V_0^\dagger X V_0. \tag{4.7.19}$$

We may take, without loss of generality,

$$V_0 = \begin{pmatrix} I_k \\ 0 \end{pmatrix} \in S_{N,k}(\mathbb{C}) \tag{4.7.20}$$

as the initial point and parametrize $X \in \mathfrak{u}(N)$, which satisfies (4.7.17), as

$$X = \begin{pmatrix} \Omega & W \\ -W^\dagger & Z \end{pmatrix} \tag{4.7.21}$$

with $W \in M(k, N-k; \mathbb{C})$ and $Z \in \mathfrak{u}(N-k)$.

Finally, we obtained a complete set for this solution (4.7.19) of the horizontal extremal equation (4.7.10) and (4.7.14). When we take the initial point V_0 as (4.7.20), the solutions are parametrized by constant matrices $\Omega \in \mathfrak{u}(k)$ and $W \in M(k, N-k; \mathbb{C})$. For definiteness, we write down the complete solution

$$V(t) = e^{tX} V_0 e^{-t\Omega}, \qquad X = \begin{pmatrix} \Omega & W \\ -W^\dagger & 0 \end{pmatrix}. \tag{4.7.22}$$

The matrix X called a *controller*. The holonomy (4.7.11) is expressed as

$$\Gamma = V^\dagger(0) V(T) = V_0^\dagger e^{TX} V_0 e^{-T\Omega} \in U(k). \tag{4.7.23}$$

The inverse problem. Once this solution (4.7.22) of the horizontal extremal equation is obtained, the remaining problem is to find the matrices Ω and W that satisfy the closed loop condition

$$V(T)V^\dagger(T) = e^{TX} V_0 V_0^\dagger e^{-TX} = V_0 V_0^\dagger \tag{4.7.24}$$

and the holonomy condition

$$V_0^\dagger V(T) = V_0^\dagger e^{TX} V_0 e^{-T\Omega} = U_{\text{gate}} \tag{4.7.25}$$

for a specific unitary gate $U_{\text{gate}} \in U(k)$. This is a kind of inverse problem. In the work by [144], a complete solution is given for this problem and is explicitly constructed for the Hadamard, the CNOT, and the two-qubit discrete Fourier transformation gates.

Chapter 5

Analytic Methods in Quantum Computation

5.1. Fuchsian System and Vector Bundle on the Riemann Sphere

Let us consider a system of linear differential equations on a $\mathbb{C}^n$-valued function f on $\mathbb{CP}^1$.

$$df = \omega f, \tag{5.1.1}$$

where ω is a $(n \times n)$ matrix-valued 1-differential form on $\mathbb{CP}^1$. Let the form be holomorphic everywhere on $\mathbb{CP}^1$ except for a finite set of points $S = \{z_1, z_2, \cdots z_k\}$. Let us consider a solution of (5.1.1) as restricted to a sectorial neighborhood centered at any $z_i \in S$. If any such solution has polynomial growth when it approaches z_i within any such sector, then the system (5.1.1) is called *regular*. If ω has only first order poles at S, then the system is called *Fuchsian*. Any Fuchsian system is linear regular, but there exist linear regular systems which are not Fuchsian.

An n order Fuchsian equation is a linear differential equation

$$f^n(z) + q_1(z) f^{n-1}(z) + \cdots + q_n(z) f(z) = 0 \tag{5.1.2}$$

such that its coefficients $q_j(z)$ have a finite set of poles $S = \{z_1, z_2, \cdots z_k\}$ and in a small neighborhood for pole z_i the coefficients of (5.1.2) have the form

$$q_j(z) = \frac{r_j(z)}{(z - z_i)^j}, \quad j = 1, \cdots, n, \tag{5.1.3}$$

where the $r_j(z)$ are holomorphic functions. Solutions of Fuchsian equations have polynomial growth when continued analytically towards a pole. This distinguishes Fuchsian differential equations from all other linear differential equations on $\mathbb{CP}^1$. Thus, for linear differential equations the notions "Fuchsian" and "regular" coincide.

Theorem 5.1.1 (see [8]). *For any Fuchsian equation on the Riemann sphere, it is possible to construct a Fuchsian system with the same singular points and the same monodromy. Converse is not true.*

Let V be the space for the solutions of a Fuchsian system or a Fuchsian equation over a point $z \in \mathbb{CP}^1 \setminus S$. Let $\pi_1(\mathbb{CP}^1 \setminus S, z)$ fundamental group of Riemann surface $\mathbb{CP}^1 \setminus S$ and $\gamma_1, ..., \gamma_n$. its generators.

Denote M_i the monodromy operator, which corresponds to the analytic continuation of solutions from V along $\gamma_i \in \pi_1(\mathbb{CP}^1 \setminus S, z)$ around $z_i \in S$ in the counterclockwise order. The (anti)representation

$$\rho : \pi_1(\mathbb{CP}^1 \setminus S, z) \to Gl(V), \rho(\gamma_j) = M_j, j = 1...n, \tag{5.1.4}$$

is called the *monodromy representation*. The monodromy matrices must satisfy the equation $M_1 \cdots M_k = Id$, because the loop $\gamma_1 \cdots \gamma_k$ is contractible.

Suppose now, a given system for a differential equation on a small disk $U \subset \mathbb{C}$ with center 0,

$$\frac{df}{dz} = A(z)f(z), \tag{5.1.5}$$

where $A(z)$ is a holomorphic matrix function on $U^* = U \backslash \{0\}$ and

$$f(z) = (f^1(z), ..., f^n(z)) \in \mathbb{C}^n$$

is unknown vector function.

Let $p : \widetilde{U^*} \to U^*$ be the universal covering of U^* and let $\tilde{z}$ and z denote the local coordinate on $\widetilde{U^*}$ and U^*, respectively.

The system (5.1.5) has n linearly independent holomorphic solutions in a small neighborhood of $z_0 \in U^*$. Denote the space of solutions by $\mathfrak{R}$. If $f \in \mathfrak{R}$, then f is a holomorphic function on $\widetilde{U^*}$. Let $E = \frac{1}{2\pi i} \ln M$,, so that, if λ_j are eigenvalues of M and μ_j of E,; then $\mu_j = \frac{1}{2\pi i} \ln \lambda_j$. Denote $\tau_j = Re\, \mu_j$ and normalize the choice of ln demanding that $0 \leq \tau_j < 1$. Then, introduce the function $\tilde{z}^E = e^{E \ln \tilde{z}}$ (which is holomorphic on $\tilde{U}^*$):

$$(\alpha \tilde{z})^E = e^{E(\ln \tilde{z} + 2\pi i)} = \tilde{z}^E M.$$

Theorem 5.1.2. *[8] 1.Let $f(z)$ be some solution of the system (5.1.5). Then $f(z)$ can be represented as follows:*

$$f(\tilde{z}) = Z(z)\tilde{z}^E,$$

where Z is holomorphic on U^*.

2. Every coordinate function $f_j(\tilde{z})$ of a solution $f(\tilde{z})$ is

$$f_j(\tilde{z}) = \sum_{p,q} \tilde{z}^{\tau_p} h_{p,q}(z) \ln^{l_q} \tilde{z}, \tag{5.1.6}$$

$$0 \leq Re\tau_q < 1, l_q \in Z, l_q \geq 0.$$

Denote then, by n^j_{pq} the order of zero of function $h_{pq}(z)$ at the point 0, and let $n^j = \min_{p,q} n^j_{p,q}$. There is the defined map

$$\varphi : \mathfrak{R} \to \mathbb{Z},\ \varphi(f) = \min_{j=1,...,n} n^j.$$

The map φ will be called *Levelt's normalization* of the solution $f(z)$. It has the following properties:

1. $\varphi(\lambda f) = \varphi(f)$, if $\lambda \in C^*$;
2. $\varphi(0) = \infty$;
3. $\varphi(f_1 + f_2) \geq \min(\varphi(f_1), \varphi(f_2))$;

The integer valued function φ defines a filtration of $\mathfrak{R}$:

$$0 \subset \mathfrak{R}^0 \subset \mathfrak{R}^1 \subset \mathfrak{R}^2 \subset \ldots \subset \mathfrak{R}^m \subset \mathfrak{R}, \tag{5.1.7}$$

such that φ is constant on the quotient space $\mathfrak{R}^j/\mathfrak{R}^{j-1}$ and if $k_j^i = \varphi(\mathfrak{R}^j/\mathfrak{R}^{j-1})$, then $k_1 > k_2 > \ldots > k_m$. Let $d_j = \dim(\mathfrak{R}^j/\mathfrak{R}^{j-1})$. We say that φ takes the value k_j with multiplicity d_j.

We shall use also the notation

$$\varphi^1 = \varphi^2 = \ldots = \varphi^{d_1} = k_1 > \varphi^{d_1+1} = \ldots = \varphi^{d_1+d_2} = k_2 > \ldots > \varphi^{d_1+d_2+\ldots+d_{m-1}}$$

$$= \varphi^{d_1+d_2+\ldots+d_m} = k_m.$$

Note then, $\varphi^1 \geq \varphi^2 \geq \ldots \geq \varphi^m$. By definition of φ it follows that $\varphi(\alpha^* f) = \varphi(f)$. Hence, it follows that α^* preserves the filtration (5.1.7) and the monodromy matrix M is the upper triangular.

A basis $f_1(z), f_2(z),\ldots,f_n(z)$ of the solution space $\mathfrak{R}$, satisfying the conditions $\varphi(f_i) = \varphi^i$ and such that the monodromy matrix M is the upper triangular, will be called a Levelt's basis.

Theorem 5.1.3. *[8] The fundamental system of solutions $\Phi(\tilde{z})$ related to a Levelt's basis is*

$$\Phi(\tilde{z}) = U(z)z^{\Psi}\tilde{z}^{E},$$

where $U(z)$ is holomorphic on U^* and $\det U(z) \neq 0$, $\Psi = diag(\varphi^1, \ldots \varphi^m)$ and $E = \frac{1}{2\pi i}M$ is upper triangular.

Remark 5.1.1. If 0 is a regular singular point, then $U(z)$ is meromorphic in U, i.e., $U(z)$ is a single valued function.

The systems $\frac{df(z)}{dz} = A(z)f(z)$ and $\frac{dg(z)}{dz} = B(z)g(z)$ are holomorphically (meromorphically) equivalent if there exists in the neighborhood of 0 a holomorphic (meromorphic) at 0 matrix function $H : V \to GL_n(\mathbb{C})$ such that the transformation $(z, f(z)) = (z, H(z)f(z))$ maps one equation to another, i.e.,

$$B(z) = \frac{dH(z)}{dz}H^{-1}(z) + H(z)A(z)H^{-1}(z). \tag{5.1.8}$$

If two systems of equations are equivalent then their monodromy groups are conjugate. Besides, if 0 is a regular singular point for the system, this system is equivalent to $\frac{df}{dz} = \frac{A}{z}f(z)$, where A is a constant matrix.

Consider system (5.1.1) and suppose $E_j = \frac{1}{2\pi i}\ln\rho(\gamma_j)$ with eigenvalues λ_j^k satisfy the conditions:

$$0 \leq Re\lambda_j^k < 1. \tag{5.1.9}$$

From this system, (5.1.1) is Fuchsian at the singular point s_j iff fundamental matrix of solution $\Phi(\tilde{z})$ admits the following representation:

$$\Phi_j(\tilde{z}) = U_j(z)(z-s_j)^{\Lambda_j}(\tilde{z}-s_j)^{E_j}, \tag{5.1.10}$$

where $\tilde{z}$ denotes the coordinates on the universal covering, U_j is holomorphic invertible at s_j, $\Lambda_j = diag(\alpha_j^1,...,\alpha_j^n), \alpha_j^1 \geq ... \geq \alpha_j^n$, is the diagonal matrix, with integer entries and E_j are upper triangular matrices as defined here, satisfying the conditions (5.1.9).

A monodromy representation is called *irreducible* if there is no proper subspace in V preserved by all the monodromy operators M_i.

The matrix $A_i = res_{z=z_i}\omega(z)$ is called the residue of a linear regular system at z_i. By the Cauchy residue theorem, $\sum_{i=1}^k A_i = 0$.

Theorem 5.1.4 (see [8]). *Any Fuchsian system has the standard form*

$$\frac{df(z)}{dz} = \sum_{i=1}^{k} \frac{A_i}{z-z_i} f(z). \tag{5.1.11}$$

The sum $\tau_j^k = \alpha_j^k + \lambda_j^k$ are the eigenvalues of matrix-residuum A_j. If the eigenvalues τ_j^k of the matrix-residue A_j have no nonzero integer differences between them, then the system (5.1.4) is called nonresonant.

Proposition 5.1.1. *For the nonresonant system (5.1.4) the monodromy matrice M_j is conjugate to $e^{2\pi i A_j}$, $j = 1,...,m$.*

Denoted by $X_m = \mathbb{CP}^1 \setminus S$. Let $\tilde{X} \to X_m$ be the universal covering map of X_m, then it is a bundle with fiber $\pi_1(X_m, z_0)$, where $z_0 \in X_m$. $\pi_1(X_m, z_0)$ is isomorphic to the group of deck transformations for this covering and therefore acts on $\tilde{X}$.

Let

$$\rho : \pi_1(X_m, z_0) \to GL_n(\mathbb{C}) \tag{5.1.12}$$

be some representation.

Consider the trivial principal bundle $\tilde{X} \times GL_n(\mathbb{C}) \to \tilde{X}$ (or vector bundle $\tilde{X} \times \mathbb{C}^n \to \tilde{X}$). The quotient space $\tilde{X} \times GL_n(\mathbb{C})/\sim$ gives the locally trivial bundle on X_m, where $\sim$ is an equivalence relation identifying the pairs $(\tilde{x}, g)$ and $(\sigma\tilde{x}, \rho(\sigma)g)$, for every $\tilde{x} \in \tilde{X}, g \in GL_n(\mathbb{C})$ (or $g \in \mathbb{C}^n$). Denote the obtained bundle by $\mathbf{P}_\rho \to X_m$ (or $\mathbf{E}_\rho \to X_m$) and call it the bundle associated with the representation ρ. Obviously, this bundle according to the transformation functions may be constructed in the following manner. Let $\{U_\alpha\}$ be a simple covering of X_m, i. e., every intersection $U_{\alpha_1} \cap U_{\alpha_2} \cap ... \cap U_{\alpha_k}$ is connected and simply connected. For each U_α, we choose a point $z_\alpha \in U_\alpha$ and join z_0 and z_α by a γ_α starting at z_0 and ending at z_α. For a point $z \in U_\alpha \cap U_\beta$, we choose a path $\tau_\alpha \subset U_\alpha$ which starts at z_α and ends at z. Consider

$$g_{\alpha\beta}(z) = \rho\left(\gamma_\alpha \tau_\alpha(z) \tau_\beta^{-1}(z) \gamma_\beta^{-1}\right). \tag{5.1.13}$$

We see that $g_{\alpha\beta}(z) = g_{\beta\alpha}(z)$ on $U_\alpha \cap U_\beta$ and $g_{\alpha\beta}g_{\beta\gamma}(z) = g_{\alpha\gamma}(z)$ on $U_\alpha \cap U_\beta \cap U_\gamma$.

The cocycle $\{g_{\alpha\beta}(z)\}$ does not depend on the choice of z. Hence, from this cocycle we obtain a flat vector (or principal) bundle, which is denoted by $\mathbf{E}'_\rho$ ($\mathbf{P}'_\rho$). Let $\{t_\alpha(z)\}$ be a trivialization of our bundle, i. e.,

$$t_\alpha : p^{-1}(U_\alpha) \to GL_n(\mathbb{C})$$

is a holomorphic mapping. Consider the matrix valued 1-form $\{\omega_\alpha\}$:

$$\omega_\alpha = -t_\alpha^{-1} dt_\alpha.$$

$\{g_{\alpha\beta}(z)\}$ are constant on the intersection $U_\alpha \cap U_\beta$ and $g_{\alpha\beta}(z)t_\beta(z) = t_\alpha(z)$, so on $U_\alpha \cap U_\beta$ the identity $\omega_\alpha = \omega_\beta$ holds. Indeed, replacing t_β by $t_\beta^{-1} g_{\alpha\beta}$ in the expression $\omega_\beta = -t_\beta^{-1} dt_\beta$, we obtain

$$\omega_\beta = -t_\alpha^{-1} g_{\alpha\beta}(z)\, dt_\alpha g_{\alpha\beta}^{-1}(z) = -t_\alpha^{-1} dt_\alpha.$$

So, $\omega = \{\omega_\alpha\}$ is a holomorphic 1-form on X_m and therefore is a connection form of the bundle $\mathbf{P}'_\rho \to X_m$. Corresponding connection is denoted by ∇'. We will extend the pair $\left(\mathbf{P}'_\rho, \nabla'\right)$ to X. As the required construction is of local character, we shall extend $\mathbf{P}'_\rho \to X_m$ to the bundle $\mathbf{P}''_\rho \to X_m \cup \{s_i\}$, where $s_i \in S$.

First, consider the extension of the principal bundle $\mathbf{P}'_\rho \to X_m$. Let neighborhood V_i for the point s_i meet $U_{\alpha_1}, U_{\alpha_2}, \ldots U_{\alpha_k}$. As we noted earlier, when constructing the bundle from transition functions (5.1.13) only one of them is different from the identity. Let us denote it by g_{1k}, then $g_{1k} = M_i$, where M_i is the monodromy which corresponds to the singular point s_i and obtained by representation (5.1.12). Mark a branch of the many valued function $(\tilde{z} - s_i)^{E_i}$ containing the point $\tilde{s}_i \in \tilde{U}_i$ (where $E_i = \frac{1}{2\pi i} \ln M_i$). Thus, the marked branch defines a function

$$g_{01} = (z - s_i)^{E_i}. \tag{5.1.14}$$

Denoted by g_{02}, the extension of g_{01} along the path which goes around s_i is counterclockwise; and, similarly for other points. At last on $U_i \cap U_{\alpha_k} \cap U_{\alpha_1}$, we shall have:

$$g_{0k}(z) = g_{01}(z) M_i = g_{01}(z) g_{0k}(z).$$

The function $g_{0k} : V_i \to GL_n(\mathbb{C})$ is the one defined at the point s_i, and takes the value which coincides with the monodromy matrix. This means, that we made an extension of the bundle to the point s_i. In the neighborhood of s_i, one will have

$$\omega_i = d g_{0k} g_{0k}^{-1} = E_i \frac{dz}{z - s_i}.$$

So, we obtained the holomorphic principal bundle $\mathbf{P}_\rho \to \mathbb{CP}^1$ on the Riemann sphere $\mathbb{CP}^1$. The vector bundle associated with $\mathbf{P}_\rho \to \mathbb{CP}^1$, which we can denote by $\mathbf{E}_\rho \to \mathbb{CP}^1$ and call canonical, is not topologically trivial. Its connection is denoted by ∇. The holomorphic sections of $\mathbf{E}_\rho$ are solutions for the equation

$$\nabla f = 0 \iff df = \omega f. \tag{5.1.15}$$

It is known that the system (5.1.15) has regular singularity at points $s_1, s_2, \ldots, s_m$ and the Chern number $c_1(\mathbf{E}_\rho)$ of $\mathbf{E}_\rho \to X$ is equal to

$$c_1(\mathbf{E}_\rho) = \sum_{i=1}^{m} tr(E_i). \tag{5.1.16}$$

Remark 5.1.2. Meromorphic connection at the point 0 is called a pair (F,∇), where F is an n-dimensional vector space over the field $K = K[\frac{1}{z}]$, whereas $\nabla : F \to F$ is an operator which satisfies the Leibnitz rule

$$\nabla(h,s) = \frac{dh}{dz}s + h\nabla s,$$

for each function $f \in K$ and $s \in F$.

Let $e_1, e_2, ..., e_n$ be a basis of F and let ∇e_i be expressed with this basis in the following manner:

$$\nabla e_i = -\sum_{j=1}^{n} \theta_{ij}(z)e_j$$

where $\theta = (\theta_{ij}(z)) \in \mathrm{End}(n,K)$, then for $s = \sum_{j=1}^{n} s_j(z)e_i$ we shall obtain

$$\nabla s = \sum_{i=1}^{n}(\frac{ds_i(z)}{dz} - \sum_{j=1}^{n}\theta_{ij}(z)e_j(z))e_i.$$

By the last formula, it follows that $\nabla s = 0$ is equivalent to $\frac{ds}{dz} = \theta s$, or $(d-\theta)s = 0$. Let us denote the matrix-valued 1-form θdz by ω, then the system will be $(d-\omega) = 0$ and the connection will be $\nabla = d - \omega$. Connections are gauge equivalent, if and only if corresponding systems of equations are equivalent; i. e., satisfy the equality (5.1.8).

Where O is the sheaf of germs for holomorphic functions on $\mathbb{CP}^1$, $\mathcal{E}$ holomorphic vector bundle of rank 2 on $\mathbb{CP}^1$, and δ^1 is the sheaf of holomorphic differentials on $\mathbb{CP}^1$. Where S is a finite non-void set of points for $\mathbb{CP}^1$ and $O(S)$ is the sheaf of germs for functions on $\mathbb{CP}^1$ holomorphic on $\mathbb{CP}^1 - S$ meromorphic in S. Let z and x be the holomorphic coordinate functions on $U_0 = \mathbb{CP}^1 \setminus \infty$ and $U_\infty = \mathbb{CP}^1 \setminus 0$, respectively; such that $z(0) = 0$, $x(\infty) = 0$ and $x = \frac{1}{z}$ on $U_0 \cap U_\infty$.

Let $e = (e_1, e_2)$ be a basis of $\mathcal{E}|_{\mathbb{CP}^1\setminus S}$. Where e is called a basis of $\mathcal{E}$ meromorphic in S if for each $s \in S$ there exists a neighborhood U for s, a basis f of $\mathcal{E}|_U$ and a $T \in GL_2(\Gamma(U, O(S)))$ such that $e = fT$ on $U \setminus S$.

Let $\bigtriangledown$ be a holomorphic connection on $\mathcal{E}$. The connection $\bigtriangledown$ induces $\mathbb{C}$-linear morphisms

$${}_z\bigtriangledown = \bigtriangledown_{\frac{d}{dz}}, {}_x\bigtriangledown = \bigtriangledown_{\frac{d}{dx}} : \mathcal{E} \to \mathcal{E}$$

One has ${}_z\bigtriangledown = (-\frac{1}{x^2})_x\bigtriangledown$. Let $e = (e_1, e_2)$ be a basis of $\mathcal{E}$ meromorhpic in $\mathbb{S}$. Then

$${}_z\bigtriangledown e_1 = \gamma_{11}e_1 + \gamma_{21}e_2,$$

$${}_z\bigtriangledown e_2 = \gamma_{12}e_1 + \gamma_{22}e_2,$$

where $\gamma_{ij} \in \Gamma(S, O(S))$.

The matrix

$$\Gamma = \Gamma({}_z\bigtriangledown, e) = \begin{pmatrix} \gamma_{11} & \gamma_{12} \\ \gamma_{21} & \gamma_{22} \end{pmatrix}$$

is called the matrix of ${}_z\bigtriangledown$ with respect to the basis e. If $\tilde{e}$ is another basis of $\mathcal{E}$ meromorphic in S, then $\tilde{e} = eT$ for some $T \in GL_2(\Gamma(\mathbb{CP}^1, O(S))$. We have

$$\tilde{\Gamma} = T^{-1}\Gamma T + T^{-1}\frac{dT}{dz}.$$

Now, let $\bigtriangledown$ be a connection on $\mathcal{E}|_{\mathbb{CP}^1\setminus S}$. $\bigtriangledown$ called a connection on $\mathcal{E}$ having regular singularities in S if for each $s \in U_0 \cap S$ ($s \in U_\infty \cap S$) there exists a neighborhood U for s and basis e of $\mathcal{E}|_U$ meromorphic in $\{s\}$ such that the coefficients of $\Gamma({}_z\bigtriangledown|_{U\setminus S}, e)$ have poles of the order at most 1 in s.

Theorem 5.1.5. *Let $S = \{s_1, .., s_m\}$ be a finite collection of points for $\mathbb{CP}^1$, $s_i \neq s_j$ if $i \neq j$, $n \geq 2$, $z(s_i) = a_i$. Let $\mathcal{E}$ be a vector bundle of rank 2 on $\mathbb{CP}^1$ and ∇ a connection on $\mathcal{E}$ having regular singularity in S. Then there exists a basis e of $\mathcal{E}$ meromorphic in S such that:*

1. if one of the points, say s_m, is ∞ then

$$\Gamma({}_z\nabla, e) = \frac{A_1}{z-a_1} + ... + \frac{A_{m-1}}{z-a_{m-1}}$$

where $A_1, ..., A_{m-1} \in M_2(\mathbb{C})$ and $M_2(\mathbb{C})$ denotes the set of 2×2-matrices with complex coefficients.

2. if $s \subset U_0$, then

$$\Gamma({}_z\nabla, e) = \frac{A_1}{z-a_1} + ... + \frac{A_m}{z-a_m}$$

where $A_1, ..., A_m \in M_2(\mathbb{C}), A_1 + ... + A_m = 0$.

See proof [8].

To study Fuchsian differential equations, K. Okubo invented what became later known as the *Okubo normal form* of a Fuchsian equation [112].

The system of differential equation

$$(z\mathbb{I} - B)\dot{X} = AX, \tag{5.1.17}$$

where X is a complex valued (n-column) vector function of complex variable z, dot denotes differentiation over z and

$$B = s_1\mathbb{I}_{n_1} \oplus ... \oplus s_p\mathbb{I}_{n_p}, \; s_i \in \mathbb{C}; s_i \neq s_j, \text{ when } i \neq j; n_1 + ... + n_p = n$$

and $A \in End(n, \mathbb{C})$ called a *system of Okubo normal form.*

Denoted by Δ the partition $n_1, ..., n_p$ of n. For the fixed partition Δ, we present the matrix A by the blocks $A = (A_{ij})_{1 \neq i, j \neq p}$.

Eigenvalues of A_{ii} denoted by Δ_i, $i = 1, ..., p$. The set of eigenvalues of A denoted by Δ_∞. It is known, that

$$\sum_{i=1}^{n} \sum_{\lambda \in \Delta_i} \lambda = \sum_{\rho \in \Lambda_\infty} \rho. \tag{5.1.18}$$

The system (5.1.17) is Fuchsian and has singular points at points $s_1, ..., s_p, \infty$. The set of characteristic exponents at s_i is Δ_i, $i = 1, ..., p$ and the set of characteristic exponents at ∞ is Δ_∞. In this definition the identity means a Fuchs's relation.

Let subgroup G_Δ of $GL_n(\mathbb{C})$ associated from partition Δ have the form $G_\Delta = GL_{n_1}(\mathbb{C}) \oplus ... \oplus GL_{n_p}(\mathbb{C})$.

Assumption 1. The matrix A is diagonalizable.

Assumption 2. For every $i = 1, ..., p$ the matrices A_{ii} are diagonalizable.

Definition 5.1.1. *Let the system (5.1.17) satisfy the previous assumption. If this system (5.1.17) is unique as determined by* $B, \Delta_1, \ldots, \Delta_p, \Delta_\infty$ *up to* G_Δ*-valued gauge transformation, this system is (5.1.17) called free from accessory parameters.*

Denoted by $L = (m_1, \ldots, m_q)$ the multiplicities of the eigenvalues of A.

Assumption 3. The entry of Δ_i are distinct, for every $i = 1, \ldots, p$.

We remark, that from the assumption 3 follows assumption 2.

Denote by integer number $N(\Delta, L)$ the following expression:

$$N(\Delta, L) = n^2 - n + 2 - \sum_{i=1}^{p} n_i^2 - \sum_{j=1}^{q} m_j^2.$$

Theorem 5.1.6. *[127] On the assumptions 1 and 3, this system (5.1.17) is free from accessory parameters if* $N(\Delta, L) = 0$.

Definition 5.1.2. *The system of differential equations*

$$\frac{dY}{dz} = A(z)Y(z)$$

is said to be reducible if there is a transformation $Y = P(z)Z$, $P(z) \in GL_n(\mathbb{C}(z))$ *such that the coefficient* $B(z)$ *transformed system*

$$\frac{dZ}{dz} = B(z)Z(z)$$

is a block triangular form. Otherwise, the system is said to be irreducible.

Yakoyama [154] gives the classification for the system of differential equations which are irreducible and free from accessory parameters. Yakoyama proved, that there exist eight classes of such a system.

The monodromy representation of that system was computed by [112], [113], [128], and [66].

Theorem 5.1.7. *Let* $\Delta = (1, 1, \ldots, 1)$ *and* $L = (n-1, 1)$. *Suppose* $B = s_1 \oplus \ldots \oplus s_p$ *in (5.1.17),* $\Lambda = (\lambda_i), i = 1, \ldots, n$ *and* $\Lambda_\infty = (\underbrace{\rho_1, \ldots, \rho_1}_{n-1}, \rho_2)$ *satisfies the condition (5.1.18):*

$$\sum_{i=1}^{n} \lambda_i = (n-1)\rho_1 + \rho_2.$$

This satisfies assumption 3 and

$$\lambda_i \neq \rho_j \text{ for every } i = 1, \ldots, n \text{ and } j = 1, 2.$$

Then the system (5.1.17) is transformed by a G_Δ*-valued transformation into*

$$(z\mathbb{I} - B)\dot{Z} = A_1 Z, \tag{5.1.19}$$

where

$$A_1 = \begin{pmatrix} \lambda_1 & \lambda_1 - \rho_1 & \ldots & \lambda_1 - \rho_1 \\ \lambda_2 - \rho_1 & \lambda_2 & \ldots & \lambda_2 - \rho_1 \\ \ldots & \ldots & \ldots & \\ \lambda_n - \rho_1 & \lambda_n - \rho_1 & \ldots & \lambda_n \end{pmatrix}.$$

It is known that this system (5.1.19) can be transformed into the Pochhammer (see [68], [71]) equation, whose monodromy is called the Burau representation, which we shall also define here.

Let β_j be the element for the group matrix with entries in ring $\mathbb{Z}[t,t^{-1}]$ and have the form:

$$\beta_i = \begin{pmatrix} \mathbb{I}_{i-1} & 0 & 0 & 0 \\ 0 & 1-t & t & 0 \\ 0 & 1 & 0 & 0 \\ 0 & 0 & 0 & \mathbb{I}_{n-k-1} \end{pmatrix}, \; 1 \leq i \leq n-1.$$

The representation Artin's braid group on n strings

$$\rho : B_n \to GL_n(\mathbb{Z}[t,t^{-1}]), \; \sigma_i \to \beta_i$$

is called the *Burau representation* for this braid group.

Proposition 5.1.2. *For β_i, we have the following identities*

$$\beta_i\beta_{i+1}\beta_i = \beta_{i+1}\beta_i\beta_{i+1}$$

and

$$\beta_i\beta_j = \beta_j\beta_i \; \textit{if} \; |i-j| > 2.$$

Consider on X_n Fuchs type Pfaffian system

$$dy = \sum_{1\leq i<j\leq n} B_{ij} d\log(z_i - z_j)\cdot y, \tag{5.1.20}$$

where

$$B_{ij} = \begin{pmatrix} 0 & \dots & \dots & \dots & 0 \\ \dots & \dots & \dots & \dots & \dots \\ i \text{ rows } \to & \lambda & \dots & -\lambda & \dots \\ & \dots & \dots & \dots & \\ j \text{ rows } \to & -\lambda & \dots & \lambda & \dots \\ \dots & \dots & \dots & \dots & \dots \\ 0 & \dots & \dots & \dots & 0 \end{pmatrix}. \tag{5.1.21}$$

Theorem 5.1.8. *The Pfassian system (5.1.20) is integrable and its monodromy representation is a Burau representation.*

◁ The proof follows from [86]. In particular, integrability of system (5.1.20) follows from what is called the *infinitesimal pure braid relations*:

$$[B_{ij}, B_{ik} + B_{jk}] = [B_{ij} + B_{ik}, B_{jk}] = 0, \text{ for } i<j<k,$$

$$[B_{ij}, B_{kl}] = 0, \text{ for distinct } i,j,k,l.$$

▷

This system (5.1.20) with a (5.1.21) is the Pochhamer system [87].

Theorem 5.1.9. *[128] The Pochhamer system (5.1.20) is the gauge equivalent to Okubo's type system (5.1.19).*

Consider the four-dimensional case of (5.1.17):

$$\left(tI - \begin{pmatrix} \lambda_1 & 0 & 0 & 0 \\ 0 & \lambda_2 & 0 & 0 \\ 0 & 0 & \lambda_3 & 0 \\ 0 & 0 & 0 & \lambda_4 \end{pmatrix}\right)\frac{dX}{dt} = AX, \tag{5.1.22}$$

where $\lambda_1 = \lambda_2$, satisfies the following assumptions:

1. $A = (a_{ij})$ is diagonalizable and has two eigenvalues ρ_1 and ρ_2 with the common multiply 2.

2. None of the quantities

$$a_{ij}(i = 1,2,3,4),\ a_{jj} - a_{kk}(j \neq k), \rho_l(l = 1,2), \rho_1 - \rho_2,$$

is an integer. (We remark here, that consequently, there is no logarithmic solution.)

This system (5.1.22) has four regular singular points $t = \lambda_1 = \lambda_2$, λ_3, λ_4, and ∞ on the $\mathbb{CP}^1$. The characteristic exponents are $(a_{11}, a_{22}, 0, 0)$ at $t = \lambda_1$, $(0,0,a_{33},0)$; at $t = \lambda_3$, $(0,0,0,a_{44})$; at $t = \lambda_4$, and $(\rho_1, \rho_1, \rho_2, \rho_2)$ at $t = \infty$..

Denoted by $e_j = exp(2\pi i a_{ii})$ and $f_k = exp(2\pi i\rho_k)$., from the Fuchs relation

$$2\rho_1 + \rho_2 = a_{11} + a_{22} + a_{33} + a_{44}$$

follows

$$(f_1 f_2)^2 = e_1 e_2 e_3 e_4.$$

It is known that the most general constant non-singular matrix T leaves B invariant, i.e., $TBT^{-1} = B$, is

$$\begin{pmatrix} T_1 & 0 \\ 0 & T_2 \end{pmatrix},$$

where T_1 is nonsingular 2×2 matrix and T_2 is a nonsingular 2×2 diagonal matrix. From this, it follows that by an appropriate choice of such a matrix T, we can make two components a_{12} and a_{21} of A to be zero and other three non-zero elements to take any preassigned values. Consider the case in which $a_{12} = a_{21} = 0$.

Proposition 5.1.3. *If the three elements a_{41}, a_{42}, and a_{43} of A are non-zero, then the non-trivial components a_{ij}, $i \neq j$ are expressible in terms of a_{41}, a_{42}, a_{43}, a_{11}, a_{22}, a_{33} and ρ_1, ρ_2*

$$a_{13} = \frac{(a_{11} - \rho_1)(a_{11} - \rho_2)}{a_{11} - a_{22}} \frac{a_{43}}{a_{41}},$$

$$a_{14} = \frac{(a_{11} - \rho_1)(a_{11} - \rho_2)}{a_{22} - a_{11}} (a_{11} + a_{33} - \rho_1 - \rho_2)\frac{1}{a_{41}},$$

$$a_{23} = \frac{(a_{22} - \rho_1)(a_{22} - \rho_2)}{a_{11} - a_{22}} \frac{a_{43}}{a_{42}},$$

$$a_{24} = \frac{(a_{11} - \rho_1)(a_{11} - \rho_2)}{a_{22} - a_{11}} (a_{11} + a_{33} - \rho_1 - \rho_2)\frac{1}{a_{41}},$$

$$a_{31} = (a_{22} + a_{33} - \rho_1 - \rho_2)\frac{a_{41}}{a_{43}},$$

$$a_{32} = (a_{11} + a_{33} - \rho_1 - \rho_2)\frac{a_{42}}{a_{43}},$$

$$a_{34} = (a_{11} + a_{33} - \rho_1 - \rho_2)(a_{22} + a_{33} - \rho_1 - \rho_2)\frac{1}{a_{43}}.$$

The system (5.1.22) has only four singular solutions defined at three finite singular points $\lambda_1 = \lambda_2$, λ_3 and λ_4, corresponding to the characteristic exponents a_{jj} at $t = \lambda$ and the normalization conditions $g_j(0) = \varepsilon_j$:

$$X_j(t) = (t - \lambda_j)^{a_{jj}} \sum_{m=0}^{\infty} g_j(m)(t - \lambda_j)^m, \ j = 1,2,3,4,$$

where the j-th element and the others of the 4-vector ε_j are 1 and 0, respectively.

In any simple connected domain contained in $\mathbb{CP}^1 \setminus \{\lambda_1, \lambda_3, \lambda_4, \infty\}$, the Wronskian of this solution is

$$detX = \frac{\sqcap_{j=1}^{4}(t - \lambda_j)^{a_{ij}}\Gamma(a_{jj}+1)}{(\Gamma(\rho_1+1)\Gamma(\rho_2+1))^2},$$

where X is the matrix (X_1, X_2, X_3, X_4), from which the linear independence of the solutions follow.

Consider the monodromy group with respect to the basis $X = (X_1, X_2, X_3, X_4)$. We fix base point $z_0 \in \mathbb{CP}^1 \setminus \{\lambda_1, \lambda_3, \lambda_4, \infty\}$. Let σ_j, $j = 1,3,4$ be a simple loop which starts at z_0, goes around λ_j once in the positive direction and returns to z_0. We may choose z_0 and σ_j such that the composition $\sigma_{infty} = \sigma_1\sigma_3\sigma_4$ is simply a loop surrounding $t = \infty$ in the negative direction, where $\sigma_k\sigma_j$ is the loop σ_j followed by σ_k. The loops σ_j, $j - 1,3,4,\infty$ generate the fundamental group $\pi(\mathbb{CP}^1 \setminus \{\lambda_1, \lambda_3, \lambda_4, \infty\}, z_0)$ with relation $\sigma_\infty^{-1}\sigma_1\sigma_3\sigma_4 = 1$. If we continue analytically the basis X along σ_j, X is transformed into XM_j where M_j is the monodromy matrix around λ_j *with respect to* X and is an element of $GL_4(\mathbb{C})$. The monodromy group G is a subgroup of $GL_4(\mathbb{C})$ generated by M_1, M_3 and M_4 with relation $M_\infty = M_1M_3M_4$. We note that, by this choice of z_0 and σ_j, XM_∞ is the result of the analytic continuation around $t = \infty$ in the negative direction.

Denoted by

$$c_{ii} = e_i - 1, c_{jk} = (e_j - 1)\gamma_{jk}, j \neq k,$$

$$L_1 = \begin{pmatrix} c_{11} & 0 & 0 & 0 \\ 0 & c_{22} & 0 & 0 \\ c_{31} & c_{32} & c_{33} & 0 \\ c_{41} & c_{42} & c_{43} & c_{44} \end{pmatrix}, L_2 = \begin{pmatrix} 0 & 0 & c_{13} & c_{14} \\ 0 & 0 & c_{23} & c_{24} \\ 0 & 0 & 0 & c_{34} \\ 0 & 0 & 0 & 0 \end{pmatrix}.$$

Obviously, we have:

1. $L_1 + L_2 = (M_1 - I) + (M_3 - I) + (M_4 - I)$;
2. $I - L_2$ is a nonsingular;
3.$M_\infty = (I - L_2)^{-1}(I + L_1)$;
4. $det(M_\infty - fI) = (f_1 - f)^2(f_2 - f)^2$ and $rank(M_\infty - f_jI) = 2$, $j = 1,2$.
5. Each f_j, $j = 1,2$ is a multiple root of the equation $det((L_1 + fL_2) - (f - 1)I) = 0$ and $rank((L_1 + f_jL_2) - (f_j - 1)I) = 2$, $j = 1,2$.

From this follows

Theorem 5.1.10. *[128]*

$$M_1 = I + \begin{pmatrix} e_1-1 & 0 & 0 & 0 \\ 0 & e_2-1 & 0 & 0 \\ 0 & 0 & 1 & 0 \\ 0 & 0 & 0 & 1 \end{pmatrix} \begin{pmatrix} 1 & 0 & \gamma_{13} & \gamma_{14} \\ 0 & 1 & \gamma_{23} & \gamma_{24} \\ 0 & 0 & 0 & 0 \\ 0 & 0 & 0 & 0 \end{pmatrix},$$

$$M_3 = I + (e_3-1) \begin{pmatrix} 0 & 0 & 0 & 0 \\ 0 & 0 & 0 & 0 \\ \gamma_{31} & \gamma_{32} & 1 & \gamma_{34} \\ 0 & 0 & 0 & 0 \end{pmatrix},$$

$$M_4 = I + (e_4-1) \begin{pmatrix} 0 & 0 & 0 & 0 \\ 0 & 0 & 0 & 0 \\ 0 & 0 & 0 & 0 \\ \gamma_{41} & \gamma_{42} & \gamma_{43} & 1 \end{pmatrix},$$

where

$$\gamma_{13} = \frac{e_2(e_1-f_1)(e_1-f_2)}{(e_1-1)(e_1-e_2)f_1f_2}\frac{\gamma_{43}}{\gamma_{41}},$$

$$\gamma_{14} = \frac{(e_1-f_1)(e_1-f_2)(e_1e_3-f_1f_2)}{e_1e_3(e_1-1)(e_4-1)(e_2-e_1)}\frac{1}{\gamma_{41}},$$

$$\gamma_{23} = \frac{e_1(e_2-f_1)(e_2-f_2)}{(e_2-1)(e_2-e_1)f_1f_2}\frac{\gamma_{43}}{\gamma_{42}},$$

$$\gamma_{24} = \frac{(e_2-f_1)(e_2-f_2)(e_2e_3-f_1f_2)}{e_2e_3(e_2-1)(e_4-1)(e_1-e_2)}\frac{1}{\gamma_{42}},$$

$$\gamma_{31} = \frac{e_2e_3-f_1f_2)}{e_2(e_3-1)}\frac{\gamma_{41}}{\gamma_{43}},$$

$$\gamma_{32} = \frac{e_1e_3-f_1f_2)}{e_1(e_3-1)}\frac{\gamma_{42}}{\gamma_{43}},$$

$$\gamma_{34} = \frac{(e_1e_3-f_1f_2)(f_1f_2-e_2e_3)}{e_1e_2e_3(e_3-1)(e_4-1)}\frac{1}{\gamma_{43}}.$$

For example, if $\gamma_{4j} = \frac{1}{e_4-1}$, $j = 1,3,4$ then the monodromy matrix M_j is represented as follows:

$$M_1 = I + \begin{pmatrix} e_1-1 & 0 & \frac{e_2(e_1-f_1)(e_1-f_2)}{(e_1-e_2)f_1f_2} & \frac{(e_1-f_1)(e_1-f_2)(e_1e_3-f_1f_2)}{e_1e_3(e_2-e_1)} \\ 0 & e_2-1 & \frac{e_1(e_2-f_1)(e_2-f_2)}{(e_2-e_1)f_1f_2} & \frac{(e_2-f_1)(e_2-f_2)(e_2e_3-f_1f_2)}{e_2e_3(e_1-e_2)} \\ 0 & 0 & 0 & 0 \\ 0 & 0 & 0 & 0 \end{pmatrix},$$

$$M_3 = I + \begin{pmatrix} 0 & 0 & 0 & 0 \\ 0 & 0 & 0 & 0 \\ \frac{e_2e_3-f_1f_2}{e_2} & \frac{e_1e_3-f_1f_2}{e_1} & e_3-1 & \frac{(e_1e_3-f_1f_2)(f_1f_2-e_2e_3)}{e_1e_2e_3} \\ 0 & 0 & 0 & 0 \end{pmatrix},$$

$$M_4 = I + \begin{pmatrix} 0 & 0 & 0 & 0 \\ 0 & 0 & 0 & 0 \\ 0 & 0 & 0 & 0 \\ 1 & 1 & 1 & e_4 - 1 \end{pmatrix}.$$

It is known that the following conditions are equivalent: [127]:

1. the system (5.1.22) is irreducible;
2. the monodromy group G is irreducible;
3. all the nontrivial components $c_{ij} = (e_i - 1)\gamma_{ij}$ are non-zero;
3. $f_1 f_2 \neq e_1 e_2$, $f_1 f_2 \neq e_1 e_2$ and $f_j \neq e_k$ $j,k = 1,2$.

Consider a four-dimensional case of (5.1.17), where B is 4×4 diagonal matrix $B = diag(\lambda_1, \lambda_1, \lambda_2, \lambda_2)$, $A = (a_{ij})$ is 4×4 constant matrix and has three distinct eigenvalues $\rho_1 = \rho_2, \rho_3, \rho_4$. We can make the following assumptions:

1. if we write $A = \begin{pmatrix} A_{11} & A_{12} \\ A_{21} & A_{22} \end{pmatrix}$ block wise, where A_{jk} are 2×2 matrices, then the diagonal blocks A_{11} and A_{22} are diagonalizable;

2. A itself is diagonalizable;

3. None the quantities a_{ij} $j = 1,2,3,4$, $a_{11} - a_{22}$, $a_{33} - a_{44}$ and $\rho_j - \rho_k$, if $\rho_j \neq rho_k$, is an integer. And further, each ρ_j is not a negative integer.

Where the system (5.1.17) in this case has the form:

$$\left(tI - \begin{pmatrix} \lambda_1 & 0 & 0 & 0 \\ 0 & \lambda_1 & 0 & 0 \\ 0 & 0 & \lambda_2 & 0 \\ 0 & 0 & 0 & \lambda_2 \end{pmatrix} \right) \frac{dX}{dt} = \begin{pmatrix} A_{11} & A_{12} \\ A_{21} & A_{22} \end{pmatrix} X. \qquad (5.1.23)$$

The properties of (5.1.23):

1.The system (5.1.23) has three regular singular points $t - \lambda_1, \lambda_2$ and ∞ in the Riemann sphere $\mathbb{CP}^1$.

2. The characteristic exponents of (5.1.23) are $(a_{11}, a_{22}, 0, 0)$, $(0, 0, a_{33}, a_{44})$ and $(\rho_1, \rho_2, \rho_3, \rho_4)$ at $\lambda_1, \lambda_2, \infty$, respectively.

3. Riemann-Fuchs relation:

$$2\rho - 1 + \rho_3 + \rho_4 = a_{11} + a_{22} + a_{33} + a_{44} \Rightarrow f_1^2 f_3 f_4 - e_1 e_2 e_3 e_4.$$

4. The system (5.1.23) is accessory parameter free.

Let λ_1, λ_2, be $0, \lambda$, respectively. Eliminating the components x_2, x_3, x_4 of $X = (x_1, x_2, x_3, x_4)$ in (5.1.23), we obtain a fourth order Fuchsian differential equation

$$A_0(t)\frac{d^4x(t)}{dt^4} + A_1(t)\frac{d^3x(t)}{dt^3} + A_2(t)\frac{d^2x(t)}{dt^2} + A_3(t)\frac{dx(t)}{dt} + A_4(t)x(t) = 0, \qquad (5.1.24)$$

where

$$A_0 = t^2(t-\lambda)^2,$$

$$A_1 = t(t-\lambda)((4 - a_{11} - a_{22})(t-\lambda) + (3 - a_{33} - a_{44})t),$$

$$A_2 = (a_{11} - 2)(a_{22} - 1)(t-\lambda)^2 + ((\rho_1 + \rho_3)(\rho_1 + \rho_4) + (\rho_3 + \rho_4 + 1)\rho_1 -$$

$$-a_{11}a_{22}-a_{33}a_{44}-3(\sum_{j=1}^{4}a_{jj})+a_{22}+7)$$

$$t(t-\lambda)+(a_{33}-1)(a_{44}-1)t^2,$$

$$A_3=(1-\rho_1)((\rho_1+\rho_3)(\rho_1+\rho_4)-(a_{11}+a_{22}-2)\rho_1+a_{11}a_{22}-a_{33}a_{44})-$$

$$-(\sum_{j=1}^{4}a_{jj}-a_{22}+1)(t-\lambda)+(1-\rho_1)((\rho_1+\rho_3)(\rho_1+\rho_4)-(a_{33}+a_{44}-1)\rho_1-a_{11}a_{22}+a_{33}a_{44}-$$

$$(\sum_{j=1}^{4}a_{jj})+a_{22}+1)t,$$

$$A_4=\rho_1(\rho_1-1)\rho_3\rho_4,$$

here x_1 is replaced by x. The characteristic exponents of (5.1.24) are $(0,1,a_{11},a_{22}+1)$, $(0,1,a_{33}+1,a_{44}+1)$ and $(-\rho_1,-\rho_1+1,-\rho_3,-\rho_4)$ at $0,\lambda,\infty$, respectively.

5.2. Schrödinger Equations and Fuchsian Systems

As already mentioned, certain physically interesting Schrödinger equations can be rewritten as Fuchsian systems [72], [50], and [51]. For us, what is especially interesting are the cases that arise for Fuchsian systems which are of an Okubo type [112] and [113].

We shall first describe the one multidimensional situation that comes from an exactly solvable model of quantum dynamics. The first to consider such models were L. Landau, N. Rosen, and C. Zener, which have been subsequently generalized by several authors (see [72] and references therein).

Consider a Schrödinger equation for the form

$$i\frac{\partial\Psi(t)}{\partial t}=H(t)\Psi(t) \tag{5.2.1}$$

with time dependent Hamiltonian $H(t)=(H_{ij}(t)),\, i,j=1,...,N$, where

$$H_{11}=\varepsilon(t), H_{12}=V_2, H_{13}=V_3,...,H_{1N}=V_N,$$

$$H_{21}=V_2, H_{31}=V_3,...,H_{N1}=V_N, \text{ and } H_{ij}=0 \text{ otherwise},$$

$\Psi(t)=(\psi_1(t),...,\psi_N(t))$ is a wave function, $V_j, j=2,\ldots,n$ are constants, and the time dependent part ε has the form $\varepsilon(t)=E_1\tanh(t/T)$ with constant E_1 and T.

Theorem 5.2.1. *Equation (5.2.1) is reducible to an N-dimensional Fuchsian system of Okubo type*

$$(zI_N-B)\frac{d\Phi(z)}{dz}=A\Phi(z) \tag{5.2.2}$$

with $B=\text{diag}(i,-i,\ldots,-i)$.

◁ Consider the following transformation for the vector function $\Psi(t) = (\psi_1(t), \dots, \psi_N(t))$:

$$g_1(t) = \psi_1(t)e^{i\int_0^t \varepsilon(\tau)d\tau}, g_j(t) = \psi_j(t), j = 2, \dots, N.$$

When taking into account

$$i\int_0^t \varepsilon(\tau)d\tau = iE_1T\log(\cosh(t/T))$$

we get a system of differential equations for functions g_i that are considered as new dependent variables:

$$g_1'(t) = T^{-1}\sum_{j=2}^{N} v_j(\cosh t/T)^{2\varepsilon_1})g_j,$$

$$g_j'(t) = (T)^{-1}v_j(\cosh(t/T))^{-2\varepsilon_1}g_1, \text{ if } 2 \le j \le N,$$

where $2\varepsilon_1 = iE_1T$, $v_j = -iV_jT$. After this change, the time variable as $z(t) = \sinh(t/T)$ for this system takes the following form

$$\begin{cases} \dfrac{dg_1(z)}{dz} = \displaystyle\sum_{j=2}^{N} v_j(1+z^2)^{\varepsilon_1 - 1/2}g_j(z), \\ \dfrac{dg_j(z)}{dz} = v_j(1+z^2)^{-\varepsilon_1-1/2}g_1(z), \\ \qquad\qquad 2 \le j \le N. \end{cases}$$

Let us take arbitrary numbers $\lambda_2, \dots, \lambda_N$ that satisfy the equality $\sum_{j=2}^{N}\lambda_j = 1$ and change the dependent variables once more by

$$\begin{cases} \phi_1(z) = (1+z^2)^{-\varepsilon_1-1/2}g_1(z), \\ \phi_j(z) = \dfrac{v_jg_j}{z+i} - \lambda(\varepsilon_1 + \dfrac{1}{2})\dfrac{z-i}{z+i}\phi_1(z), \\ \qquad\qquad 2 \le j \le N. \end{cases}$$

Finally, we obtain the following system for variables ϕ_i:

$$\begin{cases} (z-i)\dfrac{d\phi_1(z)}{dz} = -(\varepsilon_1 + \dfrac{1}{2})\phi_1(z) + \displaystyle\sum_{j=2}^{N}\phi_j(z), \\ (z+i)\dfrac{d\phi_j(z)}{dz} = \lambda_j(\varepsilon_1^2 + v_i^2 - \dfrac{1}{4})\phi_1(z) - \phi_j(z) - \lambda_j(\varepsilon_1 + \dfrac{1}{2})\displaystyle\sum_{k=2}^{N}\phi_k(z), \\ \qquad\qquad 2 \le j \le N, \end{cases}$$

which after writing in a matrix form supports (5.2.2) with $B = \mathrm{diag}(i, -i, \dots, -i)$ as was claimed. ▷

Remark 5.2.1. Hamiltonians of this type appear in the theory of nonadiabatic transition [72]. There are also Hamiltonians of other types which permit Fuchsian reduction (see, e.g., [106]) but here it does not seem appropriate to discuss this in more depth.

Remark 5.2.2. Reversing the argument used in the proof one can obtain Schrödinger realizations for certain Okubo type systems. In this way, one can construct Schrödinger equations with prescribed qualitative properties of solutions (see, e.g., [52]).

In a two-dimensional case (i.e., for $N = 2$), this result can be generalized by permitting Hamiltonians of a more general form. Namely, consider a Schrödinger equation with two-component phase function

$$i\frac{\partial f(t)}{\partial t} = H(t)f(t), \tag{5.2.3}$$

where $f(t) = (f_1(t), f_2(t))$, and time dependent Hamiltonian $H(t)$ has the form

$$H(t) = \begin{pmatrix} \varepsilon(t) & V(t) \\ V(t) & -\varepsilon(t) \end{pmatrix} \tag{5.2.4}$$

where

$$\varepsilon(t) = \frac{E_0T + E_1Ty}{1+y^2}\frac{dy}{dt}, \quad V(t) = \frac{V_0T}{\sqrt{1+y^2}}\frac{dy}{dt}$$

with some monotonically increasing differentiable function $y : \mathbb{R} \to \mathbb{R}$ satisfying $y(t) \to \pm\infty$ as $t \to \pm\infty$.

Theorem 5.2.2. *Equation (5.2.3) is reducible to a system of two hypergeometric equations of the form*

$$z(z-1)\frac{d^2g}{dz^2} + (\gamma - (1+\alpha+\beta))\frac{dg}{dz} - \alpha\beta g(z) = 0 \tag{5.2.5}$$

with appropriate constants α, β, γ.

◁ We prove the statement by considering an appropriate transformation of this equation (5.2.3). The transformation is quite standard. Rewrite (5.2.3) in the form

$$if_1'(t) = \varepsilon(t)f_1(t) + V(t)f_2(t), \tag{5.2.6}$$

$$if_2'(t) = V(t)f_1(t) - \varepsilon(t)f_2(t). \tag{5.2.7}$$

Introduce new dependent variables g_1, g_2 by the formulae

$$f_1(t) = g_1(t)e^{-i\int_0^t \varepsilon(\tau)d\tau}, \tag{5.2.8}$$

$$f_2(t) = g_2(t)e^{i\int_0^t \varepsilon(\tau)d\tau}, \tag{5.2.9}$$

then

$$g_1'(t) = f_1'(t)e^{i\int_0^t \varepsilon(\tau)d\tau} + if_1(t)\varepsilon(t)e^{i\int_0^t \varepsilon(\tau)d\tau}, \tag{5.2.10}$$

$$g_2'(t) = f_2'(t)e^{-i\int_0^t \varepsilon(\tau)d\tau} - if_2(t)\varepsilon(t)e^{i\int_0^t \varepsilon(\tau)d\tau}. \tag{5.2.11}$$

Substituting in (5.2.10) expression for $f_1'(t)$ from (5.2.6) one obtains:

$$g_1'(t) = -if_1(t)\varepsilon(t)e^{i\int_0^t \varepsilon(\tau)d\tau} - iV(t)f_2(t)e^{i\int_0^t \varepsilon(\tau)d\tau} + if_1(t)\varepsilon(t)e^{i\int_0^t \varepsilon(\tau)d\tau}.$$

Hence,

$$g_1'(t) = -iV(t)f_2(t)e^{i\int_0^t \varepsilon(\tau)d\tau}.$$

Substituting the expression for $f_2(t)$ from (5.2.9) one obtains

$$g_1' = -iV(t)g_2(t)e^{2i\int_0^t \varepsilon(\tau)d\tau}. \tag{5.2.12}$$

In a similar way, we get

$$g_2' = -iV(t)g_1(t)e^{-2i\int_0^t \varepsilon(\tau)d\tau}. \tag{5.2.13}$$

From (5.2.12), we find the second derivative of $g_1(t)$ with respect to t :

$$g_1''(t) = -iV'(t)g_2(t)e^{2i\int_0^t \varepsilon(\tau)d\tau} - iV(t)g_2'(t)e^{2i\int_0^t \varepsilon(\tau)d\tau} + 2V(t)g_2(t)\varepsilon(t)e^{2i\int_0^t \varepsilon(\tau)d\tau},$$

and substituting in this expression $g_2(t)$ from (5.2.13), we obtain

$$g_1''(t) = -iV(t)g_2(t)e^{2i\int_0^t \varepsilon(\tau)d\tau} - V^2(t)g_1(t) + 2V(t)_2(t)\varepsilon(t)e^{2i\int_0^t \varepsilon(\tau)d\tau}.$$

In order to eliminate $g_2(t)$ from the latter, we use (5.2.12). Finally, we obtain the following second order differential equation with respect to $g_1(t)$:

$$g_1''(t) - (2i\varepsilon(t) + \frac{V'(t)}{V(t)})g_1'(t) + V^2(t)g_1(t) = 0; \tag{5.2.14}$$

In a similar way, we obtain a second order equation for $g_2(t)$:

$$g_2''(t) - (\frac{V'(t)}{V(t)} - 2i\varepsilon(t))g_2'(t) + V^2(t)g_2(t) = 0; \tag{5.2.15}$$

It remains to show that these equations can be rewritten in the form (5.2.5). To this end, let us substitute expressions for ε and $V(t)$ into (5.2.14) and make a change of independent variable by the formula

$$z(t) = \frac{y(t)+i}{2i}.$$

Then, it is a matter of a simple calculation to check that the equation (5.2.14) transforms into hypergeometric equation (5.2.5), where

$$\alpha = iT(-E_1 + \sqrt{E_1^2 + V_0^2}), \beta = iT(-E_1 - \sqrt{E_1^2 + V_0^2}), \gamma = \frac{1}{2} + E_0T - iE_1T.$$

Analogously from (5.2.15), we obtain the hypergeometric equation satisfied by $g_2(z)$:

$$z(z-1)\frac{d^2g_2}{dz^2} + (\gamma' - (1 + \alpha' + \beta'))\frac{dg_2}{dz} - \alpha'\beta' g_2(z) = 0,$$

where

$$\alpha' = iT(E_1 + \sqrt{E_1^2 + V_0^2}), \beta' = iT(E_1 - \sqrt{E_1^2 + V_0^2}), \gamma' = \frac{1}{2} - E_0T + iE_1T.$$

Thus, the proof is completed. ▷

Since a single Fuchsian equation can always be rewritten as a Okubo type system [112], this theorem provides a reduction in the equation (5.2.3) to a Okubo type system. For generalized hypergeometric equations, explicit formulae for the coefficients of this resulting Okubo type system are given in [113].

Example 5.2.1. *([113]) Consider the Gauss' hypergeometric equation written as*

$$[(\delta+\alpha_1)(\delta+\alpha_2)-\frac{1}{z}\delta(\delta+\beta_1-1)]u=0, \qquad (5.2.16)$$

where $\delta = z\frac{d}{dz}$. *Introduce vector function X(z) by*

$$X_1(z)=u(z), X_2(z)=(\delta+\beta_1-1)X_1(z).$$

Then, it can be verified that $X(z)$ *satisfies a system of Okubo type with* $B=\mathrm{diag}(0,1), A=(a_{ij})$, *where* $a_{11}=1-\beta_1, a_{12}=1, a_{21}=-(\alpha_1-\beta_1+1)(\alpha_2-\beta_1+1), a_{22}=-(1-\beta_1)-(\alpha_1+\alpha_2)$.

Thus, Schrödinger equations with potentials such as this can be written in a Okubo form with explicitly known coefficient matrices, which enables one to compute the monodromy and asymptotics solutions.

Remark 5.2.3. Theorem 5.2.1 is applicable to a number of physically interesting equations. In particular, taking $y(t)=sinh(t/T)$ we obtain $\varepsilon(t)=E_0 sech(t/T)+E_1 tanh(t/T)$, $V(t)=V_0$, where E_0, E_1, T, V_0 are real constants. This Hamiltonian was studied in [106] and [72].

In order to indicate another natural source for a system with Okubo types, we reproduce one result from [128] which is formulated in terms of a so-called *weakly singular vector spaces* [112]. By definition a weakly singular vector space on a punctured Riemann sphere is nothing else but a holomorphic vector bundle permitting a logarithmic connection with prescribed nonresonant exponents at all singular points [112]. Recall that exponents are called *nonresonant* if their pairwise differences are never an integer.

Remark 5.2.4. The Hamiltonians for this type appear in the theory of nonadiabatic transition [106].

5.3. Monodromy Approach to Quantum Computing

We are going to replace a smooth bundle by a holomorphic vector bundle with the connection ∇ that have logarithmic or regular singularities along a divisor D. We will then construct topological theory of quantum computing, by which we mean that the quantum gates determined by the holonomy Γ if the connection depends only on the homotopy class of a closed path inside $X \setminus D$ [86], [87], and [73].

Assumption. *For each n, there exists a quantum system described by the Fuchsian system* $df=\omega f$ *with singularities along the subvariety D.*

Riemann-Hilbert correspondence. *Let* $U_1,...,U_N$ *be unitary matrices realizing the algorithm. There exists a Fuchsian system* $df=\omega f$ *whose monodromy matrices coincide with* $U_1,...,U_N$.

Next, we will give some distinguished special cases for this theorem.

Theorem 5.3.1. *The given collection for the unitary operators* $U_1,...,U_q$ *which realize the quantum algorithm, there exists a location of points* $s_1,....,s_q$ *on the compact Riemann surface* X, *and a Fuchsian system with monodromy matrices* $M_1,...,M_q$, *such that the following properties hold:*

1. $M_j : (\mathbb{C}^2)^{\otimes p} \to (\mathbb{C}^2)^{\otimes p}$ *are unitary operators;*

2. for the every l, M_l *acts on the elements of the type* $x_1 \otimes \ldots \otimes x_{j-1} \otimes x_j \otimes x_{j+1} \otimes \ldots \otimes x_p$ *as*

$$M_l(x_1 \otimes \ldots \otimes x_{j-1} \otimes x_j \otimes x_{j+1} \otimes \ldots \otimes x_p) =$$
$$= x_1 \otimes \ldots \otimes x_{j-1} \otimes m_l(x_j \otimes x_{j+1}) \otimes \ldots \otimes x_p;$$

where $m_l : \mathbb{C}^2 \otimes \mathbb{C}^2 \to \mathbb{C}^2 \otimes \mathbb{C}^2$ *is entangling operator;*

3. every U_j *is a product of* M_k.

From this theorem, it follows that, to realize the quantum circuit for family (C) there needs to be logical gates from $SU(2)$ with an entangling operator, and all these are obtained as monodromy matrices of a Fuchsian system.

The dynamics of the quantum system is described by parameter λ which lives on the manifold of complex structures of $X - S$, which is a $(3g + m - 3)$-dimensional complex manifold, where g is genus of the Riemann surface X. For the Riemann sphere $\mathbb{CP}^1$ it means that, if $s_1, \ldots, s_m$ any collection of distinct points on $\mathbb{CP}^1$ and $G \subset GL_n(\mathbb{C})$, then for this there does not always exist a Fuchsian system with singular points $s_1, \ldots, s_m$ and with monodromy group G, but there exist a value λ_0, such that $Im\rho_{\lambda_0} = G$ and ρ_{λ_0} is monodromy representation of some Fuchsian system.

Suppose $D = \cup_{j=1}^{m+1} D_j$ is a divisor, where D_j, $j = 1, \ldots, m+1$, are hyperplanes in $\mathbb{CP}^n$.

Let us choose a line L which intersects the divisor $D = \cup_{j=1}^{m+1} D_j$ at nonsingular points $a_j = L \cap D_j$, $j = 1, 2, \ldots, m+1$. Consider the fundamental group $\pi_1(L - \{a_1, \ldots, a_{m+1}\}, z_0)$. Let $[\gamma_1], \ldots, [\gamma_m]$ be the generators of $\pi_1(L - \{a_1, \ldots, a_{m+1}\}, z_0)$, where the loops have form $\gamma_j = \sigma_j[\alpha_j]\sigma_j^{-1}$, α is a path from z_0 to a neighborhood V_{a_j} and σ_j is small a loop in the the neighborhood V_{a_j} which generates $\pi_1(V_{a_j} - \{a_j\}) \cong \mathbb{Z}$.

It is known that $[\gamma_1], \ldots, [\gamma_m]$ are generators of $\pi_1(\mathbb{CP}^n - D, z_0)$ under some conditions.

Suppose we have a family of representations

$$\rho_\lambda : \pi_1(\mathbb{CP}^n - D, z_0) \to \mathrm{GL}_m(\mathbb{C}) \tag{5.3.1}$$

such that

$$\rho_\lambda([\gamma_j]) = 1 + \lambda M_1^j + \lambda^2 M_2^j + \ldots + \lambda^k M_k^j + \ldots, \tag{5.3.2}$$

where M_k^j are $m \times m$-matrices. The family of representations (5.3.1) satisfy this condition (5.3.2) which is called analytic.

Suppose $D_j = \{h_j = 0\}$ and consider a 1-form $\Omega_k = \sum_{j=1}^{m+1} U_k^j \frac{dh_j}{h_j}$, where U_k^j are constant matrices and $\sum_{j=1}^{m+1} U_k^j = 0$. Consider a family of meromorphic 1-forms

$$\Omega(\lambda) = \lambda\Omega_1 + \lambda^2\Omega_2 + \ldots + \lambda^k\Omega_k + \ldots. \tag{5.3.3}$$

If $U^j(\lambda) = \sum_{k=1}^{\infty} \lambda^k U_k^j$ is a converging power series, then the family $\Omega(\lambda)$ is called an analytic family of Fuchs systems.

It is known that for every λ this system satisfies the condition

$$d_z\Omega(\lambda) = 0, z \in \mathbb{C}.$$

If the analytic family of Fuchs systems satisfies the condition

$$\Omega(\lambda)\wedge\Omega(\lambda)=0, \tag{5.3.4}$$

then the family of Fuchs systems is called integrable.

Similar terminology will be used for a Pfaff system $df=\Omega(\lambda)f$.

The monodromy representation of an integrable family of Pfaffian systems

$$df=\Omega(\lambda)f \tag{5.3.5}$$

is an analytic family of representations for the fundamental group $\pi_1(\mathbb{CP}^n-D,z_0)$ [96].

Theorem 5.3.2. *[87], and [96]. 1. For every analytic family of representations (5.3.1), where for small values of λ, there exists a family of Pfaff systems (5.3.5) whose monodromy coincides with (5.3.1).*

2. Let

$$\rho:\pi_1(\mathbb{CP}^n-D,z_0)\to \mathrm{GL}_m(\mathbb{C})$$

be a representation such that $\rho([\gamma_j])$, $j=1,\dots,m$, are close to identity. Then ρ is realizable as monodromy representation of an integrable Fuchs system $df=f\Omega$, and the fundamental matrix Φ of the system represented by the exponential form $\Phi=\exp\psi$, where ψ is multi-valued function on $\mathbb{CP}^n-D$.

We now shall add some comments regarding this theorem. Representing Ψ as $\Psi=\exp\psi_\lambda$, we get

$$\Omega(\lambda)\wedge\Omega(\lambda)=\lambda^2\Omega_1\wedge\Omega_1+\lambda^3(\Omega_1\wedge\Omega_2+\Omega_2\wedge\Omega_1)+$$
$$\dots+\lambda^n(\sum_{k+l=n}\Omega_k\wedge\Omega_l)+\dots,$$

which means, that (5.3.4) is equivalent to the identity

$$\Omega_1\wedge\Omega_1=0,\Omega_1\wedge\Omega_2+\Omega_2\wedge\Omega_1=0,\ \sum_{k+l=n}\Omega_k\wedge\Omega_l=0,\dots$$

It is clear, that ψ_λ is represented by a converging power series:

$$\psi_\lambda=\lambda\psi_1+\lambda^2\psi_2+\dots+\lambda^j\psi_j+\dots,$$

and by the integrability of $\Omega(\lambda)$ one gets the following relations between ψ_j and Ω_j:

$$d\psi_1=\Omega_1,d\psi_2+h_2(\psi_1,d\psi_1)=\Omega_2,\dots,$$
$$d\psi_j+h_j(\psi_1,\dots,\psi_{j-1},d\psi_1,\dots,d\psi_{j-1})=\Omega_j,\dots,$$

where $h_2,\dots,h_j,\dots$ are 1-forms, on $P_{x_0}(\mathbb{CP}^n-D)$ and satisfy these identities

$$h_2=\frac{1}{2}[d\psi_1,\psi_1],h_3=\frac{1}{2}[d\psi_2,\psi_1]+\frac{1}{6}[[d\psi_1,\psi_1],\psi_1],\dots,$$

$$h_j=\frac{1}{2}\sum_{k+l=j}[d\psi_k,\psi_1]+\frac{1}{3!}\sum_{k+l+m}=j[[d\psi_k,\psi_l],\psi_m]+\dots$$

$$... + \frac{1}{l!} \sum_{i_1+i_2+...+i_l=j} [[d\psi_{i_1}, \psi_{i_2}], ..., \psi_{i_j} + ... + \frac{1}{j!}[...[d\psi_1, \psi_1], ..., \psi_1], ...$$

These relations can be used for finding ψ_j for a given Ω_j.

We will apply results of this section to a special divisor. Namely, suppose $D = \cup_{i<j} D_{ij} \cup D_0$, where $D_{ij} = \{(z_0, ..., z_n) \in \mathbb{CP}^n | z_i = z_j, i, j \neq 0\}$, $D_0 = \{z \in \mathbb{CP}^n | z_0 = 0\}$. Let $X_n = \mathbb{CP}^n - D = \{(z_1, ..., z_n) \in \mathbb{C}^n | z_i \neq z_j, i \neq j\}$. The fundamental group of X_n is called the pure braid group on n strings, which we denote by P_n. The symmetric group S_n acts on X_n by $g.(z_1, ..., z_n) = (z_{g(1)}, ..., z_{g(n)})$, $g \in S_n$. The fundamental group of the quotient space X_n/S_n is called the braid group on n strings and denoted by B_n. The braid group has $n-1$ generators $\sigma_1, ..., \sigma_{n-1}$ with relations

$$\sigma_i \sigma_{i+1} \sigma_i = \sigma_{i+1} \sigma_i \sigma_{i+1}, 1 \leq i \leq n-2, \tag{5.3.6}$$

$$\sigma_i \sigma_j = \sigma_j \sigma_i, |i-j| \geq 2. \tag{5.3.7}$$

Choose a point $z_0 = (1, 2, ..., n) \in X_n$ and denote by $pr : X_n \to X_n/S_n$ the natural projection. We have exact sequence of groups:

$$1 \to P_n \to B_n \to S_n \to 1$$

and generators of the pure braid group P_n are τ_{ij}, $1 \leq i < j \leq n$, where

$$\tau_{ij} = \sigma_i \sigma_{i+1} ... \sigma_{j-1} \sigma_j^2 \sigma_{j-1}^{-1} ... \sigma_i^{-1}, 1 \leq i < j \leq n, \tag{5.3.8}$$

that satisfy the relations

$$\tau_{rs}\tau_{ik} = \tau_{ik}\tau_{rs}, \text{ for } s < i \text{ or } k < s,$$

$$\tau_{ks}\tau_{ik}\tau_{ks}^{-1} = \tau_{is}^{-1}\tau_{ik}\tau_{is} \text{ for } i < k < s,$$

$$\tau_{rk}\tau_{ik}\tau_{rk}^{-1} = \tau_{ik}^{-1}\tau_{ir}^{-1}\tau_{ik}\tau_{ir}\tau_{ik} \text{ for } i < r < k,$$

$$\tau_{rs}\tau_{ik}\tau_{rs}^{-1} = \tau_{is}^{-1}\tau_{is}\tau_{ir}\tau_{ik}\tau_{ir}^{-1}\tau_{is}^{-1}\tau_{ir}\tau_{is} \text{ for } i < r < k < s.$$

Consider the matrix valued 1-form

$$\Omega = \sum_{1 \leq i < j \leq n} \Omega_{ij} d log(z_i - z_j) \tag{5.3.9}$$

holomorphic on X_n, where $\Omega_{ij} \in M_m(\mathbb{C})$, $1 \leq i < j \leq n$. Let $E \to X_n$ be a trivial holomorphic vector bundle of rank m with connection ∇ for which Ω is connection form. Holomorphic sections $f = (f^1, ..., f^m)$ of this bundle are solutions for the Fuchs system

$$df = \Omega f, \tag{5.3.10}$$

where Ω has the form (5.3.9).

The system (5.3.10) is integrable if and only if the following condition is satisfied

$$[\Omega_{ij}, \Omega_{ik} + \Omega_{jk}] = [\Omega_{ij} + \Omega_{ik}, \Omega_{ji}], i < j < k, \tag{5.3.11}$$

$$[\Omega_{ij}, \Omega_{kl}] = 0, \text{ for distinct } i, j, k, l. \tag{5.3.12}$$

From the preceding results, we obtain the solvability of Riemann-Hilbert problem. Indeed, if $\rho : P_n \to \mathrm{GL}_m(\mathbb{C})$ is such representation that $||\rho(\tau_{ij}) - 1||$ is sufficiently small for each $1 \le i < j \le n$, then there exists matrices $\Omega_{ij} \in M_m(\mathbb{C})$, $1 \le i < j \le n$ close to 0, which satisfy the conditions (5.3.11)-(5.3.12) and monodromy representation of the Fuchs system

$$df = \sum_{1 \le i < j \le n} \Omega_{ij} \frac{d(z_i - z_j)}{z_i - z_j} f$$

coincides with ρ.

From this, it follows that for a given collection of unitary matrices $U_1, \ldots, U_n$ in general there *does not exist* a gauge potential $d - \sum_{i,j} \Omega_{ij} d\log(z_i - z_j)$ on the bundle $E \to X_n$ such that monodromy matrices coincide with those in the given collection. The reason for this lies in the nonabelianness of the braid group.

Consider the case $G = SU(2)$. The Lie algebra of $SU(2)$ is generated by $J_0, J_\pm$ with relations:

$$[J_0, J_\pm] = J_\pm, [J_-, J_+] = -2J_0.$$

The group $SU(2)$ is compact, simply connected and any unitary irreducible representation of the group $SU(2)$ is determined by nonnegative integer or half-integer j.

Any elements of $SU(2)$ depends on the three parameters and represented by form $U(\alpha, \beta, \gamma) = exp\{i(\alpha\sigma_1 + \beta\sigma_2 + \gamma\sigma_3)\}$ and the general element have the form

$$\begin{pmatrix} cos\alpha & -e^{i\gamma} sin\alpha \\ e^{i(\beta-\gamma)} sin\alpha & e^{i\beta} cos\alpha \end{pmatrix},$$

where $\sigma_1 = \begin{pmatrix} 0 & 1 \\ 1 & 0 \end{pmatrix}$, $\sigma_2 = \begin{pmatrix} 0 & -i \\ i & 0 \end{pmatrix}$, $\sigma_3 = \begin{pmatrix} 1 & 0 \\ 0 & -1 \end{pmatrix}$.

Let $\Gamma_A(\gamma) = Pexp\{i \int_0^\tau A(t)dt\}$ holonomy operator and $Hol(A)$ be the holonomy group. If $Hol(A) = U(n)$, then A is called irreducible. It is known that for $U(2)$, there exists a irreducible connection.

Let M_1, M_2, M_3, M_4 for such matrices, such that $E_j = (1/2\pi i) ln M_j$, $j = 1, \ldots, 4$ and M_1, M_2, M_3 make a basis for the Lie algebra of $SU(2)$ and satisfy the condition $M_1 M_2 M_3 M_4 = 1$. Let $s_1, s_2, s_3, s_4 \in \mathbb{CP}^1$ be any points. Denoted by $X_4 = \mathbb{CP}^1 - \{s_1, s_2, s_3, s_4\}$ and by $\gamma, \gamma_2, \gamma_3, \gamma_4$ the generators of $\pi(X_4, z_0)$ with relation $\gamma_1\gamma_2\gamma_3\gamma_4 = 1$. Consider the representation ρ of fundamental group $\pi(X_4, z_0)$ in $SU(2)$ defined by the correspondence

$$\rho : \pi(X_4, z_0) \to SU(2), \gamma_j \mapsto M_j.$$

Then for ρ there exits the system of differential equations for the Fuchs type

$$dF(z) = (\frac{A_1}{z - s_1} dz + \frac{A_2}{z - s_2} dz + \frac{A_3}{z - s_3} dz + \frac{A_4}{z - s_4} dz) F(z)$$

where monodromy representation coincides with ρ. The representation ρ induces two-dimensional vector bundle $E \to \mathbb{CP}^1$ with meromorphic connection form

$$\omega = \sum_{j=1}^{4} \frac{A_j}{z - s_j} dz$$

and Chern number $c_1(E) = \sum_{j=1}^{4} E_j$.

The solution space of Υ in the system is a two-dimensional vector space and $\pi(X_4, z_0)$ act on the Υ and any unitary operator shell that is obtained in this way.

Therefore, we have proved the following result (see [53], [54], and [55]).

Theorem 5.3.3. *The connection form* ω *generates the basis for the local gates.*

For ending this theorem proof we shall rely on, as proof, the necessary well known facts.

Proposition 5.3.1. *[82] Any unitary operator* $U : C^p \to C^p$ *can be submitted as product* $p(p-1)$ *time following quasitriangle operators*

$$diag(\mathbf{1}, \begin{pmatrix} a & b \\ c & d \end{pmatrix}, \mathbf{1})$$

where

$$\begin{pmatrix} a & b \\ c & d \end{pmatrix} \in U(2)$$

Proposition 5.3.2. *[82] Basis, consists of all operators acting on pairs of qubits given to all extended operators.*

In particular, any operator, which acts on the pairs of qubits is obtained by the product for at most 6 operator in the form:

$$\begin{pmatrix} 1 & 0 & 0 & 0 \\ 0 & a & b & 0 \\ 0 & c & d & 0 \\ 0 & 0 & 0 & 1 \end{pmatrix}$$

For example, in two-dimensional spin-$\frac{1}{2}$ XXZ Heisenberg models the two qubit gates have the form:

$$e^{i\tau J_z} \begin{pmatrix} e^{-2i\tau J_z} & 0 & 0 & 0 \\ 0 & cos(2\tau J_{xy}) & -isin(2\tau J_{xy}) & 0 \\ 0 & -sin(2\tau J_{xy}) & cos(\tau J_{xy}) & 0 \\ 0 & 0 & 0 & e^{-2i\tau J_z} \end{pmatrix}$$

Remark 5.3.1. The complex structure of the manifold X_m depend on the 1-dimension family of parameters. The holomorphic type of the bundle $E \to \mathbb{CP}^1$ is not dependent on the complex structure of X_m. So, the holomorphic type of E is defined by the integer value vector $(k_1, k_2) \in Z$, $k_1 \geq k_1$, which are expressed by Chern number bundle E and its conjective bundle E^* following way: $k_1 = \frac{1}{2}(c_1(E) + c_1(E^*))$,$k_1 = \frac{1}{2}(c_1(E) - c_1(E^*))$ [56].

The vector bundle, induced by the representation ρ will not appear stable; whereas the necessary and sufficient condition for stability in the sense of Gohberg-Krein-Bojarski is that the inequalities $|k_i - k_j| \leq 1$ hold for all pairs i, j, but at small indignation it will proceed as a stable bundle, holomorphic type which will be $(c_1(E)/2, c_1(e)/2)$, if $c_1(E)$ is even, and will be $([c_1(E)/2]+1, c_1(E)/2)$, if $c_1(E)$ is odd, where $[c_1(E)/2]$ denoted integer part of the rational number $[c_1(E)/2]$.

Denoted by $\Omega_{z_0}X_4$ as the space of closed loops. Let $\gamma_j : [0,1] \to U(2)$ be any element of this space, when going around s_j, $j = 1,...,4$, i.e., $\gamma(0) = \gamma(1) = z_0$. Then the connection form ω defines the local representation $\Gamma : \Omega_{z_0}X_4 \to U(2)$ in the following way:

$$\Gamma_\gamma(\omega) = \frac{1}{2\pi i}\int_0^1 \frac{A_j}{\gamma(t)-s_j}\gamma'(t)dt.$$

Now, let us consider this general case. Let X be a compact Riemann surface genus $g \geq 2$ and let $E \to X$ flat holomorphic vector bundle. Then, by the Weil theorem there exits the unitary representation ρ of the fundamental group $\pi_1(X)$ for surface X, where $\alpha_1, \beta_1, ..., \alpha_g, \beta_g$ generators of $\pi_1(X)$ with relations $\prod_{j=1}^{g}(\alpha_j\beta_j\alpha_j^{-1}\beta_j^{-1}) = 1$, that the bundle $E \to X$ is induced by ρ. Let $E \to X$ be the two-dimension flat holomorphic vector bundle, then there exits the holomorphic connection of $E \to X$, which holonomy coincides with ρ. Therefore, we proof the following theorem:

Theorem 5.3.4. *Let X be a compact Riemann surface genus $g \geq 2$ and let $E \to X$ stable flat holomorphic vector bundle with holomorphic connection ∇. Then ∇ gives the basis for the quantum computation.*

For example, stable flat two-dimensional vector bundles on Riemann surface genus 2 that makes a so-called moduli space and this is the complex manifold of dimension 5.

By these reasons, it holds for 2-level quantum systems. Also, it does not present any difficulty to generalize these results for the d-level systems. If we want to construct d-level quantum computing, it is necessary to take d-dimension vector bundle with unitary irreducible connection. In general, for any holomorphic vector bundles such a connection does not exist. For example, for the compact Riemann surface genus $g \geq 2$, it is possible to consider a stable bundle. For a Riemann sphere, to find the pair (E, ∇) this is possible by advanced methods in theory, which is devoted to research in the Riemann-Hilbert monodromy problem [57].

Consider a 1-form Ω for a special type defined by (5.3.9). Let $V_1, ..., V_m$ be sl_2-modules. Put $V = V_1 \otimes ... \otimes V_m$. Let $\{I_j\}$ be an orthogonal basis of $\mathfrak{sl}_2$ and $\mathbf{c} = \sum_j I_j I_j \in U(\mathfrak{sl}_2)$ be the Casimir element in the universal enveloping algebra $U(\mathfrak{sl}_2)$. Let $\Delta : U(\mathfrak{sl}_2) \to U(\mathfrak{sl}_2) \otimes U(\mathfrak{sl}_2)$ be the diagonal homomorphism determined by $\Delta(x) = x \otimes 1 + 1 \otimes x$, $x \in \mathfrak{sl}_2$. Set $\Omega = \frac{1}{2}(\Delta\mathbf{c} - \mathbf{c} \otimes 1 - 1 \otimes \mathbf{c})$. Consider a family $\rho_i : \mathfrak{sl}_2 \to \mathrm{End}(V_i)$, $i = 1, ..., n$ of irreducible representations and define the representations

$$\rho_i \otimes \rho_j : \mathfrak{sl}_2 \to \mathrm{End}(V_1 \otimes ... \otimes V_n) \tag{5.3.13}$$

by the formula

$$(\rho_i \otimes \rho_j)(x) = I_1 \otimes ... \otimes \rho_i(x) \otimes ... \otimes I_j \otimes ... \otimes I_n + I_1 \otimes ... \otimes I_i \otimes ... \otimes \rho_j(x) \otimes ... \otimes I_n,$$

where I_k denotes the identity operator acting on V_k.

The representations (5.3.13) extend to the universal enveloping algebra $U(\mathfrak{sl}_2)$, these representations we denote again by $\rho_i \otimes \rho_j$; thus we have the representations $\rho_i \otimes \rho_j$:

$U(\mathfrak{sl}_2) \to \mathrm{End}(V_1 \otimes \ldots \otimes V_n)$. Let $\Omega_{ij} = (\rho_i \otimes \rho_j)\Omega$. The linear operators $\Omega_{ij} : V \to V$, $i < j$, act as $\Omega \otimes 1 + 1 \otimes \Omega$ on $V_i \otimes V_j$ and trivially on all of the other factors. The Fuchs type Pfaff system

$$\frac{\partial \Psi(z_1,\ldots,z_n)}{\partial z_i} = \frac{1}{\lambda} \sum_{j=1, i\neq j}^{n} \frac{\Omega_{ij}}{z_i - z_j} d(z_i - z_j)\Psi, \quad i = 1,\ldots,n, \tag{5.3.14}$$

where $\Psi(z_1,\ldots,z_n)$ is a V-valued function on X_n, is called the *Knizhnik-Zamolodchikov equation* [85]. Here, λ is a complex parameter. Solutions of (5.3.14) are covariant constant sections of the trivial bundle $X_n \times V \to X_n$ with flat connection

$$\sum_{\substack{j=1 \\ i\neq j}}^{n} \frac{\Omega_{ij}}{z_i - z_j} d(z_i - z_j).$$

It follows from previous the subsections that (5.3.14) is integrable if and only if the conditions (5.3.11)-(5.3.12) are satisfied. The monodromy representation of (5.3.14) can be extended to a representation of the braid group [86]. Let $V_1 = V_2 = \ldots = V_n \cong \mathbb{C}^2$, then we can obtain that the braid group B_n acts on $V^{\otimes n} \cong (\mathbb{C}^2)^{\otimes n}$. For quantum computing, an appropriate choice of action is necessary (see [39], and [40]).

The Knizhnik-Zamolodchikov equation was invented in conformal field theory. Its solutions describe $(n+1)$-point correlation functions on the Riemann sphere in the Wess-Zumino-Witten model. This model is uniquely determined by the choice of a simple Lie algebra $\mathfrak{g}$ for every $k > 0$, $k \in \mathbb{Z}$. Any model of conformal field theory has a certain set of primary fields $\{\phi_j\}$. They are in one-to-one correspondence with irreducible representations $\{\rho_j\}$ of the Kac-Moody algebra $\widehat{\mathfrak{g}}$. Let $\mathfrak{g} = \mathfrak{sl}_2(\mathbb{C})$. For each nonnegative half integer j there exists a unique irreducible $\mathfrak{g}$-module V_j, called the module of spin j, with highest weight $j\alpha$ and $\dim V_j = 2j+1$.

Theorem 5.3.5. *[85], [87]. The n-point function $\Psi = \langle \phi_1(z_1)\ldots\phi_n(z_n)\rangle$, where ϕ_j are primary fields, satisfies the system of differential equations (5.3.14) and the monodromy representation of this system is unitarizable.*

For example, consider the case $n = 2$ with

$$\Omega = \Omega_{12} = \Omega_{21} = \begin{pmatrix} 1/2 & 0 & 0 & 0 \\ 0 & -1/2 & 1 & 0 \\ 0 & 1 & -1/2 & 0 \\ 0 & 0 & 0 & 1/2 \end{pmatrix},$$

then (5.3.14) have the form

$$\frac{\partial F}{\partial z_1} = \frac{1}{\lambda} \frac{\Omega_{12}}{z_1 - z_2} F, \tag{5.3.15}$$

$$\frac{\partial F}{\partial z_2} = \frac{1}{\lambda} \frac{\Omega_{21}}{z_2 - z_1} F, \tag{5.3.16}$$

where F is a function on $X_2 = \mathbb{C}^2 - \{z_1 = z_2\}$ with values in $V^{\otimes 2}$, $dim_{\mathbb{C}} V = 2$. The solutions to this system are given by $F(z) = e^{\frac{1}{\lambda}\ln(z_1 - z_2)\Omega} C$, where C is a constant 4-vector. The image of the generator $\sigma_1 \in B_2$ by the monodromy representation of the system (5.3.15)-(19.16) is $e^{\frac{-\pi i}{\lambda}\Omega}$ [93].

Proposition 5.3.3. *The operator* $U = e^{-\frac{\pi i}{\lambda}\Omega}$ *is an entangling operator.*

◁ This proof will use an online program, see `www.physics.uq.edu.au/gqc`. In the canonical decomposition from proposition 1.4.4, we set

$$A_1 = \begin{pmatrix} -i & 0 \\ 0 & i \end{pmatrix}, B_1 = \begin{pmatrix} 1 & 0 \\ 0 & i \end{pmatrix}, A_2 = \begin{pmatrix} i & 0 \\ 0 & i \end{pmatrix}, B_2 = \begin{pmatrix} 1 & 0 \\ 0 & -1 \end{pmatrix},$$

$$\theta_x = \theta_y \cong 1,07, \theta = 0,5.$$

In this approximate calculation U has the form

$$U = \begin{pmatrix} 1 & 0 & 0 & 0 \\ 0 & 0,21 - 0,45i & 0,71 + 0,45i & 0 \\ 0 & 0,71 + 0,45i & 0,29 - 0,45i & 0 \\ 0 & 0 & 0 & 1 \end{pmatrix} \triangleright$$

Remark 5.3.2. In this calculation the numerical parameter $-\frac{\pi i}{\lambda}$ has not been used at all. Moreover, it was not needed that U belong to the special unitary group. These requirements do not change anything.

The Knizhnik-Zamolodchikov equation is generalized to those Fuchsian systems which are invariant under the action of a Weyl group on a root system in $\mathbb{R}^n$. Singularities of this equation constitute the union of those hyperplanes which are orthogonal to the given root system. Fundamental groups complement the divisor of singular points for this obtained Fuchsian system. These are generalized braid groups from the series of classical groups B_n, D_n, etc., (see [62]).

Following the work by [95], we now present an example of monodromy representation of a generalized braid group for type B_n or D_n which seems to fit the monodromic quantum computation ideology.

Let $\mathfrak{U}\mathfrak{g}$ be the universal enveloping algebra of a finite dimensional complex Lie algebra $\mathfrak{g}$, $(,)$ be its Killing form, and H be a Cartian subalgebra of $\mathfrak{g}$. Denoted by $\{e_\alpha\}$, the corresponding root system and by $\{h_i = h_{\alpha_i}\}_{i=1}^d$ a basis of H over $\mathbb{C}$. Put $H_{ij} = (h_i, h_j)$, $G = H^{-1}$.

Introduce then

$$\xi = \sum_{\alpha \neq 0} e_\alpha \otimes e_{-\alpha} + \sum_{i,j=1}^{d} G_{ij} h_i \otimes h_j \in \mathfrak{U}\mathfrak{g},$$

$$\mu = \sum_{\alpha \neq 0} e_\alpha \otimes e_{-\alpha} - \sum_{i,j=1}^{d} G_{ij} h_i \otimes h_j \in \mathfrak{U}\mathfrak{g},$$

$$\nu = \frac{1}{2}\left(\sum_{\alpha \neq 0} e_\alpha^2 + \sum_{\alpha \neq 0} e_\alpha e_{-\alpha}\right) \in \mathfrak{U}\mathfrak{g}.$$

Let

$$\chi_i : \mathfrak{U}\mathfrak{g} \to \mathfrak{U}\mathfrak{g}^{\otimes n}, \chi_i(u) = 1^{\otimes(i-1)} \otimes u \otimes 1^{\otimes(n-i)};$$

$$\chi_{ij} : \mathfrak{U}\mathfrak{g} \otimes \mathfrak{U}\mathfrak{g} \to \mathfrak{U}\mathfrak{g}^{\otimes n}, \chi_{ij}(u \otimes v) = 1^{\otimes(i-1)} \otimes u \otimes 1^{\otimes(j-i-1)} \otimes v \otimes 1^{\otimes(n-j)};$$

$$\xi_{ij} = \chi_{ij}(\xi), \mu_{ij} = \chi_{ij}(\mu), \nu_i = \chi_i(\nu) \in \mathfrak{U}\mathfrak{g}^{\otimes n}.$$

Let $\widehat{D}$ be the divisor given by the equations $z_i - z_j = 0, z_i + z_j = 0, z_i = 0$. The generalized braid group $\widehat{B_n}$ has generators $a_1, \ldots, a_{n-1}, c_1$ with relations

$$a_i a_j = a_j a_i, |i-j| \geq 2,$$

$$a_i a_{i+1} a_i = a_{i+1} a_i a_{i+1},$$

$$c_1 a_i = a_i c_1, i \geq 2,$$

$$(c_1 a_1)^2 = (a_1 c_1)^2;$$

Consider the following 1-form on $\widehat{X_n} = \mathbb{C}^n - \widehat{D}$

$$\widehat{\omega} = \lambda(\sum_{1 \leq i < j \leq n} \rho_n(\xi_{ij}) d\log(z_i - z_j) +$$

$$+ \sum_{1 \leq i < j \leq n} \rho_n(\mu_{ij}) d\log(z_i + z_j) + \sum_{1 \leq i < j \leq n} \rho_n(\nu_{ij}) d\log z_i)$$

where λ is complex parameter. Then the monodromy representation θ induced by the differential equation $df = \widehat{\omega} f$ is

$$\theta(a_i) = \frac{1}{\sqrt{q}} \mathbf{1}^{\otimes(i-1)} \otimes \begin{pmatrix} 1 & 0 & 0 & 0 \\ 0 & 0 & q & 0 \\ 0 & q & 1-q^2 & 0 \\ 0 & 0 & 0 & 1 \end{pmatrix} \otimes \mathbf{1}^{\otimes(n-i-1)},$$

$$\theta(c_1) = q^{-r} \begin{pmatrix} 0 & 1 \\ 1 & q^{-2s} - q^{2s} \end{pmatrix} \otimes \mathbf{1}^{\otimes(n-1)}.$$

Let divisor $\widetilde{D}$ be given by equations $z_i - z_j = 0, z_i + z_j = 0$. Analogically, the generalized braid group $\widetilde{B_n}$ have generators $a_1, \ldots, a_{n-1}, b_1$ with relations

$$a_i a_j = a_j a_i, |i-j| \geq 2,$$

$$a_i a_{i+1} a_i = a_{i+1} a_i a_{i+1},$$

$$b_1 a_i = a_i b_1, i \neq 2,$$

$$b_1 a_2 b_1 = a_2 b_1 a_2);$$

Consider on $\widetilde{X_n} = \mathbb{C}^n - \widetilde{D}$ the holomorphic 1-form

$$\widetilde{\omega} = \lambda(\sum_{1 \leq i < j \leq n} \rho_n(\xi_{ij}) d\log(z_i - z_j) + \sum_{1 \leq i < j \leq n} \rho_n(\mu_{ij}) d\log(z_i + z_j))$$

and the system of differential equations $df = \widetilde{\omega} f$. The monodromy matrices are then:

$$a_i \to \frac{1}{\sqrt{q}} \mathbf{1}^{\otimes(i-1)} \otimes \begin{pmatrix} 1 & 0 & 0 & 0 \\ 0 & 0 & q & 0 \\ 0 & q & 1-q^2 & 0 \\ 0 & 0 & 0 & 1 \end{pmatrix} \otimes \mathbf{1}^{\otimes(n-i-1)},$$

$$b_1 \to \frac{1}{\sqrt{q}} \begin{pmatrix} 0 & 0 & 0 & q \\ 0 & 1 & 0 & 0 \\ 0 & 0 & 1 & 0 \\ q & 0 & 0 & 1-q^2 \end{pmatrix} \otimes \mathbf{1}^{\otimes(n-2)}.$$

The Weil group $\widehat{W_n}$ acts on the trivial bundles $V^{\otimes n} \times \widehat{X}_n \longrightarrow \widehat{X}_n$ and after factorizing, we obtain a nontrivial bundle over $\widehat{X}_n/\widehat{W}_n$ with the fiber $V^{\otimes n}$ and flat connection $\widehat{\nabla} = d - \widehat{\omega}$; whose monodromy representation coincides with θ, since the fundamental group of manifold $\widehat{X}_n/\widehat{W}_n$ is $\widehat{B}_n$.

An analogous argument is applicable to the connection $\widetilde{\nabla} = d - \widetilde{\omega}$ and we obtain the desired bundle in a similar way.

It is known that every finitely generated irreducible subgroup of $GL_n(\mathbb{C})$ is a monodromy group of a Fuchsian system on $\mathbb{CP}^1$ with prescribed poles. The generators are assumed to correspond to the monodromy matrices. Therefore, regular system with irreducible monodromy group is an equivalent to a Fuchsian system. Besides, for the given n and q there exists a constant $N(n,q) \leq 2n(n-2)(p+3)+p+1$ such that every finitely generated subgroup of $GL_n(\mathbb{C})$ which is the monodromy group of a regular system on $\mathbb{CP}^1$ with given poles that is Fuchsian at all the poles with exception, possibly, of one. The order of this pole is not greater than $N(n,q)$.

The Fuchsian connection on the bundle E over Riemann sphere have the form $\omega = \sum_{j=1}^{q} \frac{A_j}{z-s_j} dz$ which satisfies the condition: $A_1 + \ldots + A_q = 0$ if $\infty \overline{\in} \{s_1, \ldots, s_q\}$. The $n \times n$-matrices called matrices-residua. Not every Fuchsian system gives be irreducible monodromy group. This is connected to the Deligne-Simpson problem: give necessary and sufficient conditions for the choice of the cojugacy classes $[C_j] \subset GL_n(\mathbb{C})$ (resp. $[c_j] \subset gl_n(\mathbb{C})$ so that, there exists irreducible $q+1$-types of matrices $M_j \in [C_j]$ (resp. $A_j \in [c_j]$) such that $M_1 \ldots M_{q+1} = 1$ (resp. $A_1 + \ldots + A_q = 0$). The matrices M_j and are interpreted as monodromy matrices and as matrices-residua of Fuchsian systems on Riemann sphere.

5.4. Scattering Matrices as the Gates for Quantum Computer

Quantum field theory provides the framework for the most fundamental physical theories to be confirmed experimentally, and has enabled predictions of unprecedented precision. However, calculations defining this physical observable often require great computational complexity and can generally be performed only when the interaction strength is weak. A full understanding of the foundations and rich consequences of quantum field theory remains an outstanding challenge. Presently, a quantum algorithm to compute relativistic scattering amplitudes in massive ϕ^4 theory in spacetime of four and fewer dimensions is developing. The algorithm runs in a time that is a polynomial in the number of particles, their energy, and the desired precision, and applies at both weak and strong coupling. Thus, it offers exponential speedup over existing classical methods at high precision or strong coupling.

In this section, we propose the theory of quantum scattering and notion of unitary scattering matrix to formulate quantum input-output relations. This differs from the standard approach to this subject in which the quantum gates for quantum computing are considered to be unitary time evolution operators for a given, fixed time, and that the Hamiltonian can describe the dynamics.

In our understanding, most of the experimental designs for quantum gates are described in scattering experiments. Thus, the representation of quantum gates as scattering matrices (S-matrices) may have the physical realization. In this section, we study one-dimensional scattering problem in the electro-magnetic field. For a varying electro-magnetic field and momentum of the electron we get a 1-parametric family of S-matrices. We prove that by adjusting the electro-magnetic field, we can deliver relevant 2-order and 4-order unitary matrix. It is well known that the collection of such matrices can be used as the set of universal gates for quantum computing [24].

Another kind of scattering problem emerges in two-level quantum systems controlled by electric pulse. Using time-depended Hamiltonian, we also achieve the representation of desired gates as scattering matrices. In both approaches, we use the well known solution of the inverse problem to construct the quantum gates.

The application of quantum scattering process to quantum computing was studied in the works by [88], [94], and [101]. In previous sections, we considered the monodromic approach to the quantum computing. In this section, we construct a set of universal gates by monodromy matrices with a Fuchsian connection of holomorphic vector bundles. Unlike the monodromy matrices, the scattering matrices are encountered in physical experiments. Moreover, there exists a one-to-one correspondence between the monodromy and scattering matrices [10] in one-dimensional case. Here, we describe the potentials and corresponding S-matrices which generate the set of universal gates.

5.4.1. Scattering on the Line and Universal Gates

In this subsection, we discuss the possible applications of $1-D$ electron scattering in an electro-magnetic potential to the problem of design with universal quantum gates for quantum computation. Using the methodology of an inverse scattering problem, we construct genuine potentials to obtain some kind of universal gates as a scattering matrix [58], and [145].

Let us consider the stationary Schrödinger equation on the line

$$\frac{d^2}{dx^2}\psi(x) + (k^2 + Q(x))\psi(x) = 0, \; x \in (-\infty, \infty) \tag{5.4.1}$$

where $Q(x)$ is continuous potential function vanishing at infinity, i.e., $Q(x) \to 0$, as $|x| \to \infty$.

Let $\varphi(x)$ be a solution for (5.4.1), which coincides with e^{-ikx} for $x \to -\infty$. Its complex conjugate function $\overline{\varphi}(x)$ also satisfies the equation (5.4.1) that coincides with e^{ikx} as $x \to -\infty$. Moreover, we denote by $\psi(x)$ and $\overline{\psi}(x)$ the solution of (5.4.1) which coincides with e^{ikx} and e^{-ikx}, respectively as $x \to \infty$. It is clear that there exists 2×2 matrix $M(k)$ such that

$$(\varphi(x), \bar{\varphi}(x)) = (\bar{\psi}(x), \psi(x))M(k). \tag{5.4.2}$$

It is known that for $M(k) \in SU(1,1)$, this matrix is called the monodromy matrix [10]. The matrix $M(k)$ can be represented as

$$M(k) = \begin{pmatrix} a(k) & \bar{b}(k) \\ b(k) & \bar{a}(k) \end{pmatrix}, |a(k)|^2 - |b(k)|^2 = 1.$$

as an element of $SU(1,1)$. Therefore, we have

$$\begin{cases} \varphi = a\bar{\psi} + b\psi \\ \bar{\varphi} = \bar{b}\bar{\psi} + \bar{a}\psi \end{cases} \Rightarrow \begin{cases} \bar{\psi} = \frac{1}{a}\varphi - \frac{b}{a}\psi \\ \bar{\varphi} = \bar{b}(\frac{1}{a}\varphi - \frac{b}{a}\psi) + \bar{a}\psi = \frac{\bar{b}}{a}\varphi + \frac{1}{a}\psi \end{cases}.$$

Thus $(\bar{\psi}, \bar{\varphi}) = (\varphi, \psi)\widetilde{S}(k)$, where $\widetilde{S}(k) = \begin{pmatrix} \frac{1}{a} & \frac{\bar{b}}{a} \\ -\frac{b}{a} & \frac{1}{a} \end{pmatrix}$. Similarly, $(\bar{\varphi}, \bar{\psi}) = (\varphi, \psi)S(k)$, where $S(k) = \begin{pmatrix} \frac{\bar{b}}{a} & \frac{1}{a} \\ \frac{1}{a} & -\frac{b}{a} \end{pmatrix}$. The matrices $\widetilde{S}(k), S(k)$ are called [109], and [37] the scattering matrices in various contexts.

We will use the notations $T(k) = \frac{1}{a(k)}$, $R(k) = \frac{b(k)}{a(k)}$ for the interpretation of transmission and reflection amplitudes, respectively. The particle comes by left-to-right with impulse $k > 0$ goes to the barrier with transmission probability $|T(k)|^2$ and goes back with reflection probability $|R(k)|^2$,, then output will be $e^{ikx} + be^{-ikx}$ on left from barrier and ae^{ikx} on the right from barrier.

We consider here only the $S(k)$ case. Such a matrix is symmetric and unitary simultaneously (but it does not belong to $SU(2)$ in general) and many interesting gates can be represented in such form. Let us consider the mapping

$$\tau : SU(1,1) \to U(2), \begin{pmatrix} a & \bar{b} \\ b & \bar{a} \end{pmatrix} \mapsto \begin{pmatrix} \frac{\bar{b}}{a} & \frac{1}{a} \\ \frac{1}{a} & -\frac{b}{a} \end{pmatrix}.$$

The image of mapping $\tau(SU(1,1))$ as the subset of the matrix in $U(2)$, can be represented as a scattering matrix.

Remark 5.4.1. Let initial states be denoted by $|0\rangle = e^{ikx}$, $|1\rangle = e^{-ikx}$. If the scattering matrix has the form $H = \frac{1}{\sqrt{2}}\begin{pmatrix} 1 & 1 \\ 1 & -1 \end{pmatrix}$ then we obtain the states: $(|0\rangle + |1\rangle)\sqrt{2}$ and $(|0\rangle - |1\rangle)\sqrt{2}$.

To write the quantum, it registers a natural number from 1 to x, then n-qubit is needed. We take n-th tensor product of Hadamard gates and obtain:

$$(H \otimes \ldots \otimes H)|0\rangle \otimes \ldots \otimes |0\rangle = \frac{1}{\sqrt{2^n}} H|0\rangle \otimes H|0\rangle \otimes \ldots \otimes H|0\rangle =$$

$$= \frac{1}{\sqrt{2^n}}(|0\rangle + |1\rangle) \otimes \ldots \otimes (|0\rangle + |1\rangle) = \frac{1}{\sqrt{2^n}} \sum_{x=0}^{2^n-1} |x\rangle.$$

If $f : B^n \to B^n$ is the Boolean function, then it is well known that there exists a unitary operator U_f, which computes all values of f in following way:

$$U_f(\frac{1}{\sqrt{2^n}} \sum_{x=0}^{2^n-1} |x,0\rangle) = \frac{1}{\sqrt{2^n}} \sum_{x=0}^{2^n-1} U_f(|x,0\rangle) = \frac{1}{\sqrt{2^n}} \sum_{x=0}^{2^n-1} |x, f(x)\rangle).$$

On the other hand, through well known results (see [24]) there exists unitary matrices $U_2 \in U(2)$ and $U_4 \in U(4)$ such that each unitary matrix from $U(2^n)$ can be represented as product of matrices $\mathbb{I}\otimes\ldots\otimes U_2\otimes\ldots\otimes\mathbb{I}$ and $\mathbb{I}\otimes\ldots\otimes U_4\otimes\ldots\otimes\mathbb{I}$. Such U_2, U_4 are called the universal gates. Our aim is to represent the universal gates as S-matrices.

Example 1. The so-called Hadamard matrix $H = \frac{1}{\sqrt{2}}\begin{pmatrix} 1 & 1 \\ 1 & -1 \end{pmatrix}$ can be represented as $\tau\begin{pmatrix} \sqrt{2} & 1 \\ 1 & \sqrt{2} \end{pmatrix}$. Similarly, the matrix $\begin{pmatrix} 0 & 1 \\ 1 & 0 \end{pmatrix}$ has a representation $\tau\begin{pmatrix} 1 & 0 \\ 0 & 1 \end{pmatrix}$.

Example 2. The complex numbers $a = \sqrt{n^2+1}, b = n$ defines

$$\tau\begin{pmatrix} \sqrt{n^2+1} & n \\ n & \sqrt{n^2+1} \end{pmatrix} = \frac{1}{\sqrt{n^2+1}}\begin{pmatrix} n & 1 \\ 1 & -n \end{pmatrix}.$$

As $n \to \infty$, this matrix approximates the universal gate $NOT = \begin{pmatrix} 1 & 0 \\ 0 & -1 \end{pmatrix}$. Similarly, we can approximate a more general gate $\begin{pmatrix} 1 & 0 \\ 0 & e^{i\phi} \end{pmatrix}$, if we take $a = \sqrt{n^2+1}, b = -ne^{i\phi/2}$. Indeed, we have

$$T = \frac{1}{a} = \frac{1}{\sqrt{n^2+1}}e^{i\phi/2}, \ \frac{b}{a} = \frac{n}{\sqrt{n^2+1}}e^{i\phi}, \ -\frac{\overline{b}}{a} = \frac{n}{\sqrt{n^2+1}}$$

and

$$\begin{pmatrix} \frac{1}{\sqrt{n^2+1}} & \frac{1}{\sqrt{n^2+1}}e^{i\phi/2} \\ \frac{1}{\sqrt{n^2+1}}e^{i\phi/2} & \frac{n}{\sqrt{n^2+1}}e^{i\phi} \end{pmatrix} \to \begin{pmatrix} 1 & 0 \\ 0 & e^{i\phi} \end{pmatrix}, \ \textit{as } n \to \infty.$$

□

It is known that the Hadamard matrix H and controlled phase gate $\begin{pmatrix} 1 & 0 & 0 & 0 \\ 0 & 1 & 0 & 0 \\ 0 & 0 & 1 & 0 \\ 0 & 0 & 0 & e^{i\phi} \end{pmatrix}$ can be used as the universal set of gates, as well as H and the controlled NOT $\begin{pmatrix} 1 & 0 & 0 & 0 \\ 0 & 1 & 0 & 0 \\ 0 & 0 & 0 & 1 \\ 0 & 0 & 1 & 0 \end{pmatrix}$.

Remark 5.4.2. There exists the mapping $SU(1,1) \to SU(2)$

$$\begin{pmatrix} a & \overline{b} \\ \overline{b} & a \end{pmatrix} \mapsto \begin{pmatrix} \frac{1}{a} & \frac{b}{a} \\ -\frac{\overline{b}}{a} & \frac{1}{\overline{a}} \end{pmatrix},$$

which is used in some subjects for differential geometry [111] but neither S nor $\widetilde{S}$ could not be generated by this mapping.

Now, we show why the universal gates can be constructed via potential $Q(x)$ and particle's momentum k.

Scattering data of equation (5.4.1) is defined as a set $\mathcal{S}^+ = \{R(k) = \frac{b(k)}{a(k)}, (k_j, b_j)_{j=1}^N\}$, where $R(k)$ is meromorphic function in upper half plane, $k_j = i\eta_j, j = 1, ..., N$ with a finite set of poles for $R(k)$ and $b_j = \frac{\varphi(x,i\eta_j)}{\psi(x,i\eta_j)}$. From this scattering data $\mathcal{S}^+$, we can recover the scattering matrix; i.e., its unknown element $T(k) = \frac{1}{a(k)}$ in following way

$$-\ln T(k) = \ln a(k) = \sum_{j=1}^{N} \ln \frac{k - i\eta_j}{k + i\eta_j} - \frac{1}{2\pi i} \int_{\mathbb{R}} \frac{\ln(1 - |R(\zeta)|^2)}{\zeta - k - i0} d\zeta$$

or

$$T(k) = \prod_{j=1}^{N} \frac{k + i\eta_j}{k - i\eta_j} e^{\frac{1}{2\pi i} \int_{\mathbb{R}} \frac{\ln(1-|R(\zeta)|^2)}{\zeta - k - i0} d\zeta}$$

Since by the Plemelj formula $\frac{1}{\zeta - k - i0} = \text{v.p.} \frac{1}{\zeta - k} + i\pi\delta(\zeta - k)$ ([37]. 50 p.), then

$$-\ln T(k) = -\frac{1}{2} \ln(1 - |R(k)|^2) + \sum_{j=1}^{N} \ln \frac{k - i\eta_j}{k + i\eta_j} - \frac{1}{2\pi i} \text{v.p.} \int_{\mathbb{R}} \frac{\ln(1 - |R(\zeta)|^2)}{\zeta - k} d\zeta$$

and finally, we have

$$T(k) = \sqrt{1 - |R(k)|^2} \prod_{j=1}^{N} \frac{k + i\eta_j}{k - i\eta_j} \exp\left(\frac{1}{2\pi i} \text{v.p.} \int_{\mathbb{R}} \frac{\ln(1 - |R(\zeta)|^2)}{\zeta - k} d\zeta\right)$$

Evidently $|T(k)| = \sqrt{1 - R(k)^2}$, since $|k + i\eta| = |k - i\eta|$ and $\frac{1}{2\pi i} \text{v.p.} \int_{\mathbb{R}} \frac{\ln(1-|R(\zeta)|^2)}{\zeta - k} d\zeta$ belongs to $[0, 1]$. Thus, we obtain that the phase of $T(k)$ was defined by the Hilbert transform of function $\ln(1 - R(\zeta)|^2)$. Suppose now the sets of numbers $\{k_1, ..., k_N\}, \{\alpha_1, ..., \alpha_N\}$ are $\{\beta_1, ..., \beta_N\}$ are given, and for simplicity assume $k_j > 0$. One can find continuous function $f(k), k > 0$ such that $f(k_j) = \alpha_j$ and $\gamma_j \doteq \frac{1}{\pi} \int_0^\infty \frac{f(\zeta)}{k_j - \zeta} d\zeta > \beta_j$. Also, we can choose a continuous $g(k), k \leq 0$ such that $\frac{1}{\pi} \int_{-\infty}^0 \frac{g(\zeta)}{k_j - \zeta} d\zeta = \beta_j - \gamma_j$. Thus, the function

$$F(k) = \begin{cases} f(k), k > 0 \\ g(k), k \leq 0 \end{cases}$$

acquires the property

$$F(k_j) = \alpha_j, \quad \frac{1}{\pi} \int_{\mathbb{R}} \frac{F(\zeta)}{k_j - \zeta} d\zeta = \beta_j.$$

Using these relations for a given sequence $k_j, t_j, r_j, j = 1, .., N$, such that $|t_j|^2 = |r_j|^2$ we can construct a pair function for $(T(k), R(k))$; such that $T(k_j) = t_j, R(k_j) = r_j$, since for $\alpha_j = t_j, \beta_j = \ln(1 - |r_j|^2)$, we find $F(k)$ and $T(k) = \frac{1}{\pi} \text{v.p.} \int_{\mathbb{R}} \frac{F(\zeta)}{k_j - \zeta} d\zeta = \beta_j$ and further define $R(k)$ by the equation $|R(k)|^2 = 1 - e^{-F(k)}$.

The potential $Q(x)$ can be recovered in the following way (see [109]). If

$$C(z) = -i \sum_{j=1}^{N} \gamma_j e^{-\eta_j} + \frac{1}{2\pi} \int_{\mathbb{R}} R(\xi) e^{i\xi x} d\xi,$$

where $\gamma_j = \frac{b_j}{a'(i\eta_j)}$, then Q is defined by the solution of Gelfand-Levitan integral equation

$$K(x,y) + C(x+y) + \int_x^\infty K(x,s) C(s+y) ds = 0, \ \ y > x$$

as $Q(x) = 2\frac{d}{dx}K(x,x)$.

Therefore, we have proven the following proposition.

Proposition 5.4.1. *Let $k_1, k_2, \ldots, k_m$ be a sequence of pairwise distinct numbers and $S_1, S_2, \ldots, S_m$ are matrices from the set $\tau(SU(1,1))$. Then there exists potential $Q(x)$ such that the corresponding scattering matrix satisfies the conditions $S(k_1) = S_1, S(k_2) = S_2, \ldots, S(k_m) = S_m$.*

Now, consider the particle with spin moving on the line in presence of electromagnetic field $(E(x), B(x)), x \in \mathbb{R}$, where electric field $E(x)$ is directed along the line and magnetic field $B(x)$ is directed across the line.

Let $A(x)$ be the potential for magnetic field of $B(x)$, and $Q(x)$ is potential for the electric fields, i.e., $A'(x) = B(x)$, $Q'(x) = -E(x)$.

The Schrödinger operator (as Hamiltonian) of such system has the form

$$\mathcal{H} = -(\frac{d^2}{dx^2} - A^2(x) + Q(x))\mathbb{I} - B(x)\sigma_3, \tag{5.4.3}$$

where $\mathbb{I} = \begin{pmatrix} 1 & 0 \\ 0 & 1 \end{pmatrix}$, and $\sigma_3 = \begin{pmatrix} 1 & 0 \\ 0 & -1 \end{pmatrix}$ is third Pauli matrix.

Here, for $\psi(x) = \begin{pmatrix} \psi_1(x) \\ \psi_2(x) \end{pmatrix} \in L^2(\mathbb{R}) \otimes \mathbb{C}$ we have

$$\mathcal{H}\psi = -\begin{pmatrix} \psi_1''(x) + (Q(x) + A'(x) - A(x)^2)\psi_1(x) \\ \psi_2''(x) + (Q(x) - A'(x) - A(x)^2)\psi_2(x) \end{pmatrix}.$$

Maybe it is more plausible to consider the behavior of the particle on the plane. In this case, the potential of magnetic field is of the form $a(x,y) = (a_1(x,y), a_2(x,y))$ and in the gauge $a_1 = 0$ we can take $a_2'(x) = B(x)$. Then, Hamiltonian can be written as

$$\mathcal{H} = (-\frac{\partial^2}{\partial x^2} - \frac{\partial^2}{\partial y^2} + ia_2(x)\frac{\partial}{\partial y} - Q(x) + a_2^2(x))\mathbb{I} - B(x)\sigma_3$$

If we take $A(x) = a_2(x)$ and consider the wave functions independent on y-variable, then we get the Hamiltonian (5.4.3).

If, in advance, we take two arbitrary potential $U(x)$ and $V(x)$, then $A(x)$ and $Q(x)$ are defined by relation

$$A'(x) = \frac{1}{2}(U(x) - V(x)),\ Q(x) = A^2(x) + \frac{1}{2}(U(x) + V(x))$$

satisfy the equations

$$Q(x) + A'(x) - A^2(x) = U(x),\ Q(x) - A'(x) - A^2(x) = V(x).$$

Hence, we have

$$-\mathcal{H}\psi(x) = \frac{d^2}{dx^2}\psi(x) + \begin{pmatrix} U(x) & 0 \\ 0 & V(x) \end{pmatrix}\psi(x) \equiv -\begin{pmatrix} H_U\psi_1(x) \\ H_V\psi_2(x) \end{pmatrix}.$$

By definition, the scattering operator follows that the scattering operator for Hamiltonian $\mathcal{H} = \begin{pmatrix} H_U & 0 \\ 0 & H_V \end{pmatrix}$ has the form $S = \begin{pmatrix} S_U & 0 \\ 0 & S_V \end{pmatrix}$, where S_U and S_V are scattering operator for H_U and H_V, respectively.

Therefore, k^2-energy level in the scattering matrix of this system (5.4.3), which is 4×4-matrix, can be represented as

$$S_H(k) = \begin{pmatrix} S_U(k) & 0 \\ 0 & S_V(k) \end{pmatrix}, \tag{5.4.4}$$

where $S_U(k)$ and $S_V(k)$ are scattering matrices corresponding to the potentials of U and V, respectively. Evidently, they are 2×2-matrices from $\tau(SU(1,1))$.

Therefore, this proposition is proven.

Proposition 5.4.2. *Let $U(x)$ and $V(x)$ be two arbitrary potentials. Then there exists electromagnetic potential $(A(x), Q(x))$ such that the Hamiltonian of (5.4.3) takes the form $\mathcal{H} = \begin{pmatrix} H_U & 0 \\ 0 & H_V \end{pmatrix}$, where $H_U = -\frac{d^2}{dx^2} - U(x)$ and $H_V = -\frac{d^2}{dx^2} - V(x)$. Moreover, the scattering matrix for the stationary Schrödinger equation $\mathcal{H}\psi = k^2\psi$ can be represented as matrix (5.4.4).*

As a corollary, we get that 4×4-matrix from the example 2

$$\begin{pmatrix} 1 & 0 & 0 & 0 \\ 0 & 1 & 0 & 0 \\ 0 & 0 & 1 & 0 \\ 0 & 0 & 0 & e^{i\phi} \end{pmatrix} \text{ and } \begin{pmatrix} 1 & 0 & 0 & 0 \\ 0 & 1 & 0 & 0 \\ 0 & 0 & 0 & 1 \\ 0 & 0 & 1 & 0 \end{pmatrix}$$

can be represented as the scattering matrix of the system governed by (5.4.3). Moreover, we can take potential $U(x) = 0$ in both cases and $V(x)$ as indicated in examples 1 and 2.

5.4.2. Two-Level System in Electric Field

In this subsection, we consider the electric field that interacts with two-level (or four-level) quantum systems. The transition from initial state to final state are used as gates. The well known inverse scattering problem procedure allows to obtain potentials that maintain some universal gates. Finally, we consider some potentials which allow to compute S-matrix as monodromy of corresponding Fushian system.

Let us consider behavior of two-level system in electric field. It can be described by the equation

$$i\frac{d}{dt}\begin{pmatrix} A(t) \\ B(t) \end{pmatrix} = \begin{pmatrix} \zeta & E(t) \\ \overline{E(t)} & -\zeta \end{pmatrix}\begin{pmatrix} A(t) \\ B(t) \end{pmatrix} \tag{5.4.5}$$

where A and B denote the amplitudes of the first and second levels, respectively; $E(t)$ is complex envelope of electric field and ζ is difference between support frequency and proper frequency. The free Hamiltonian here is $H_0 = \zeta\sigma_3$ and free dynamics is given by unitary matrix $e^{-i\zeta t\sigma_3} = \begin{pmatrix} e^{-i\zeta t} & 0 \\ 0 & e^{i\zeta t} \end{pmatrix}$.

Let $\Psi(t)$ and $\Phi(t)$ be matrix solution of (5.4.5) with condition

$$\Psi(t) \sim \begin{pmatrix} e^{-i\zeta t} & 0 \\ 0 & e^{i\zeta t} \end{pmatrix}, t \to \infty \text{ and } \Phi(t) \sim \begin{pmatrix} e^{-i\zeta t} & 0 \\ 0 & e^{i\zeta t} \end{pmatrix}, t \to -\infty \tag{5.4.6}$$

respectively. They are unitary unimodular matrices, since $\begin{pmatrix} \zeta & E(t) \\ \overline{E(t)} & -\zeta \end{pmatrix}$ is Hermitian traceless matrix. Thus, this so-called monodromy matrix $M(\zeta) = \Psi^{-1}(t)\Phi(t)$ belongs to $SU(2)$ and can be represented as $M(\zeta) = \begin{pmatrix} a(\zeta) & -\overline{b}(\zeta) \\ b(\zeta) & \overline{a}(\zeta) \end{pmatrix}$ with $|a(\zeta)|^2 + |b(\zeta)|^2 = 1$.

Let $U(t,\tau)$ be fundamental matrix of (5.4.5), i.e.,

$$U(\tau,\tau) = I,\ i\frac{\partial}{\partial t}U(t,\tau) = \begin{pmatrix} \zeta & E(t) \\ \overline{E(t)} & -\zeta \end{pmatrix} U(t,\tau),\ t > \tau.$$

Evidently there exists the matrices W_φ and W_ψ such that $\Psi(t) = U(t,0)W_\Psi$, $\Phi(t) = U(t,0)W_\Phi$. The relation (5.4.6) can be rewritten as

$$\lim_{t\to\infty} \Psi(t)^\star e^{-\zeta t\sigma_3} = \mathbb{I} \text{ and } \lim_{t\to-\infty} \Phi(t)^\star e^{-\zeta t\sigma_3} = \mathbb{I}.$$

On the other hand, by the definition of wave operators, we have [124]

$$W_\pm = \lim_{t\to\mp\infty} U(t,0)^* e^{-\zeta t\sigma_3}.$$

Hence,

$$\begin{aligned} W_+ = \lim_{t\to-\infty} U(t,0)^* e^{-\zeta t\sigma_3} &= \lim_{t\to-\infty} W_\Phi (U(t,0)W_\Phi)^* e^{-\zeta t\sigma_3} \\ &= \lim_{t\to-\infty} W_\Phi \Phi(t)^* e^{-\zeta t\sigma_3} = W_\Phi. \end{aligned}$$

Similarly, we get $W_- = W_\Psi$. Thus, we have

$$\begin{aligned} S(\zeta) &= W_-^{-1}W_+ = W_\Psi^{-1}W_\Phi \\ &= \Psi(t)^{-1}U(t,0)U(t,0)^{-1}\Phi(t) = \Psi(t)^{-1}\Phi(t) = M(\zeta) \end{aligned}$$

i.e.,

$$S(\zeta) = M(\zeta) = \begin{pmatrix} a(\zeta) & -\overline{b}(\zeta) \\ b(\zeta) & \overline{a}(\zeta) \end{pmatrix},\ |a(\zeta)|^2 + |b(\zeta)|^2 = 1.$$

The function $a(\zeta)$ is a boundary value of holomorphic function on the upper half plane with zeros ζ_j, $Im\zeta_j > 0$, $j = 1,\ldots,N$. For $\zeta = \zeta_j$, there exists the solution $\chi(t)$ of (5.4.5) with condition $\chi(t) \sim \begin{pmatrix} 0 \\ e^{-k_j t} \end{pmatrix}$, $t \to -\infty$, $\chi(t) \sim \begin{pmatrix} e^{ik_j t} \\ 0 \end{pmatrix} d_j$, $t \to \infty$. The scattering dates for the system (5.4.5) is the set $\{b(\zeta), \zeta_j, d_j, j = 1,\ldots,N\}$. The function $a(\zeta)$ is defined as

$$a(\zeta) = e^{-\frac{1}{2\pi i}\int_{\mathbb{R}} \frac{\ln(1-|b(\zeta')|^2)}{\zeta-\zeta'-i0} d\zeta'} \prod_{l=1}^{N} \frac{\zeta - \zeta_l}{\zeta - \overline{\zeta}_l}.$$

The field $E(t)$ can be restored by scattering data (see [38]); $E(t) = -2iK(t,t)$, where $K(t) = \begin{pmatrix} K_{11}(t) & K_{12}(t) \\ K_{21}(t) & K_{22}(t) \end{pmatrix}$ is a solution of the Gelfand-Levitan integral matrix equation

$$K(t,s) + \widehat{F}(t,s) + \int_t^{\infty} K(t,s)\widehat{F}(z+s)dz, \ t < s$$

with

$$\widehat{F}(t) = \begin{pmatrix} 0 & F(t) \\ \overline{F}(t) & 0 \end{pmatrix}, F(t) = \frac{1}{2\pi}\int_{\mathbb{R}} r(k)e^{ikt}dk + \sum_{j=1}^{N} m_j e^{ik_j t},$$

$$r(\zeta) = \frac{b(\zeta)}{a(\zeta)}, \ m_j = \frac{d_j}{a'(\zeta_j)}.$$

As in the previous case, we define the function $b(\zeta)$ to obtain a desired scattering matrix $S(\zeta)$. Hence, we have the next proposition.

Proposition 5.4.3. *For the finite set of numbers $\{\zeta_1, \zeta_2, ..., \zeta_m\}$ and matrices $S_1, S_2, ..., S_m$ from the set $SU(2)$ there exists potential $Q(x)$ such that the corresponding scattering matrix satisfies the conditions $S(\zeta_1) = S_1, S(\zeta_2) = S_2, ..., S(\zeta_m) = S_m$.*

In particular cases, we can construct the matrices of example 1 up to constant: $iH = \frac{i}{\sqrt{2}}\begin{pmatrix} 1 & 1 \\ 1 & -1 \end{pmatrix}$, and $e^{-i\varphi/2}\begin{pmatrix} 1 & 0 \\ 0 & e^{i\varphi} \end{pmatrix}$.

Now, we suppose that electric field acts on the pair of 2-level particle performing dipol-dipol interaction between dipolar particle states. We assume that dipols are oriented along electric fields and therefore the Hamiltonian of two-particle system is presented as $\mathcal{H}_{AB}(E(t), \Phi_{AB})$ [94], where

$$\mathcal{H}_{AB}(x,y) = (\mathcal{H}_A + x\hat{d}_A) \otimes I + I \otimes (\mathcal{H}_B + x\hat{d}_B) + y\hat{d}_A \otimes \hat{d}_B$$

and

$$\mathcal{H}_C = \begin{pmatrix} W_+^C & 0 \\ 0 & W_-^C \end{pmatrix} \ \hat{d}_C = \begin{pmatrix} 0 & d_C \\ \bar{d}_C & 0 \end{pmatrix} \ C = A, B$$

In the case of $\Phi_{AB} = y$, the Hamiltonian can be rewritten as 4×4 matrix

$$\mathcal{H}_{AB}(E(t), y) = \begin{pmatrix} W_+^A + W_+^B & d_A E(t) & d_A E(t) & y d_A d_B \\ \bar{d}_A E(t) & W_+^A + W_-^B & y d_A \bar{d}_B & d_B E(t) \\ \bar{d}_A E(t) & y \bar{d}_A d_B & W_-^A + W_+^B & d_B E(t) \\ y \bar{d}_A \bar{d}_B & \bar{d}_B E(t) & \bar{d}_B E(t) & W_-^A + W_-^B \end{pmatrix}.$$

Proposition 5.4.4. *Let $d_A, d_B, W_+^A, W_+^B, W_-^A, W_-^B$ be arbitrary complex numbers. Then there exists continuous electric pulse $E(t)$ and interacting potential $\Phi_{AB}(t)$ vanished at infinity and such that the corresponding scattering matrix does not belong in $SU(2) \otimes SU(2)$; i.e., it is an entangled operator.*

Proof. It is easy to see that the Hamiltonian $\hat{d}_A \otimes \hat{d}_B$ is not represented as a matrix $M \otimes I + I \otimes N$ for some $M, N \in su(2)$. Hence, the matrix

$$F(T) = \exp\{iT\mathcal{H}_{AB}(0,0)\}\exp\{-2iT\mathcal{H}_{AB}(x,y)\}\exp\{iT\mathcal{H}_{AB}(0,0)\} \quad (5.4.7)$$

for some T and $y \neq 0$ cannot be represented as an element of $SU(2) \otimes SU(2)$, as

$$F'(0) = i\mathcal{H}_{AB}(0,0) - 2i\mathcal{H}_{AB}(x,y) + i\mathcal{H}_{AB}(0,0)$$

also does not represent a matrix $M \otimes I + I \otimes N$ for some $M, N \in su(2)$. It is easy to see that the S-matrix in this system

$$i\frac{d}{dt}Y(t) = \mathcal{H}_{AB}(E(t), \Phi_{AB}(t))Y(t)$$

for the electric pulse $E(t) = x1_{[-T,T]}(t)$ and $\Phi_{AB}(t) = y1_{[-T,T]}(t)$ coincides to matrix of (5.4.7).

Since $SU(2) \otimes SU(2)$ is a closed subset of $U(4)$, then for sufficiently small nonzero x the matrix (5.4.7) is also considered an entangled one; i.e., where it does not belong to $SU(2) \otimes SU(2)$ then the each continuous pulse $E_\varepsilon(t)$ $\Phi_{AB,\varepsilon}(t)$ being close enough to $E(t), \Phi_{AB}(t)$ also generated the entangled operator. □

This proposition and results of [24] allow us to tune the electric pulse to construct universal gates' system.

5.4.3. The Fuchsian Systems Monodromy and S-Matrix

Let $\{s_1, ..., s_n\}$ be the complex numbers on $\mathbb{C}$ and $s_i \neq s_j$, $i \neq j$. Consider the Fuchsian system

$$df = \omega f, \tag{5.4.8}$$

where ω is meromorphic 1-form on $\mathbb{C}$

$$\omega = \left(\frac{A_1}{z - s_1} + \frac{A_2}{z - s_2} + ... + \frac{A_n}{z - s_n}\right) dz,$$

and A_j, $j = 1, ..., n$ are constant 2×2-matrices. Let (f_1, f_2) the solutions for (5.4.8) in the neighborhood of $z_0 \in \mathbb{C} \setminus \{s_1, ..., s_n\}$ and $\gamma_1, ..., \gamma_n$ are the generators of $\pi_1(\mathbb{C} \setminus \{s_1, ..., s_n\}, z_0)$. After the extension (f_1, f_2) along the γ_j, we obtain other solutions (f_1^j, f_2^j) for (5.4.8) and $(f_1, f_2) = M_j(f_1^j, f_2^j)$, where M_j, $j = 1, ..., n$ are monodromy matrices. Conversely, for fixed data $(s_1, ..., s_n, M_1, ..., M_n)$ there exists a Fuchsian system (5.4.8) with given singular points s_j, $j = 1, ..., n$ and monodromy matrices M_j, $j = 1, ..., n$. We take monodromy matrices from $SU(2)$ and consider them as scattering matrices. The singular points $s_1, ..., s_n$ consider as the source of energy and ω is gauge potential induces from the given date $(s_1, ..., s_n, M_1, ..., M_n)$.

Now, we intend to show how a S-matrix system may be expressed as a monodromy matrix for a corresponding Fushian system. First, we recall that two differential equations $i\frac{d}{dt}\psi_1(t) = \mathcal{H}_1(t)\psi_1(t)$ and $i\frac{d}{dt}\psi_2(t) = \mathcal{H}_2(t)\psi_2(t)$ are called gauge equivalent if there exists differentiable 1-parameter family of invertible matrices such that $\mathcal{H}_2(t) = iU'(t)U(t)^{-1} + U(t)\mathcal{H}_1(t)U(t)^{-1}$. It is easy to see that if $\psi_1(t)$ is a solution for the first equation then $\psi_2(t) = U(t)\psi_1(t)$ is the solution for the second equation. By this definition, the Schrödinger equation of type $i\frac{d}{dt}\psi_0(t) = (\mathcal{H}_0 + V(t))\psi_0(t)$ is the gauge equivalent to the so-called interacting representation $i\frac{d}{dt}\psi_1(t) = V_1(t)\psi_1(t)$ via 1-parameter family $U(t) = e^{it\mathcal{H}_0}$, where $V_1(t) = e^{it\mathcal{H}_0}V(t)e^{-it\mathcal{H}_0}$.

Now, we consider the field of type $E(t)e^{-i\omega t}$. Then system (5.4.5) recasts

$$i\frac{d}{dt}\begin{pmatrix} A(t) \\ B(t) \end{pmatrix} = \zeta\sigma_3 + \begin{pmatrix} 0 & E(t)e^{-i\omega t} \\ E(t)e^{i\omega t} & 0 \end{pmatrix}\begin{pmatrix} A(t) \\ B(t) \end{pmatrix}. \tag{5.4.9}$$

Since

$$\begin{pmatrix} e^{i\zeta t} & 0 \\ 0 & e^{-i\zeta t} \end{pmatrix}\begin{pmatrix} 0 & e^{-i\omega t} \\ e^{i\omega t} & 0 \end{pmatrix}\begin{pmatrix} e^{-i\zeta t} & 0 \\ 0 & e^{i\zeta t} \end{pmatrix} = \begin{pmatrix} 0 & e^{-i(\omega-2\zeta)t} \\ e^{i(\omega t-2\zeta)} & 0 \end{pmatrix}.$$

then the interacting representation for the system (5.4.9) takes the form

$$i\frac{d}{dt}\Phi(t) = \begin{pmatrix} 0 & E(t)e^{-i(\omega-2\zeta)t} \\ E(t)e^{i(\omega-2\zeta)t} & 0 \end{pmatrix}\Phi(t) \tag{5.4.10}$$

Let $\Phi(t,\tau)$ denotes the fundamental matrix of (5.4.10). Then the S-matrix for system (5.4.9) can be represented as

$$S(\zeta) = \lim_{\substack{t\to+\infty, \\ \tau\to-\infty}} \Phi(t,\tau). \tag{5.4.11}$$

In this particular case, the analytical function of $E(t)$ and the S-matrix coincides with the monodromy matrix for the corresponding Fushian system.

Example 3. Consider the case $\omega = 2\zeta$, $E(t,a,b) = \frac{2ab}{t^2+a^2} \equiv \frac{ib}{t+ia} - \frac{ib}{t-ia}$, $a,b \in \mathbb{R}$. By the equality $\sigma_1 = H\sigma_3 H^{-1}$, system (5.4.10) is a gauge equivalent to

$$i\frac{d}{dt}\Psi(t) = \frac{2ab}{t^2+a^2}\sigma_3\Psi(t).$$

The solution of the last equation is

$$\Psi(t) = \begin{pmatrix} \exp(-ib\int\frac{2adt}{t^2+a^2}) & 0 \\ 0 & \exp(ib\int\frac{2adt}{t^2+a^2}) \end{pmatrix}$$

and thanks to the formula (5.4.11) S-matrix is $e^{-2\pi b\int\frac{2adt}{t^2+a^2}i\sigma_3} = e^{-bi\sigma_3}$. Since $\Phi(z) = H\Psi(z)$ then

$$S(\zeta) = M(\gamma) = H\begin{pmatrix} e^{2\pi bit} & 0 \\ 0 & e^{-2\pi bit} \end{pmatrix}H^{-1} = e^{-2i\pi b\sigma_1}.$$

Now, extend the range of variables for t to the complex plane $\mathbb{C}$ and compute monodromy matrix for the obtained Fushian system. The system (5.4.10) looks as

$$i\frac{d}{dw}\Phi(w) = \frac{2ab\sigma_1}{w^2+a^2}\Phi(w),$$

where $\sigma_1 = \begin{pmatrix} 0 & 1 \\ 1 & 0 \end{pmatrix}$. By the equation $\sigma_1 = H\sigma_3 H^{-1}$, this system is a gauge equivalent to

$$i\frac{d}{dw}\Psi(w) = \frac{2ab\sigma_3}{w^2+a^2}\Psi(w).$$

The line $Imw = 0$ on the Riemann sphere $P_1(\mathbb{C}) \supset \mathbb{C}$ becomes a loop. In order to express this loop explicitly, we change the coordinate $z = \frac{1}{w-i}$, which is holomorphic map $P_1(\mathbb{C}) \to P_1(\mathbb{C})$ with inverse $w = \frac{1}{z} + i$. This 1-form mapping for the $\frac{2ab}{w^2+a^2}dw$ carries into 1-form $-\frac{2ab}{a^2z^2+(iz+1)^2}dz$. In view

$$-\frac{2ab}{a^2z^2+(iz+1)^2} = \frac{ib}{z-\frac{i}{a+1}} - \frac{ib}{z+\frac{i}{a-1}}$$

we get the Fushian system

$$\frac{d}{dz}\Psi(z) = \left(\frac{b\sigma_3}{z-\frac{i}{a+1}} - \frac{b\sigma_3}{z+\frac{i}{a-1}}\right)\Psi(z), \tag{5.4.12}$$

which has the solution

$$\Psi(z) = \begin{pmatrix} \exp\left(b\int\frac{dz}{z-\frac{i}{a+1}} - b\int\frac{dz}{z+\frac{i}{a-1}}\right) & 0 \\ 0 & \exp\left(-b\int\frac{dz}{z-\frac{i}{a+1}} + b\int\frac{dz}{z+\frac{i}{a-1}}\right) \end{pmatrix}.$$

$$Bynewcoordinates, theline$$

w=t*becomesthecurve*z=1$\frac{}{t-i\equiv\frac{t}{1+t^2}+i\frac{1}{1+t^2}}$, which is the circle $\gamma: x^2+(y-\frac{1}{2})^2 = \frac{1}{4}$. The poles of system (5.4.12) $\frac{i}{a+1}$ and $\frac{i}{1-a}$ lie inside and outside of the circle, respectively. Hence, we can consider the circle γ as a loop around pole $\frac{i}{a+1}$ and monodromy matrix of (5.4.12) coincides to S-matrix given by (5.4.11) Since the solution of (5.4.12) $\Phi(z) = H\Psi(z) = H^{-1}\Psi(z)$ then

$$S(\zeta) = M(\gamma) = H\begin{pmatrix} e^{2\pi bi} & 0 \\ 0 & e^{-2\pi bi} \end{pmatrix}H^{-1} = e^{-2\pi bi\sigma_1}.$$

□

If $E(t) = \sum\frac{2a_kb_k}{t^2+a_k^2}$, then $S(\zeta) = \prod M(\gamma_k)$, where γ_k is a loop around $\frac{1}{1+a_k}$ and $M(\gamma_k)$ corresponding monodromy matrix.

Example 4. If we consider the field $E(t) = \frac{2t}{t^2+a^2}$, then the integral $\int_{-\infty}^{\infty}\frac{2tdt}{t^2+a^2}$ exists only in the sense of a principle value. Extending the field in the complex plane, after the changed variable, we obtain

$$\frac{d}{dz}\Phi(z) = \frac{1}{a}\left(\frac{1}{z}+i\right)\left(\frac{\sigma_1}{z-\frac{i}{a+1}} - \frac{\sigma_1}{z+\frac{i}{a-1}}\right)\Phi(z). \tag{5.4.13}$$

Hence, the complementary pole $z = 0$ appeared, which lies in γ. We pass to the gauge equivalent differential equation of type

$$\frac{d}{dz}\Psi(z) = \left(\frac{b_1}{z} + \frac{b_2}{z-p} + \frac{b_3}{z-q}\right)\sigma_3\Psi(z), \; p = \frac{i}{1+a}, q = \frac{i}{1-a}$$

for some $b_1, b_2, b_3 \in \mathbb{C}$. The monodromy matrix of such a system around γ is matrix

$$\begin{pmatrix} e^{\int_\gamma\left(\frac{b_1}{z}+\frac{b_2}{z-p}+\frac{b_3}{z-q}\right)dz} & 0 \\ 0 & e^{-\int_\gamma\left(\frac{b_1}{z}+\frac{b_2}{z-p}+\frac{b_3}{z-q}\right)dz} \end{pmatrix}$$

which contains Caushy type integral v.p. $\int_\gamma \frac{b_1}{z}dz$. Since 0 lies on γ and p does inside γ then this matrix is equal to

$$\begin{pmatrix} \exp(i\pi b_1 + 2\pi i b_2) & 0 \\ 0 & \exp(-i\pi b_1 - 2\pi i b_2) \end{pmatrix} = \exp(i\pi(b_1+2b_2)\sigma_3).$$

Hence, the system (5.4.13) has the monodromy matrix $\exp(i\pi(b_1+2b_2)\sigma_1)$ similarly to previous example.

Remark 5.4.3. In the case of $\omega \neq 2\zeta$, a more complicated irregular singularities emerged.

5.5. Quantum Monodromy

As pointed out earlier, the rôle of local and entangling operators in monodromic quantum computing is played by monodromy matrices in connection with logarithmic singularities acting on the fiber of a holomorphic vector bundle on the space of qubits.

In the previous section, we considered Fuchsian systems for which the divisor for the fundamental group complement the singularities in the generalized braid group. Next, we will give Hamiltonians from known quantum models.

We consider Fuchsian differential equations on Riemann sphere as completely integrable systems. These systems are also known in mathematical physics as the *Calogero-Moser-Sutherland* systems [115]. For example, in these particular cases, the trigonometric Calogero-Moser-Sutherland system describes the dynamics of a quantum mechanical system as n particles moving on the real line under the influence of a pair potential that is proportional to the inverse square by the hyperbolic sine *sh* distance of the particles. It is also known, that the trigonometric Calogero-Moser-Sutherland system is obtained as a limit for the elliptic Calogero-Moser-Sutherland system.

In [114], the authors classified integrable models of quantum mechanics which are invariant under the action of a Weyl group with certain assumptions. For the case of B_N ($N \geq 3$), the generic model coincides with the BC_N *Inozemtsev* model. In [116], the eigenvalue problem for the Hamiltonian of the BC_1 (one particle) Inozemtsev model is transformed to the *Heun* equation. The Heun equation is a second-order Fuchsian differential equation with four regular singular points. Note, that the Lame equation is a special case of the Heun equation.

In [99], a certain equation is given on coefficients for the Heun equation which transforms the latter into a hypergeometric equation. The author calls this transformation a Heun with a hypergeometric reduction. This property reducibility of the Heun equation to the hypergeometric one we will call *PHH*.

Consider the class of linear second-order differential equations on the Riemann sphere $\mathbb{CP}^1$ which are Fuchsian, i.e., have only regular singular points. Any such equation with exactly three singular points can be transformed to the hypergeometric equation by appropriate changes of the independent and dependent variables. The Gauss hypergeometric equation is

$$\frac{d^2y}{dz^2} + \left(\frac{c}{z} + \frac{a+b-c+1}{z-1}\right)\frac{dy}{dz} + \frac{ab}{z(z-1)}y = 0, \tag{5.5.1}$$

where $a, b, c \in \mathbb{C}$ are parameters. This equation and its solution space are specified by the Riemann P-symbol

$$P\left\{\begin{matrix} 0 & 1 & \infty & \\ 0 & 0 & a & ;z \\ 1-c & c-a-b & b & \end{matrix}\right\},$$

where each column, except the last, refers to a regular singular point. The first entry is its location, and the final two are the characteristic exponents of the solutions there. The exponents at each singular point are obtained by solving an indicial equation. In general, each finite singular point z_0 has ζ as an exponent if and only if the equation has a local (Frobenius) solution for the form $(z-z_0)^\zeta h(z)$ in a neighborhood of $z = z_0$, where h is analytic and nonzero at $z = z_0$. If the exponents at $z = z_0$ differ by an integer, this statement must be modified: the solution corresponding to the smaller exponent may have a logarithmic singularity at $z = z_0$. The definition extends in a straightforward way to $z_0 = \infty$, and also to ordinary points, each of which has exponents $0, 1$.

There are $2 \times 3 = 6$ local solutions of ($\mathfrak{h}$) in all: two per singular point. If c is not a nonpositive integer, the solution at $z = 0$ belonging to the exponent zero will be analytic. When normalized to unity at $z = 0$, it will be the Gauss hypergeometric function ${}_2F_1(a, b; c; z)$. This is the sum of a hypergeometric series, which converges in a neighborhood of $z = 0$. In general, ${}_2F_1(a, b; c; z)$ is not defined when c is a nonpositive integer.

Similarly, second-order Fuchsian differential equations on the Riemann sphere $\mathbb{CP}^1$ with exactly four singular points can be transformed to the Heun equation.

The Heun equation is usually written in the form

$$\frac{d^2u}{dt^2} + \left(\frac{\gamma}{t} + \frac{\delta}{t-1} + \frac{\varepsilon}{t-d}\right)\frac{du}{dt} + \frac{\alpha\beta t - q}{t(t-1)(t-d)}u = 0. \tag{5.5.2}$$

Here, $d \in \mathbb{C}$ is the location of the fourth singular point, a parameter ($d \neq 0, 1$), and $\alpha, \beta, \gamma, \delta, \varepsilon \in \mathbb{C}$ are exponent-related parameters. The P-symbol is

$$P\left\{\begin{matrix} 0 & 1 & d & \infty & \\ 0 & 0 & 0 & \alpha & ;t \\ 1-\gamma & 1-\delta & 1-\varepsilon & \beta & \end{matrix}\right\}.$$

This does not uniquely specify the equation and its solutions, since it omits the accessory parameter $q \in \mathbb{C}$. The exponents are constrained by

$$\alpha + \beta - \gamma - \delta - \varepsilon + 1 = 0. \tag{5.5.3}$$

This is a special case of Fuchs's relation, according to which the sum of the $2n$ characteristic exponents of any second-order Fuchsian equation on $\mathbb{CP}^1$ with n singular points must equal $n-2$.

There are $2\times 4=8$ local solutions of $(\mathfrak{H})$ in all: two per singular point. If γ is not a nonpositive integer, the solution at $t=0$ that belong to the exponent zero will be analytic. When normalized to unity at $t=0$, it is called the local Heun function, and is denoted as $Hl(d,q;\alpha,\beta,\gamma,\delta;t)$. It is the sum of a Heun series, which converges in a neighborhood of $t=0$. In general, $Hl(d,q;\alpha,\beta,\gamma,\delta;t)$ is not defined when γ is a nonpositive integer.

If $\varepsilon=0$ and $q=\alpha\beta d$, the Heun equation loses a singular point and becomes a hypergeometric equation. Similar losses occur if $\delta=0$, $q=\alpha\beta$, or $\gamma=0$, $q=0$. If $\alpha\beta=0$ and $q=0$, the Heun equation is said to be trivial. Triviality implies that one of the exponents at $t=\infty$ is zero (i.e., $\alpha\beta=0$), and is implied by absence of the singular point at $t=\infty$ (i.e., $\alpha\beta=0$, $\alpha+\beta=1$, $q=0$).

The transformation to $(\mathfrak{H})$ or $(\mathfrak{h})$ of a linear second-order Fuchsian differential equation with singular points at $t=0,1,d,\infty$ (resp. $z=0,1,\infty$), and with arbitrary exponents, is accomplished by certain linear changes of the dependent variable, called F-*homotopies*. If an equation with singular points at $t=0,1,d,\infty$ has dependent variable u, carrying out the substitution $\tilde{u}(t)=t^{-\rho}(t-1)^{-\sigma}(t-d)^{-\tau}u(t)$ will convert the equation to a new one, with the exponents at $t=0,1,d$ reduced by ρ,σ,τ respectively, and those at $t=\infty$ increased by $\rho+\sigma+\tau$. By this technique, one exponent at each finite singular point can be shifted to zero.

In fact, the Heun equation has a group of F-homotopic automorphisms isomorphic to $(\mathbb{Z}_2)^3$, since at each of $t=0,1,d$, the exponents $0,\zeta$ can be shifted to $-\zeta,0$, i.e., to $0,-\zeta$. Similarly, the hypergeometric equation has a group of F-homotopic automorphisms isomorphic to $(\mathbb{Z}_2)^2$. These groups act on the 6 and 3-dimensional parameter spaces, respectively. For example, one of the latter actions is $(a,b;c)\mapsto(c-a,c-b;c)$, which is induced by an F-homotopy at $z=1$. From this F-homotopy follows Euler's transformation

$$ {}_2F_1(a,b;c;z)=(1-z)^{c-a-b}{}_2F_1(c-a,c-b;c;z), \tag{5.5.4}$$

which holds because ${}_2F_1$ is a local solution at $z=0$, rather than at $z=1$.

If the singular points of the differential equation are arbitrarily placed, transforming it to the Heun or hypergeometric equation will require a Möbius (i.e., projective linear or homographic) transformation, which repositions the singular points to the standard locations. A unique Möbius transformation maps any three distinct points in $\mathbb{CP}^1$ to any other three; but the same is not true of four points, which is why $(\mathfrak{H})$ has the singular point d as a free parameter.

The characterization of Heun equations that can be reduced to the hypergeometric equation will employ the cross-ratio orbit of $\{0,1,d,\infty\}$, defined as follows. If $A,B,C,D\in\mathbb{CP}^1$ are distinct, their cross-ratio is

$$(A,B;C,D)=\frac{(C-A)(D-B)}{(D-A)(C-B)}\in\mathbb{CP}^1\setminus\{0,1,\infty\}, \tag{5.5.5}$$

which is invariant under Möbius transformations. Permuting A,B,C,D yields an action of the symmetric group S_4 on $\mathbb{CP}^1\setminus\{0,1,\infty\}$. The cross-ratio is invariant under interchange of A,B and C,D, and also under simultaneous interchange of the two points in each pair. So each orbit contains no more than $4!/4=6$ cross-ratios. The possible actions of S_4 on $s\in\mathbb{CP}^1\setminus\{0,1,\infty\}$ are generated by $s\mapsto 1-s$ and $s\mapsto 1/s$, and the orbit of s comprises

$$s,\quad 1-s,\quad 1/s,\quad 1/(1-s),\quad s/(s-1),\quad (s-1)/s, \tag{5.5.6}$$

which may not be distinct. This is called the *cross-ratio orbit* of s; or, if $s = (A,B;C,D)$, the cross-ratio orbit of the unordered set $\{A,B,C,D\} \subset \mathbb{CP}^1$. Two sets of distinct points $\{A_i,B_i,C_i,D_i\}$ $(i = 1,2)$ have the same cross-ratio orbit iff they are related by a Möbius transformation.

Cross-ratio orbits generically contain six values, but there are two exceptional orbits: one with three and the other with two. If $(A,B;C,D) = -1$, the cross-ratio orbit of $\{A,B,C,D\}$ will be $\{-1,\frac{1}{2},2\}$. The value -1 for $(A,B;C,D)$ defines a so-called *harmonic* configuration: A,B and C,D are said to be harmonic pairs. More generally, if $\{A,B,C,D\}$ has cross-ratio orbit $\{-1,\frac{1}{2},2\}$, it is said to be a *harmonic quadruple.*

The cross-ratio orbit containing exactly two values is $\{\frac{1}{2} \pm i\frac{\sqrt{3}}{2}\}$. Any set $\{A,B,C,D\}$ with this as the cross-ratio orbit is said to be an *equianharmonic quadruple.* $\{A,B,\infty,D\}$ will be an equianharmonic quadruple iff A,B,D are the vertices of an equilateral triangle in $\mathbb{C}$.

According to theory for the Riemann P-function, any Möbius transformation M of the independent variable will preserve characteristic exponents. For the hypergeometric equation (5.5.1), this implies that if M is one of the 3! Möbius transformations that permute the singular points $z = 0,1,\infty$, the exponents of the transformed equation $(\tilde{\mathfrak{h}})$ at its singular points $M(0),M(1),M(\infty)$ will be those of (5.5.1) at $0,1,\infty$. But if M is not affine, i.e., $M(\infty) \neq \infty$, then $(\tilde{\mathfrak{h}})$ will not in general be a hypergeometric equation, since its exponents at $M(\infty)$ may both be nonzero. To convert $(\tilde{\mathfrak{h}})$ to a hypergeometric equation, the permutation must in this case be followed by an F-homotopic transformation of the form $\tilde{y}(z) = [z - M(\infty)]^{-a}y(z)$ or $\tilde{y}(z) = [z - M(\infty)]^{-b}y(z)$.

The *automorphism group* $\mathrm{Aut}(\mathfrak{h})$ *for the hypergeometric equation* (5.5.1), is the group of changes for a variable (Möbius of the independent variable, linear of the dependent) which leave (5.5.1) invariant, up to parameter changes. Similarly, $\mathrm{Aut}(\mathfrak{H})$ is the automorphism group of $(\mathfrak{H})$. $\mathrm{Aut}(\mathfrak{h})$ acts on the 3-dimensional parameter space of (5.5.1). It contains the symmetric group S_3 of permutations for singular points as a subgroup, and the group $(\mathbb{Z}_2)^2$ of F-homotopies as a normal subgroup. So $\mathrm{Aut}(\mathfrak{h}) \times (\mathbb{Z}_2)^2 \times S_3$ is a semidirect product. It is isomorphic to S_4, the octahedral group.

The Möbius automorphism subgroup is the group $\mathcal{M}(\mathfrak{h}) \stackrel{\text{def}}{=} \{1\} \times S_3$, which permutes the singular points $z = 0,1,\infty$. The subgroup of affine automorphisms is $\mathcal{A}(\mathfrak{h}) \stackrel{\text{def}}{=} \{1\} \times S_2$, which permutes the *finite* singular points $z = 0,1$, and fixes ∞. (It is generated by the involution $z \mapsto 1 - z$.) The F-homotopic automorphism subgroup is $(\mathbb{Z}_2)^2 \times \{1\}$.

The Möbius automorphism subgroup is the group $\mathcal{M}(\mathfrak{H}) \stackrel{\text{def}}{=} \{1\} \times S_4$, which maps between sets of singular points for the form $\{0,1,d',\infty\}$. The subgroup of affine automorphisms is $\mathcal{A}(\mathfrak{H}) \stackrel{\text{def}}{=} \{1\} \times S_3$, which maps between sets of *finite* singular points for the form $\{0,1,d'\}$, and fixes ∞. The F-homotopic automorphism subgroup is $(\mathbb{Z}_2)^3 \times \{1\}$.

The theorem will characterize when a homomorphism of rational substitution type from the Heun equation (5.5.2) to the hypergeometric equation (5.5.1) exists. It will list the possible substitutions, up to affine automorphisms of the two equations. It is really a characterization of the $\mathcal{A}(\mathfrak{H})$-orbits that can be mapped by homomorphisms of this type to $\mathcal{A}(\mathfrak{h})$-orbits. The possible substitutions, it turns out, are all polynomial.

In the statement of the theorem and what follows, $S = 1 - R$.

Theorem 5.5.1. *[99] A Heun equation(5.5.2), which has four singular points and is non-*

trivial (i.e., $\alpha\beta \neq 0$ or $q \neq 0$), can be transformed to the hypergeometric equation ($\mathfrak{h}$) by a rational substitution $z = R(t)$ if and only if R is a polynomial, $\alpha\beta \neq 0$, and one of the following two conditions are satisfied.

(1) *$\triangle 01d$ is similar to $\triangle 01D$, for one of the values of D listed in subcases* 1a–1c*; each of which is real, so the triangle must be degenerate. Also, the normalized accessory parameter $q/\alpha\beta$ must equal one of $0,1,d$, which may be denoted d_0. Each subcase lists the value of d_0 when $d = D$.*

(2) *$\triangle 01d$ is similar to $\triangle 01D$, for one of the values of D listed in subcases* 2a–2d*; each of which is non-real and has real part equal to $\frac{1}{2}$, so the triangle must be isosceles. Each subcase lists the value of $q/\alpha\beta$ when $d = D$.*

Besides specifying D and the value of $q/\alpha\beta$ when $d = D$, each subcase imposes restrictions on the characteristic exponent parameters at the singular points $0,1,d$. The subcases of the 'real' case 1 *are the following.*

(1a) *$D = 2$. Suppose $d = D$. Then d_0 must equal 1, and $t = 0,d$ must have the same characteristic exponents, i.e., $\gamma = \varepsilon$. In general, either R or S will be the degree-2 polynomial $t(2-t)$, which maps $t = 0,d$ to $z = 0$ and $t = 1$ to $z = 1$ (with double multiplicity). There are special circumstances in which R may be quartic, which are listed separately, as subcase* 1c.

(1b) *$D = 4$. Suppose $d = D$. Then d_0 must equal 1, the point $t = 1$ must have characteristic exponents that are double those of $t = d$, i.e., $1-\delta = 2(1-\varepsilon)$, and $t = 0$ must have exponents $0,1/2$, i.e., $\gamma = 1/2$. Either R or S will be the degree-3 polynomial $(t-1)^2(1-t/4)$, which maps $t = 0$ to $z = 1$ and $t = 1,d$ to $z = 0$ (the former with double multiplicity).*

(1c) *$D = 2$. Suppose $d = D$. Then d_0 must equal 1, and $t = 0,d$ must have the same characteristic exponents, i.e., $\gamma = \varepsilon$. Moreover, the exponents of $t = 1$ must be twice those of $t = 0,d$, i.e., $1-\delta = 2(1-\gamma) = 2(1-\varepsilon)$. Either R or S will be the degree-4 polynomial $4[t(2-t)-\frac{1}{2}]^2$, which maps $t = 0,1,d$ to $z = 1$ ($t = 1$ with double multiplicity).*

The subcases of the 'non-real' case 2 *are the following.*

(2a) *$D = \frac{1}{2} + \mathrm{i}\frac{\sqrt{3}}{2}$. $q/\alpha\beta$ must equal the mean of $0,1,d$, and $t = 0,1,d$ must have the same characteristic exponents, i.e., $\gamma = \delta = \varepsilon$. Suppose $d = D$. Then $q/\alpha\beta$ must equal $\frac{1}{2} + \mathrm{i}\frac{\sqrt{3}}{6}$. In general, either R or S will be the degree-3 polynomial $\left[1 - t/(\frac{1}{2} + \mathrm{i}\frac{\sqrt{3}}{6})\right]^3$, which maps $t = 0,1,d$ to $z = 1$ and $t = q/\alpha\beta$ to $z = 0$ (with triple multiplicity). There are special circumstances in which R may be sextic, which are listed separately, as subcase* 2d.

(2b) *$D = \frac{1}{2} + \mathrm{i}\frac{5\sqrt{2}}{4}$. Suppose $d = D$. Then $q/\alpha\beta$ must equal $\frac{1}{2} + \mathrm{i}\frac{\sqrt{2}}{4}$, $t = d$ must have characteristic exponents $0,1/3$, i.e., $\varepsilon = 2/3$, and $t = 0,1$ must have exponents $0,1/2$, i.e., $\gamma = \delta = 1/2$. Either R or S will be the degree-4 polynomial*

$\left[1-t/(\frac{1}{2}+\mathrm{i}\frac{5\sqrt{2}}{4})\right]\left[1-t/(\frac{1}{2}+\mathrm{i}\frac{\sqrt{2}}{4})\right]^3$, *which maps* $t=d,q/\alpha\beta$ *to* $z=0$ *(the latter with triple multiplicity) and* $t=0,1$ *to* $z=1$.

(2c) $D=\frac{1}{2}+\mathrm{i}\frac{11\sqrt{15}}{90}$. *Suppose* $d=D$. *Then* $q/\alpha\beta$ *must equal* $\frac{1}{2}+\mathrm{i}\frac{\sqrt{15}}{18}$, $t=d$ *must have characteristic exponents* $0,1/2$, *i.e.,* $\varepsilon=1/2$, *and* $t=0,1$ *must have exponents* $0,1/3$, *i.e.,* $\gamma=\delta=2/3$. *Either* R *or* S *will be the degree-5 polynomial* $At(t-1)\left[t-(\frac{1}{2}+\mathrm{i}\frac{\sqrt{15}}{18})\right]^3$, *which maps* $t=0,1,q/\alpha\beta$ *to* $z=0$ *(the last with triple multiplicity). The factor* A *is chosen so that it maps* $t=d$ *to* $z=1$, *as well; explicitly,* $A=-\mathrm{i}\frac{2025\sqrt{15}}{64}$.

(2d) $D=\frac{1}{2}+\mathrm{i}\frac{\sqrt{3}}{2}$. $q/\alpha\beta$ *must equal the mean of* $0,1,d$, *and* $t=0,1,d$ *must have characteristic exponents* $0,1/3$, *i.e.,* $\gamma=\delta=\varepsilon=2/3$. *Suppose* $d=D$. *Then* $q/\alpha\beta$ *must equal* $\frac{1}{2}+\mathrm{i}\frac{\sqrt{3}}{6}$. *Either* R *or* S *will be the degree-6 polynomial* $4\left\{[1-t/(\frac{1}{2}+\mathrm{i}\frac{\sqrt{3}}{6})]^3-\frac{1}{2}\right\}^2$, *which maps* $t=0,1,d,q/\alpha\beta$ *to* $z=1$ *(the last with triple multiplicity).*

Definition 5.5.1. *A rational map* $R:\mathbb{CP}^1\to\mathbb{CP}^1$ *is said to map the characteristic exponents of the Heun equation (5.5.2) to the characteristic exponents of the hypergeometric equation (*$\mathfrak{h}$*). If, for all* $t_0\in\mathbb{CP}^1$, *the exponents of* $t=t_0$ *according to the Heun equation, divided by the multiplicity of* $t_0\mapsto z_0\stackrel{\mathrm{def}}{=}R(t_0)$, *equal the exponents of* $z=z_0$ *according to the hypergeometric equation.*

Properties:

1.A Heun equation for this (5.5.2) form can be reduced to a hypergeometric equation as a (5.5.1) form by a rational substitution $z=R(t)$ by its independent variable only if R maps exponents-to-exponents.

2. A Heun equation for this (5.5.2) form is trivial (i.e., $\alpha\beta=0$ and $q=0$); that will be reduced to a hypergeometric equation (5.5.1) form by a specified rational substitution $z=R(t)$, where its independent variable if and only if R maps exponents-to-exponents.

3. A Heun equation for this 5.5.2) form has four singular points and is nontrivial (i.e., $\alpha\beta\neq 0$ or $q\neq 0$), that will be reduced to a hypergeometric equation (5.5.1) form by a specified rational substitution $z=R(t)$, where its independent variable if and only if R maps exponents-to-exponents. Moreover, R is a polynomial, $\alpha\beta\neq 0$, and one of the following two conditions on the normalized accessory parameter $p\equiv q/\alpha\beta$ is satisfied.

(1) p equals one of $0,1,d$. Call this point d_0, and the other two singular points d_1 and d_2. In this case, d_0 must be a double zero of R or S, and each of d_1,d_2 must be a simple zero of R or S.

(2) p does not equal any of $0,1,d$. In this case, each of $0,1,d$ must be a simple zero of R or S, and p must be a triple zero of either R or S.

In both cases, R and S must have no additional simple zeroes or zeroes of order greater than two. Also, if $1-c$ (the nonzero exponent at $z=0$) does not equal $1/2$, then R must have no additional double zeroes; and if $c-a-b$ (the nonzero exponent at $z=1$) does not equal $1/2$, then S must have no additional double zeroes. Moreover, in both cases no additional double zero, if any, must be mapped by R to the point (out of $z = 0, 1$) to which p is mapped.

Suppose a Heun equation has four singular points and is nontrivial ($\alpha\beta \neq 0$ or $q \neq 0$). Then the only reductions of its local Heun function Hl to ${}_2F_1$ that can be performed by a rational transformation of the independent variable involve polynomial transformations of degrees 2, 3, 4, 5, and 6. There are seven distinct types, each of which can exist only if d lies on an appropriate cross-ratio orbit. The following list includes a representative reduction of each type. The ones with real d (and $\deg R = 2, 3, 4$) include

$$Hl(2, \alpha\beta; \alpha, \beta, \gamma, \alpha+\beta-2\gamma+1; t) = {}_2F_1\left(\tfrac{\alpha}{2}, \tfrac{\beta}{2}; \gamma; t(2-t)\right),$$

$$Hl\left(4, \alpha\beta; \alpha, \beta, \tfrac{1}{2}, \tfrac{2(\alpha+\beta)}{3}; t\right) = {}_2F_1\left(\tfrac{\alpha}{3}, \tfrac{\beta}{3}; \tfrac{1}{2}; 1-(t-1)^2(1-t/4)\right),$$

$$Hl\left(2, \alpha\beta; \alpha, \beta, \tfrac{\alpha+\beta+2}{4}, \tfrac{\alpha+\beta}{2}; t\right) = {}_2F_1\left(\tfrac{\alpha}{4}, \tfrac{\beta}{4}; \tfrac{\alpha+\beta+2}{4}; 1-4\left[t(2-t)-\tfrac{1}{2}\right]^2\right),$$

and the ones with non-real d (and $\deg R = 3, 4, 5, 6$) include

$$Hl\left(\tfrac{1}{2} \pm i\tfrac{\sqrt{3}}{2}, \alpha\beta(\tfrac{1}{2} \pm i\tfrac{\sqrt{3}}{6}); \alpha, \beta, \tfrac{\alpha+\beta+1}{3}, \tfrac{\alpha+\beta+1}{3}; t\right) =$$

$$= {}_2F_1\left(\tfrac{\alpha}{3}, \tfrac{\beta}{3}; \tfrac{\alpha+\beta+1}{3}; 1-\left[1-t/(\tfrac{1}{2} \pm i\tfrac{\sqrt{3}}{6})\right]^3\right),$$

$$Hl\left(\tfrac{1}{2} \pm i\tfrac{5\sqrt{2}}{4}, \alpha(\tfrac{2}{3}-\alpha)(\tfrac{1}{2} \pm i\tfrac{\sqrt{2}}{4}); \alpha, \tfrac{2}{3}-\alpha, \tfrac{1}{2}, \tfrac{1}{2}; t\right) =$$

$$= {}_2F_1\left(\tfrac{\alpha}{4}, \tfrac{1}{6}-\tfrac{\alpha}{4}; \tfrac{1}{2}; 1-\left[1-t/(\tfrac{1}{2} \pm i\tfrac{5\sqrt{2}}{4})\right]\left[1-t/(\tfrac{1}{2} \pm i\tfrac{\sqrt{2}}{4})\right]^3\right),$$

$$Hl\left(\tfrac{1}{2} \pm i\tfrac{11\sqrt{15}}{90}, \alpha(\tfrac{5}{6}-\alpha)(\tfrac{1}{2} \pm i\tfrac{\sqrt{15}}{18}); \alpha, \tfrac{5}{6}-\alpha, \tfrac{2}{3}, \tfrac{2}{3}, t\right) =$$

$${}_2F_1\left(\tfrac{\alpha}{5}, \tfrac{1}{6}-\tfrac{\alpha}{5}; \tfrac{2}{3}; (\mp i\tfrac{2025\sqrt{15}}{64})\, t(t-1)\left[t-(\tfrac{1}{2} \pm i\tfrac{\sqrt{15}}{18})\right]^3\right),$$

$$Hl\left(\tfrac{1}{2} \pm i\tfrac{\sqrt{3}}{2}, \alpha(1-\alpha)(\tfrac{1}{2} \pm i\tfrac{\sqrt{3}}{6}); \alpha, 1-\alpha, \tfrac{2}{3}, \tfrac{2}{3}, t\right) =$$

$$= {}_2F_1\left(\tfrac{\alpha}{6}, \tfrac{1}{6}-\tfrac{\alpha}{6}; \tfrac{2}{3}; 1-4\left\{\left[1-t/(\tfrac{1}{2} \pm i\tfrac{\sqrt{3}}{6})\right]^3-\tfrac{1}{2}\right\}^2\right).$$

In the preceding reductions, α, β, γ are free parameters. Each of these equalities holds in a neighborhood of $t = 0$ whenever the two sides are defined, e.g., whenever the fifth argument of Hl and the third argument of ${}_2F_1$ are not equal to a nonpositive integer.

Recall that the *Weierstrass function* $\wp(u) \equiv \wp(u; g_2, g_3)$ with invariants $g_2, g_3 \in \mathbb{C}$, which cannot both equal zero, has a double pole at $u = 0$ and satisfies

$$\wp'^2 = 4\wp^3 - g_2\wp - g_3 = 4(\wp - e_1)(\wp - e_2)(\wp - e_3).$$

Where e_1, e_2, e_3, the zeroes of the defining cubic polynomial are the finite critical values of $\wp$, the sum of which is zero; they are required to be distinct. $\wp$ is doubly periodic on $\mathbb{C}$, with periods denoted $2\omega, 2\omega'$. So, it can be viewed as a function on the torus $\mathbb{T} \overset{\text{def}}{=} \mathbb{C}/\mathcal{L}$, where $\mathcal{L} = 2\omega\mathbb{Z} \oplus 2\omega'\mathbb{Z}$ is the period lattice. It turns out that the half-lattice $\{0, \omega, \omega', \omega +$

$\omega'\} + \mathcal{L}$ comprises the critical points of $\wp$. The map $\wp : \mathbb{T} \to \mathbb{CP}^1$ is a double branched cover of the Riemann sphere, but $\mathbb{T}$ is uniquely coordinatized by the pair $(\wp, \wp')$.

The modular discriminant $\Delta \stackrel{\text{def}}{=} g_2^3 - 27g_3^2 \neq 0$ is familiar from the elliptic function theory. If $g_2, g_3 \in \mathbb{R}$ and $\Delta > 0$, ω, ω' can be chosen to be real and imaginary, respectively. If $\Delta < 0$, they can be chosen to be complex conjugates, so that the third critical point $\omega_2 \stackrel{\text{def}}{=} \omega + \omega'$ is real.

Clarkson and Olver [28] considered the Weierstrass-form Lamé equation

$$\frac{d^2\psi}{du^2} - [\ell(\ell+1)\wp(u) + B]\,\psi = 0, \tag{5.5.7}$$

which is a Fuchsian equation on $\mathbb{T}$ with exactly one singular point (at $(\wp, \wp') = (\infty, \infty)$) and a single accessory parameter, B. In particular, they considered the case $g_2 = 0$, $g_3 \neq 0$, $B = 0$. Clarkson and Olver mapped $u \in \mathbb{T}$ to $z \in \mathbb{CP}^1$ by the formal substitution

$$u = \frac{\mathrm{i}}{(16g_3)^{1/6}} \int^{(1-z)^{1/3}} \frac{d\tau}{\sqrt{1-\tau^3}},$$

and showed that the Lamé equation is reduced to

$$z(1-z)\frac{d^2\psi}{dz^2} + \left(\frac{1}{2} - \frac{7}{6}z\right)\frac{d\psi}{dz} + \frac{\ell(\ell+1)}{36}\psi = 0. \tag{5.5.8}$$

This is a hypergeometric equation with $(a, b; c) = (-\ell/6, (\ell+1)/6; 1/2)$. It has exponents $0, 1/2$ at $z = 0$; $0, 1/3$ at $z = 1$; and $-\ell/6, (\ell+1)/6$ at $z = \infty$.

For the Lamé equation (5.5.7) to be viewed as a Heun equation on $\mathbb{CP}^1$, it must be transformed by $s = \wp(u)$ to its algebraic form. The algebraic form is

$$\frac{d^2\psi}{ds^2} + \left(\frac{1/2}{s-e_1} + \frac{1/2}{s-e_2} + \frac{1/2}{s-e_3}\right)\frac{d\psi}{ds} + \frac{[-\ell(\ell+1)/4]s - B/4}{(s-e_1)(s-e_2)(s-e_3)}\psi = 0. \tag{5.5.9}$$

This is a special case in the general Heun equation, the canonical version of the natural general-form Heun equation, with distinct finite singular points $d_1, d_2, d_3 = e_1, e_2, e_3$. Also, $\alpha, \beta = -\ell/2, (\ell+1)/2$, $\gamma = \delta = \varepsilon = 1/2$, and $q' = B/4$. It has characteristic exponents $0, 1/2$ at $s = e_1, e_2, e_3$, and $-\ell/2, (\ell+1)/2$ at $s = \infty$.

Theorem 5.5.2. *[99] The algebraic-form Lamé equation (5.5.9) for a the equianharmonic case $g_2 = 0$, $g_3 \neq 0$, can be reduced by $\ell(\ell+1) \neq 0$ to a hypergeometric equation for this (5.5.1) form by a rational transformation of $z = R(s)$ iff the accessory parameter B equals zero. If this is the case, R will necessarily be a cubic polynomial;*

$$z = 4s^3/g_3, \qquad z = 1 - 4s^3/g_3$$

will both work, and they are the only possibilities.

The Weierstrass-form Lamé equation (5.5.7), in the equianharmonic case $g_2 = 0$, $g_3 \neq 0$, can be reduced by $\ell(\ell+1) \neq 0$ to a hypergeometric equation for this (5.5.1) form through a substitution of $z = R(\wp(u))$, where R is rational, iff the accessory parameter B equals zero. In this case,

$$z = 4\wp(u)^3/g_3, \qquad z = 1 - 4\wp(u)^3/g_3 \tag{5.5.10}$$

will both work, and they are the only such substitutions.

If the Heun equation (5.5.2) is trivial it may be solved by quadratures. The basis for these solutions were

$$u_1(t) = 1, \quad u_2(t) = \int^t \exp\left[-\int^v \left(\frac{\gamma}{w} + \frac{\delta}{w-1} + \frac{\varepsilon}{w-d}\right) dw\right] dv. \tag{5.5.11}$$

The canonical polynomial substitutions give rise to many *nonpolynomial* rational reductions of trivial Heun equations for hypergeometric equations, by composing certain Möbius transformations. In the context of *nontrivial* Heun equations, Möbius automorphisms that are not affine could not be employed; essentially because, moving the point at infinity would require a compensating F-homotopy. But in the trivial case no such issue can arise: the Heun equation is reduced to a hypergeometric equation by a rational substitution of its independent variable, $z = R(t)$, iff the substitution maps exponents-to-exponents. Möbius transformations that are not affine certainly preserve exponents.

Theorem 5.5.3. *[99] A Heun equation of this (5.5.2) form that has four singular points and is trivial (i.e.,* $\alpha\beta = 0$ *and* $q = 0$*), can be reduced to a hypergeometric equation (5.5.1) form by any rational substitution for the form* $z = M_2(R_1(M_1(t)))$*, where* $z = R_1(t)$ *is a polynomial that maps* $\{0, 1, D\}$ *to* $\{0, 1\}$*, listed (along with D) in one of the seven subcases of Theorem* 5.5.1*, and where* $M_1 \in \mathcal{M}(\mathfrak{H})$*, and* $M_1 \in \mathcal{M}(\mathfrak{h})$*. That is,* M_1 *maps* $\{0, 1, d, \infty\}$ *onto* $\{0, 1, D, \infty\}$*, and* M_2 *maps* $\{0, 1, \infty\}$ *onto* $\{0, 1, \infty\}$*. The necessary conditions on characteristic exponents stated in Theorem* 5.5.1 *must be satisfied, the conditions on exponents at specified values of t being taken to refer to the exponents at the preimages of these points under* M_1.

Remark 5.5.1. It is known that *Lamé* operator

$$L_m = \frac{d^2}{dx^2} - m(m+1)\wp - B,$$

is a nontrivial example of an operator for the algebraic type (i.e., its eigenvalue problem is explicitly solvable in quadratures.)

To be more exact, $L_{m,B}$ is the algebraic type if and only if m is a nonnegative integer. The equation

$$\frac{d^2}{dx^2}u - (m(m+1)\wp + B)u = 0,$$

is a Schrödinger equation with an elliptic potentials extended to the complex domain.

In [143], K. Takemura established the relationship between the BC_1 Inozemtsev model and the Heun equation. Here, we shall give Takemura's result.

The Hamiltonian of the BC_1 Inozemtsev model is given as follows:

$$H := -\frac{d^2}{dx^2} + \sum_{i=0}^{3} l_i(l_i+1)\wp(x+\omega_i), \tag{5.5.12}$$

where $\wp(x)$ is the Weierstrass $\wp$-function with periods $(2\omega_1, 2\omega_3)$, $\omega_0 = 0$, $\omega_2 = -(\omega_1 + \omega_3)$, and l_i $(i = 0,1,2,3)$ are coupling constants. Let $f(x)$ be an eigenfunction of H with an eigenvalue E, i.e.,

$$(H-E)f(x) = \left(-\frac{d^2}{dx^2} + \sum_{i=0}^{3} l_i(l_i+1)\wp(x+\omega_i) - E\right) f(x) = 0, \tag{5.5.13}$$

We will make a correspondence between equation (5.5.13) and the Heun equation as was done in [116]. Set $z = \wp(x)$ and $e_i = \wp(\omega_i)$ $(i = 1,2,3)$. By a straightforward calculation, we have

$$\frac{d^2}{dx^2} = \wp'(x)^2 \left(\frac{d^2}{dz^2} + \frac{1}{2}\left(\frac{1}{z-e_1} + \frac{1}{z-e_2} + \frac{1}{z-e_3}\right)\frac{d}{dz}\right).$$

Hence, equation (5.5.13) is equivalent to the equation

$$\left\{\frac{d^2}{dz^2} + \frac{1}{2}\left(\frac{1}{z-e_1} + \frac{1}{z-e_2} + \frac{1}{z-e_3}\right)\frac{d}{dz} - \frac{1}{4(z-e_1)(z-e_2)(z-e_3)}\cdot \right. \tag{5.5.14}$$
$$\left. \cdot\left(\tilde{C} + l_0(l_0+1)z + \sum_{i=1}^{3} l_i(l_i+1)\frac{(e_i-e_{i'})(e_i-e_{i''})}{z-e_i}\right)\right\}\tilde{f}(z) = 0,$$

where $i', i'' \in \{1,2,3\}$, $i' < i''$, i' and i'' are different from i, $\tilde{f}(\wp(x)) = f(x)$ and $\tilde{C} = -E + \sum_{i=1}^{3} l_i(l_i+1)e_i$. Note that $e_1 + e_2 + e_3 = 0$.

Equation (5.5.14) has four singular points e_1, e_2, e_3 and ∞ on the Riemann sphere and all singular points are regular. Therefore, this equation is a Heun kind of equation. The exponents at $z = e_i$ $(i = 1,2,3)$ are $(l_i+1)/2$ and $-l_i/2$, and the exponents at ∞ are $(l_0+1)/2$ and $-l_0/2$. The parameter $\tilde{C}$ corresponds to the accessory parameter.

In the theory of holonomic quantum computing, [156], [117], and [144] established that for the controllability of the correspondence quantum system necessary conditions is the irreducibility of the connection, in this case holonomy group of connections coincides with $SU(2)$.

The following theorem imposes on the hypergeometrical equation those conditions which ensure construction of local gates.

Theorem 5.5.4. *The physical system described by Schrödinger equation (5.2.3) with Hamiltonian (5.2.4) defines a universal set of gates, iff satisfying the following conditions:*

$$1) iT(-E_1 + \sqrt{E_1^2+V_0^2}) \overline{\in} \mathbb{Z};$$

$$2) iT(-E_1 - \sqrt{E_1^2+V_0^2}) \overline{\in} \mathbb{Z};$$

$$3) \frac{1}{2} + E_0T + iT\sqrt{E_1^2+V_0^2} \overline{\in} \mathbb{Z};$$

$$4) \frac{1}{2} + E_0T - 2iE_1T + iT\sqrt{E_1^2+V_0^2} \overline{\in} \mathbb{Z};$$

$$5) \frac{1}{2} - E_0T + iT\sqrt{E_1^2+V_0^2} \overline{\in} \mathbb{Z};$$

$$6) \frac{1}{2} - E_0T - iT\sqrt{E_1^2+V_0^2} \overline{\in} \mathbb{Z};$$

◁ It is known, that the equation (5.2.5) defines the two dimension holomorphic vector bundle on $\mathbb{CP}^1$ with holomorphic connection with logarithmic singularities [57]. This fiber bundle considers the space of qubits. The monodromy matrices for this equation (5.2.5) acts on the fiber in natural way. It is also known, that associated connection gives a universal set of gates, iff monodromy representation is irreducible [156]. For the equation (5.2.5) irreducibility condition means, that in Riemann P function

$$P\begin{pmatrix} 0 & 1 & \infty \\ 0 & 0 & \alpha \\ 1-\gamma & \gamma-\alpha-\beta & \beta \end{pmatrix} \tag{5.5.15}$$

the constants α, β, γ satisfies the conditions

$$\alpha, \beta, \gamma-\alpha, \gamma-\beta, 1-\gamma+\alpha, 1-\gamma+\beta \overline{\in} \mathbb{Z}.$$

From this definition, the universal set of gates with irreducibility for the monodromy representation in a hypergeometric equation it then presents the following: iff the representation associated from equation (5.2.5) is irreducible, then the system is completely controllable. From this condition of irreducibility it then follows, that the constants from (5.2.4) can be satisfied for conditions 1),2),3),4), 5), and 6) in the theorem.▷

If $\alpha = 1$, $q = 1$, or $\alpha = 0$, $q = 0$ the equation (5.5.2) reduces to a hypergeometric equation.

Theorem 5.5.5. *If Heun equation satisfies the PHH properties, then it defines a completely controllable quantum system.*

◁ If the conditions of theorem are satisfied, then the property of complete controllability coincides with irreducibility of monodromy representation for the hypergeometric equation which was obtained from PHH-reduction. The irreducibility condition of the monodromy representation for the hypergeometric equation is given by the exponents of equation. ▷

Theorem 5.5.6. *The Hamiltonian of the BC_1 Inozemtsev model*

$$H = -\frac{d^2}{dx^2} + \sum_{i=0}^{3} l_i(l_i+1)\wp(x+\omega_i) \tag{5.5.16}$$

for appropriate values of parameters (the conditions for the parameters are given in the proof) defines completely controllable quantum system.

◁ We follow the discussion from [143] as given earlier. From (5.5.16), one obtains Fuchsian equation (5.5.2) with 4 singular points and therefore obtains the Heun equation with singular points $(0, 1, t, \infty)$. Denoted by ∇, the connection of the two-dimensional vector bundle induces from the Heun equation. Let eigenvalues for residues of ∇ be:

$$Res_0\nabla = (\lambda_0, -\lambda_0), Res_1\nabla = (\lambda_1, -\lambda_1),$$

$$Res_t\nabla = (\lambda_t, -\lambda_t), Res_\infty\nabla = (\lambda_\infty, -\lambda_\infty).$$

It is known, that if $\sum \varepsilon_i\lambda_i \overline{\in} \mathbb{Z}$, where $(\varepsilon_1, \varepsilon_2, \varepsilon_3, \varepsilon_4) \in (\mathbb{Z}/2\mathbb{Z})^4$, then monodromy representation of the ∇ is irreducible and two-dimensional vector bundle is stable. From this it follows the controllability of the Hamiltonian (5.5.16).▷

Remark 5.5.2. The sixth Painleve equation P_{VI} is the nonlinear differential equation in the order of two on $\mathbb{CP}^1$ with at most four regular singularities and without movable branch points $0, 1, t, \infty$:

$$\frac{d^2x}{dt^2} = \frac{1}{2}\left(\frac{1}{x} + \frac{1}{x-1} + \frac{1}{x-t}\right)\left(\frac{dx}{dt}\right)^2 - \left(\frac{1}{t} + \frac{1}{t-1} + \frac{1}{x-t}\right)\frac{dx}{dt} +$$

$$+ \frac{x(x-1)(x-t)}{t^2(t-1)^2}\left(\alpha - \beta\frac{t}{x^2} + \gamma\frac{t-1}{(x-1)^2} + \left(\frac{1}{2} - \delta\right)\frac{t(t-1)}{(x-t)^2}\right).$$

This is isomonodromic condition of the Heun equation. The parameters $\alpha, \beta, \gamma, \delta$ represent the eigenvalues $\delta_0, \delta_1, \delta_t, \delta_\infty$ of the residues of logarithmic connection Δ as before:

$$\alpha = \frac{1}{2}\lambda_\infty^2,\ \beta = \frac{1}{2}\lambda_0^2,\ \gamma = \frac{1}{2}\lambda_1^2,\ \delta = \frac{1}{2}\delta_t^2.$$

The Painleve P_{VI} is Hamilton type equation

$$\frac{dx}{dt} = \frac{\partial H}{\partial p},\ \frac{dp}{dt} = -\frac{\partial H}{\partial x},$$

with a Hamiltonian

$$H = \frac{1}{t(t-1)}[x(x-1)(x-t)p^2 - \{\lambda_0(x-1)(x-t) +$$

$$+\lambda_1 x(x-t) + (\lambda_t - 1)x(x-1)\}p + \frac{1}{4}((\lambda_0 + \lambda_1 + \lambda_t - 1)^2 - \lambda_\infty^2)(x-t)].$$

Remark 5.5.3. A second order differential equation on the Riemann sphere with coefficients in the field of rational functions induces a two-dimensional vector bundle on $\mathbb{CP}^1$. In the Fuchsian case, with $n \geq 3$ singular points, the equation will be determined up to isomorphism by *location of singular points*, a *difference of characteristic exponents,* and $n-3$ *accessory parameters* which is more global than the local significance.

Remark 5.5.4. The equation (5.5.1) is the two-dimensional case of (5.1.17) in which

$$A = \begin{pmatrix} 1-c & 1 \\ (a-c+1)(c-b-1) & c-a-b-1 \end{pmatrix} \text{ and } B = \begin{pmatrix} 0 & 0 \\ 0 & 1 \end{pmatrix}.$$

It is known that for every hypergeometric equation (5.5.1) there exists a Fuchsian system, with the same singular points and the some monodromy. This system has the form:

$$\frac{dF(z)}{dz} = \begin{pmatrix} 0 & 0 \\ -ab & c \end{pmatrix}\frac{F(z)}{z} + \begin{pmatrix} 0 & 1 \\ 0 & c-(a+b) \end{pmatrix}\frac{F(z)}{z-1}$$

where $F(z)$ is 2-dimensional vector function.

References

[1] Agrachev, A.A.; Sachkov, Y.L., Control theory from the geometric viewpoint. *Encyclopedia of Mathematical Sciences*, 87. Control Theory and Optimization, II. Springer-Verlag, Berlin, 412 pp., 2004.

[2] Aharonov, D.; Win van Dam, Kempe, J.; Landau, Z.; Lloyd, S.; Regev, O., *Adiabatic quantum computation is equivalent to standard quantum computation*. ArXiv:quant-ph/0405098

[3] Akhtarshenas, S.J., Canonical coset parametrization and the Bures metric of the three-level quantum systems. *J. Math. Phys.* 48, no. 1, 012102, 12 pp., 2007.

[4] Akhtarshenas, S.J.; Jafarizadeh, M.A., Lewenstein-Sanpera decomposition of a generic $2 \otimes 2$ density matrix by using Wootters's basis. *Quantum Inf. Comput.* 3, 229-248, 2003.

[5] Allen, L.; Eberly, J.H., *Optical Resonance and Two–Level Atoms*, Wiley, New York, 1975.

[6] Altafini, C., Controllability of quantum mechanical systems by root space decomposition of $\mathfrak{su}(N)$. *J. Math. Phys.* 13(5), pp 2051-2062, 2002.

[7] Ambrose, W.; Singer, I.M., A theorem on holonomy. *Trans. Amer. Math. Soc.,* 75, 428-443, 1953.

[8] Anosov, D.V.; Bolibruch, A.A., The Riemann-Hilbert problem, *Aspects of Mathematics,* Vieweg, Braunschweig, Wiesbaden, 1994.

[9] Arestova, M.L.; Bikovski, A.Iu., The method of realization of the optoelectronic scheme in many valued logics. *Quantum electronics.* vol. 10, no. 2, 1995.

[10] Arnold, V.I., Complementary Chapters of Differential Equations, Nauka, 302p, 1978 (Russian).

[11] Arvind, Mallesh, K.S.; Mukunda, N., A generalized Pancharatnam geometric phase formula for three-level quantum systems. *J. Phys. A* 30 (1997), no. 7, 2417-2431.

[12] Barenco, A.; Bennett, C.; Cleve, R.; DiVincenzo, D.; Margulus, N.; Shor, P.; Sleator, T.; Smolin, J.; Weinfurter, H., Elementary gates for quantum computation. *Phys. Rev. A* 2, 3457-3467, 1995.

[13] Belavkin,V.; Ohya, M., Quantum entropy and information in discrete entangled states. *Infin. Dimens. Anal. Quantum Probab. Relat. Top.* 4, no. 2, 137-160, 2001.

[14] Benioff, P., Models of quantum Turing Machines. ArXiv: quant-ph/9708054.

[15] Berry, M., Quantal phase factors accompanying adiabatatic changes. *Proc. Royal Soc. London, A* 392, 45-57, 1984.

[16] Bernstein, E.; Vazirani, U.; Quantum complexity theory. SIAM J. Comput. 26, no. 5, 1411-1473, 1997.

[17] Bloore, F.J., Geometrical description of the convex sets of states for systems with spin- $1/2$ and spin-1. *J. Phys. A: Math. Gen.* 9, 2059-2067, 1976.

[18] Boscain, U.Th.; Chambrion, Jean-Paul Gauthier, On the $K+P$ problem for a three-level quantum system: Optimality implies resonance. *J. Dynam. Control system,* Vol.8, pp.547-572, 2002.

[19] Boykin, P.O.; Mor, T.; Pulver, M.; Roychowdhury, V.; Vatan, F., On universal and Fault-Tolerant quantum computing: A novel basis and a new constructive proof of universality for Shor's basis. In *Proc. 40th Ann. Symp. Found. Comp. Sci.*IEEE Comp.Soc. Press, Los Alamitos. pp.486-494, 1999.

[20] Byrd, M.S.; Slater, P.; Bures measures over the spaces of two- and three-dimensional density matrices. *Phys. Lett. A* 283, 152-156, 2001.

[21] Bures, D.J.C., An extension of Kakutani's theorem on infinite product measures to the tensor product of semifinite w^*-algebras. *Trans. Am. Math. Phys.* 135, 199-212, 1969.

[22] Braunstein, S.L.; Lo, H.-K., (special issue editors), Kok, P. (assistant editor). Experimental Proposals for Quantum Computation. *Fortschritte der Physik (Progress of Physics)* 48, 767-1138, 2000.

[23] Breuillard, E.; Gelander, T., *On Dense Free Subgroups of Lie Groups.* arXiv:math.GR/0206236 v2.

[24] Brylinski, J.L.; Brylinski, R., Universal quantum gates. *In Mathematics of Quantum Computation,* Chapman & Hall/CRC Press, Boca Raton, Florida, 2002.

[25] Cirac, J.I.; Zoller, P., *Phys.Rev.Lett.* 74, 4091, 1995.

[26] Chaturvedi, S.; Ercolessi, E.; Marmo, G.; Morandi, G.; Mukunda, N.; Simon, R., Geometric phase for mixed states: A differential geometric approach. *Eur. Phys. J. C Part. Fields* 35, no. 3, 413-423, 2004.

[27] Chen, K.-T., Iterated path integrals. *Bull. Amer. Math. Soc.* Vol. 83, No. 5, 1977.

[28] Clarkson, P. A.; Olver, P. J., Symmetry and the Chazy equation, *J. Differential Equations* 124 (1), 225–246, 1996.

[29] D'Alessandro, D., *The Lie Algebra Rank Condition for Non-Bilinear Quantum Systems.* arXiv: quant-ph/0301144 v2, 2003

[30] D'Alessandro, D., *Uniform Finite Generation of Compact Lie Groups and Universal Quantum Gates.* arXiv: quant-ph/0111133 v1.

[31] Deutsch, D., Quantum theory, the Church-Turing principle and the universal quantum computer. *Proc. Royal Soc. London. Ser.A* 400, pp.97-117, 1985.

[32] Diţă, P., Finite-level systems, Hermitian operators, isometries and a novel parametrization of Stiefel andGrassmann manifolds. *J. Phys. A: Math. Gen.* 38, 2657-2668, 2005.

[33] Diţă, P., Factorization of unitary matrices. *J. Phys. A: Math. Gen.* 36, 11, 2781-2789, 2003.

[34] Dye, H., *Unitary solution to the Yang-Baxter equation in dimension four.* ArXiv: quant-ph/0211050.

[35] Ekert, A.; Ericsson, M.; Hayden, P.; Inamori, H.; Jones, J. A.; Oi, D.K.L.; Vedral, V., Geometric quantum computation. *J. Mod. Opt.* 47(14-15): 2501-2513, 2000.

[36] Ercolessi, E.; Marmo, G.; Morandi, G.; Mukunda, N. *Geometry of mixed states and degeneracy structure of geometric phases for multi-valued quantum systems.* ArXiv: quant-ph/0105007.

[37] Faddeev L.D. and Jakubovski O.A., *Lectures in Quantum Mechatins,* Leningrad, 198p,1980 (russian).

[38] Faddeev, L.D.; Takhtajan, L.A., *Hamilton Approach to the Theory of Solitons*, Nauka 528p. 1986 (russian).

[39] Freedman, M., *P/NP* and the quantum field computer. *Proc. Nat. Acad. Sci. USA,* Vol. 95, pp. 98-100, 1998.

[40] Freedman, M.H.; Kitaev, A.; Wang, Z., *Simulation of topological field theories by quantum computer.* ArXiv: quant-ph/0001071

[41] Feynman, R., *Quantum mechanical computations.* Optics News, N 11, p.11, February, 1985.

[42] Fujii, K., *More on optical holonomic quantum computer.* ArXiv: quant-ph/0005129.

[43] Fujii, K., Note on coherent states and adiabatic connections, curvatures. *J. Math. Phys.*41, 4406-4412, 2000.

[44] Fujii, K., *Mathematical Foundations of Holonomic Quantum Computer.* quant-ph/0004102.

[45] Fujii, K., *Mathematical Foundations of Holonomic Quantum Computer II.* quant-ph/0101102.

[46] Fujii, K., *Introduction to Grassmann Manifolds and Quantum Computations.* quant-ph/0103011.

[47] Gauthier, J-P.; Guérin, S.; Jauslin, H-R., Optimal control in laser-induced population transfer for two- and three-level quantum systems. *J. Math. Phys.* 43, no. 5, 2107-2132, 2002.

[48] Gilmore, R., *Lie Groups, Lie Algebras, and Some of Their Applications*, John-Wiley Publishing Co., 1974.

[49] Giorgadze, G., Monodromic approach to quantum computing. *Intl. J. Modern Physics B* 16 No. 30, 4593-4605, 2002.

[50] Giorgadze, G., On the Hamiltonians induced from a Fuchsian system. *Mem. Diff. Eq. and math. Phys.* vol.31, 69-82, 2004.

[51] Giorgadze, G.; Khimshiashvili, G., On Shrödinger equations of Okubo type. *J. Dyn. Cont. Syst.* 10, N. 2, 171-186, 2004.

[52] Giorgadze, G., *G*-systems and holomorphic principal bundles on Riemann surfaces. *J. Dyn. Contr. Syst.* 8, N. 2, 245-291, 2002.

[53] Giorgadze, G., Holomorphic quantum computing bull.*Georgian Acad. Sci.* vol. 166, N.2, 228-230, 2002.

[54] Giorgadze, G., On the existence of a irreducible connections and its application. *Bull.Georgian Acad. Sci.* vol.168, N.2, 227-229, 2003.

[55] Giorgadze, G., On representation of connections by exponential iterated integral bull.*Georgian Acad. Sci.* vol.168, N.3, 413-416, 2003.

[56] Giorgadze, G., Holomorphic structures and connections on low dimensional vector bundles on the Riemann sphere. *Bulletin de la Societe des Sciences et des Lettres de Lodz.* Vol. LII, pp.93-100, 2002.

[57] Giorgadze, G., Regular systems on Riemann surfaces.*J. Math. Sci.* 118 (5), 5347-5399, 2003.

[58] Giorgadze, G.; Tevzadze, R., Scattering matrices as the gates for quantum computer. *Bull. Georgian Natl. Acad. Sci.* 173, no. 1, 7-10, 2006.

[59] Giunashvili, Z., Methods of geometric control theory for quantum computations.*J. Math. Sci. (N. Y.)* 141, no. 2, 1113-1133, 2007.

[60] Giunashvili, Z., Geometric methods for constructing elements of quantum circuits.*J. Math. Sci. (N. Y.)* 153, no. 2, 120-158, 2008.

[61] Gisin, N.; Bechmann-Pasquinucci, H., Bell inequality, Bell states and maximally entangled states for n qubits. *Phys. Lett. A* 246, no. 1-2, 1-6, 1998.

[62] Golubeva, V.A.; Leksin, V.P., On two types of representations of braid group associated with the Knizhnik-Zamolodchikov equation of the B_n type. *J. Dynamic. Contr. Syst.* Vol. 5, No. 4, pp. 565-596, 1999.

[63] Große, D.; Wille, R.; Dueck, G.; Drechsler, R., Exact synthesis of elementary quantum gate circuits. *J. Mult.-Valued Logic Soft Comput.* 15, no. 4, 283-300, 2009.

[64] Hall, M.J W., Random quantum correlations and density operator distributions. *Phys. Lett. A* 242, 123-129, 1998.

[65] Haraoka, Y., Canonical forms of diffrential equations free from accessory parameters. *SIAM J. Math. Anal.* 25, 1203-1226, 1994.

[66] Haraoka, Y., Monodromy representation of the system of differential equation free from accessory parameters. *SIAM J. Math. Anal.* 25, 1595-1621, 1994.

[67] Iablonskiĭ, S.V., *Introduction in Discrete Mathematics.* Moscow, 384 p. 2003.

[68] Ince, E.L., *Ordinary Differential Equations*, Dover, 1956.

[69] Jurdjevic, V., *Geometric Control Theory*, Cambridge Univ. Press, Cambridge, UK, 1996.

[70] Jurdjevich, V.; Sussmann, H., Control Systems on Lie Groups. *J. Differential Equations* 12,pp 313-329, 1972.

[71] Kamke E. *Differental gleichungen. Lüsungsmethoden und Lösungen,* Leipzig, 1959.

[72] Kato, T.; Nakamura, K.; Lakshmanan, M., Analytic calculation of nonadiabatic transition probalities from monodromy of differetial equations. *J. Phys. A: Math. Gen.* 36, pp.5803-5815, 2003.

[73] Tsuchiya, A.; Kanie, Y., Vertex operators in conformal field theory on CP^1 and monodromy representations of braid group.*Adv. Stud. Pure Mathematics* 16. Conformal Field Theory and Solvable Lattice Models. pp. 292–372, 1988.

[74] Kauffman, L.; Lomonaco, S., *Braiding Operators are Universal Quantum Gates.* ArXiv:quant-ph/0401090.

[75] Kauffman, L.H.; Lomonaco, S.J., Quantum entanglement and topological entanglement. *New Journal of Physics* 4, 73.1-73.18, 2002.

[76] Kieu, T., *A Reformulation of Hilbert's Tenth Problem Through Quantum Mechanics.* AtXiv: quant-ph/0111063, v3, 2003.

[77] Kieu, T., Quantum algorithm for Hilbert's tenth problem. *Intl. J. Theor. Phys.* 42 (2003), no. 7, 1461-1478.

[78] Kieu, T., A reformulation of Hilbert's tenth problem through quantum mechanics. *Proc. R. Soc. Lond. Ser. A Math. Phys. Eng. Sci.* 460, no. 2045, 1535-1545. 2004.

[79] Kieu, T., Quantum algorithm for Hilbert's tenth problem. *Intl. J. Theor. Phys.* 42 (2003) 1461-1478.

[80] Kieu, T., A *Reformulation of Hilbert's Tenth Problem Through Quantum Mechanics.* arXiv:quant-ph/0111062 v2.

[81] Kieu, T., *Quantum Principles and Mathematical Computability.* arXiv:quant-ph/0205093 v2.

[82] Kitaev, A., Quantum computations: Algorithms and error correction. *Russ. Math. Surv.* 52 (6): 1191-1249, 1995.

[83] Khanna, G.; Mukhopadhyay, S.; Simon, R.; Mukunda, N., Geometric phases for SU(3) representations and three level quantum systems. *Ann. Phys.* 253, no. 1, 55-82, 1997.

[84] Cohen-Tannoudji, C.; Dupont-Roc, J.; Grynberg, G. *Atom-Photon Interactions.* - Weinheim: Wiley-VCH, 656 p., 2004.

[85] Knizhnik, V.G.; Zamolodchikov, A.B., Current algebrand Wess-Zumino models in two dimensions. *Nucl. Phys. B* 247, 83-103, 1984.

[86] Kohno, T., Monodromy representations of braid groups and Yang-Baxter equation. *Ann. Inst. Fourier, Grenoble,* 37, 4, pp.139–160, 1987.

[87] Kohno, T., Hecke algebra representations of braid groups and classical Yang-Baxter equation. *Adv. Stud. Pure Mathematics.* 28, pp. 255–269, 1988.

[88] Kostrykin, V.; Schrader, R., Kirchoff's rule for quantum wires, ll *The Inverse Problem with Possible Applications to Quantum Computers,* arXiv:quant-ph/99/0053 v1.

[89] Kuiken, K., Heun's equation and the hypergeometric equation, *SIAM J. Math. Anal.*10 (3), pp.655-657, 1979.

[90] Kraus, B.; Cirac, J.I., Optimal creation of the entanglement using a two-qubit gate. *Phys. Rev. A,* 63, 062309,2000.

[91] Kuranishi, M., On everywhere dense embedding of free groups in Lie groups. *Nagoya Math. J,* 2, pp. 63-71, 1951.

[92] Kuś, M.; Zyczkowski, K., Geometry of entangled states. *Phys. Rev. A* (3) 63 no. 3, 032307, 2001.

[93] Lawrence R. *Homology representations of braid group.* Ph.Thesis, Oxford University, 1989.

[94] Lee, C.; Ostrovskaya, E.A., *Quantum Computation with Diatomic Bits in Optical Lattices*, arXiv:quant-ph/0502154 v1.

[95] Leibman, A., Some monodromy representations of generalized braid groups. *Commun. Math. Phys.* 164, pp.293-304, 1994.

[96] Leksin, V.P., The Riemann-Hilbert problem for analytic families of representations, *Math. USSR-Zam.* Vol.50, No.2, pp.89-97, 1991.

[97] Lobry, C., Dynamical polysystems and control theory, geometric methods in system theory. *Proc. NATO Adv. Study Institute, London,* pp.1-42, 1973.

[98] Liu, L., Optical implementation of parallel fuzzy logic. *Opt.Commun.* 73(3), p.183-187, 1989.

[99] Maier, R., Transforming the Heun equations to the hypergeometric equation: I. *Polynomial Transformations.* ArXiv: math.CA/0203264.

[100] Melikishvili, Z.G.; Giorgadze, G., Atom-photon interactions in connection with quantum computations: A three-level atom in a two-mode field. *J. Math. Sci.(N.Y.)* 153, no. 2, 167-185, 2008.

[101] Masalas, M.; Flischauer, M., *Scattering of Dark-state Polaritons in Optical Lattices and Quantum Phase Gates for photons,* arXiv:quant-ph/0402142 v1.

[102] Matiyasevich, Yu.V., The Diophantineness of enumerable sets. *Dokl. Akad. Nauk SSSR,* 191, 279-282 , 1970.

[103] Matiyasevich, Y., *Hilbert's Tenth Problem.* MIT Press, Cambridge, Massachussetts, 1993.

[104] Milnor, J.; Stasheff, I.D.,*Characteristic Classes.* Princeton, N.T.: Princeton University Press, 1974.

[105] Montgomery, R., Isoholonomic problems and some applications. *Commun. Math. Phys.* 128, N 3, 565-592, 1991.

[106] Nakamura, H., *Nonadiabatic Transition: Concepts, Basic Theories, and Applications.* World Scientific, Singapore, 2002.

[107] Niskanen, A.; Nakahara, M.; Salomaa, M., *Realization of Arbitrary Gates in Holonomic Quantum Computation.* ArXiv: quant-ph/0209015

[108] Nielsen, M.; Chuang, I., *Quantum Computation and Quantum Information.* Cambridge Unversity Press, New York, 676 p., 2000.

[109] Newell, A.C., Solitons in Mathematical Physics,*SIAM*, 324p, 1989.

[110] Nomizu, K., *Lie Groups and Differential Geometry.* The Mathematical Society of Japan, 1956.

[111] Dubrovin, B.A.; Novikov, S.P.; Fomenko, A.T., Modern Geometry, *Nauka*, vol.1, 736 p., 1979 (russian).

[112] Okubo, M., *On the group of Fuchsian equation. Seminar Notes*, Tokyo Metropolitan University, 1987.

[113] Okubo, K.; Takano, K.; Yoshida, S., A connection problem for the generalized hypergeometric equation. *Funkcial. Ekvac.* 31, 483-495,1988.

[114] Ochiai, H.; Oshima, T.; Sekiguchi, H., Commuting families of symmetric differential operators, *Proc. Japan. Acad.* Vol. 70, pp.62-66, 1994.

[115] Olshanetsky, M.; Perelomov, A., Quantum integrable systems related to Lie algebras. *Phys. Rep.* Vol. 94, pp. 313-404, 1983.

[116] Oshima, T.; Sekiguchi, H., Commuting families of differential operators invariant under the action of a Weyl group, *J. Math. Sci. Univ. Tokyo.* Vol. 2 pp.1-75, 1995.

[117] Pachos, J.; Zanardi, P.; Rasetti, M., *Non-Abelian Berry connections for quantum computation.* ArXiv: quant-ph/9907103.

[118] Pachos, J.; Zanardi, P.; Rasetti, M., Non-Abelian Berry Connections for quantum computation. *Phys. Rev.* A61, 010305, 2000.

[119] Pachos, J.; Zanardi, P., Quantum holonomies for quantum computing. *Int. J. Mod. Phys.* B15, 1257-1286, 2001.

[120] Pachos, J.; Chountasis, S., Optical holonomic quantum computer *Phys. Rev. A* 62, 2318, 2000.

[121] Petz, D.; Sudár, C., Geometries of quantum states. *J. Math. Phys.* 37, 2662-2673, 1996.

[122] Puri, R.R., *Mathematical methods of quantum optics. - Berlin:* Springer, 285 pp., 2001.

[123] Ramakrishna, V.; Flores, K.L.; Rabitz, H.; Ober, R., Quantum control by decomposition of $su(2)$, *Phys. Rev. A* 62, 2000.

[124] Reed, M.; Simon, B., Methods of Modern Mathematical Physics, vol. lll. *Scattering Theory*, Academic Press, New York, 1979.

[125] Reinhard, H., Comment on "Phase Change during a Cyclic Quantum Evolution". *Phys. Rev. Lett.* 59, 2823, 1987.

[126] Ronveaux, A.,(ed.) Heun's differential equations. *Oxford Sci.Pub.* Oxford University Press, Oxford, 1995.

[127] Sasai, T., On the monodromy group and irreducibility conditions of a fourth order Fuchsian differential system of Okubo type. *J. Reine Angew. Math.*, 299/300, 1978, 38-50.

[128] Sasai, T.; Tsuchiya, S., On a fourth order Fuchsian differential equation of Okubo type. Funkcial. *Ekvac.* 34, 211-221, 1991.

[129] Scully, M.O.; Zubairy, M.S., *Quantum Optics. - Cambridge:* University Press, 630 p., 1997.

[130] Shapere, A.; Wilczek, F., (Eds.) *Geometric Phases in Physics,* World Scientific, 1987.

[131] Schirmer, S.G.; Pullen, I.C.H.; Solomon, A.I., Controllability of Quantum Systems. *Hamiltonian and Lagrangian Methods in Nonlinear Control* (Elsevier Science Ltd, 2003).

[132] Sovolay, R., Lie groups and quantum circuits. *MSRI Conf. Mathematics of the Quantum Computation,* Berkeley, CA, 2000.

[133] Sommers, H.J.; Zyczkowski, K., Bures volume of the set of mixed quantum states. *J. Phys. A: Math. Gen.* 36, 10083-10100, 2003.

[134] Sussmann, H., Orbits of families of vector fields and integrability of distributions. *Trans. Amer. Math. Soc.* 180,171-188, 1973.

[135] Slater, P., Bures geometry of the three-level quantum systems. *J. Geom. Phys.* 39, 207-216, 2001.

[136] Slater, P.B., Apriori probability that two qubits are unentangled. *Quantum Inf. Process.* 1, 387-408, 2002.

[137] Slater, P.B., Silver mean conjectures for 15-dimensional volumes and 14-dimensional hyperareas of the separable two-qubit systems. *J. Geom. Phys.* 53, 74 , 2005.

[138] Slater, P.B., Qubit-qutrit separability-probability ratios. *Phys. Rev. A* 71, 052319, 2005.

[139] Schleich, W.P., *Quantum Optics in Phase Space.* Wiley-VCH, Berlin, 2001.

[140] Solovarov, N.K., Quantum coherence dynamics of a three-level atom in a two-mode field. *J. Exper. and Theor. Phys.*, Vol. 106, No. 6, pp. 1033-1044, 2008.

[141] Suzko, A.A., Exactly solvable models with time-dependent potentials. *Phys.Lett. A* 308, 267, 2003.

[142] Suzko A.A.; Giorgadze, G., Quantum computers based on exactly solvable models, and geometric phases. *J. Math. Sci. (N.Y.)* 153, no. 2, 186-196, 2008.

[143] Takemura, K., The Heun equation and the Calogero-Moser-Sutherland system I: The Bethe Ansatz method. *Comm. Math. Phys.* Vol. 235, pp. 467-494, 2003.

[144] Tanimura, S.; Hayashi, D.; Nakahara, M., Exact solutions of holonomic quantum computation. *Phys. Lett. A* 325, No.3-4,199-205, 2004.

[145] Tevzadze, R.; Giorgadze, G.; Quantum computations with scattering matrices. *J. Math. Sci. (N.Y.)*153, no. 2, 197-209, 2008.

[146] Tilma, T.; Byrd, M.; Sudarshan, E.C.G., A parametrization of bipartite systems based on $SU(4)$ Euler angles. *J. Phys. A: Math. Gen.* 35, 10445, 2002.

[147] Tilma, T.; Sudarshan, E.C.G., Generalized Euler angle parametrization for SU(N). *J. Phys. A: Math. Gen.* 35, 10467-10501, 2002.

[148] Tilma, T.; Sudarshan, E.C.G., Generalized Euler angle parameterization for $U(N)$ with applications to $SU(N)$ coset volume measures. *J. Geom. Phys.* 52, 263-283, 2004.

[149] Velicheva, E.P.; Suzko, A.A., Exact solutions of nonstationary Schrödinger equations. *Theor. Math. Phys* 115, 687, 1998.

[150] Wang, S.J., Nonadiabatic Berry's phase for a psi particle in a rotating magnetic field. Phys. Rev. A 42 (1990) 5107; *Nonadiabatic Berry's phase for a quantum system with a dynamical semisimple Lie group.* Ibid p. 5103.

[151] Wilczek, F.; Zee, A., Appearance of gauge structure in simple dynamical systems. *Phys. Rev. Lett.* 52, 2111, 1984.

[152] Williamson, M.S.; Vedral, V., Composite geometric phase for multipartite entangled states. *Phys. Rev. A* (3) 76, no. 3, 032115, 4 pp., 2007.

[153] Wu, Jian-da; Chen, Jian-lan; Zhang, Yong-de, *The Generation and Exchange of Entanglement via Zeeman Effect.* arXiv:quant-ph/0609074 v1.

[154] Yakoyama, T., On an irreducibility condition for hypergeometric systems, *Funkcial. Ekvac.*, 38,11-19, 1995.

[155] Yamabe, H., On an arcwise connected subgroup of a Lie group. *Osaka Math. J.,* 13-14, 1950.

[156] Zanardi, P.; Rasetti, M., Holonomic quantum computation. *Phys. Lett.* A264, 94, 1999.

[157] Zhang, I.; Liu, L.; et al., Fuzzy reasoning morphological operators and their optical emplementation. *App. Opt.* v.36, no. 11, 2328, 1997.

[158] Duan, L. M.; Cirac, J.I.; Zoller, P., Geometric manipulation of trapped ions for quantum computation. *Science* 292, 1695, 2001.

[159] Życzkowski, K.; Horodecki, P.; Sanpera, A.; Lewenstein, M., Volume of the set of separable states. *J. Phys. Rev. A* 58, 883-892, 1998.

[160] Życzkowski, K., Volume of the set of separable states. II. *Phys. Rev. A* 60, 3496-3507, 1999.

[161] Życzkowski, K.; Słomczyński, W., The Monge metric on the sphere and geometry of quantum states. *J. Phys. A: Math. Gen.* 34, 6689-6722, 2001.

Index